RISK-BASED RELIABILITY ANALYSIS AND GENERIC PRINCIPLES FOR RISK REDUCTION

RISK-BASED RELIABILITY ANALYSIS AND GENERIC PRINCIPLES FOR RISK REDUCTION

M.T. Todinov
Cranfield University, Cranfield, UK

ELSEVIER

AMSTERDAM • BOSTON • HEIDELBERG • LONDON • NEW YORK • OXFORD
PARIS • SAN DIEGO • SAN FRANCISCO • SINGAPORE • SYDNEY • TOKYO

ELSEVIER B.V.
Radarweg 29
P.O. Box 211, 1000 AE Amsterdam
The Netherlands

ELSEVIER Inc.
525 B Street, Suite 1900
San Diego, CA 92101-4495
USA

ELSEVIER Ltd
The Boulevard, Langford Lane
Kidlington, Oxford OX5 1GB
UK

ELSEVIER Ltd
84 Theobald's Road
London WC1X 8RR
UK

First edition 2007

ISBN-13: 978-0080-44728-5

ISBN-10: 0080-44728-7

Typeset by Charon Tec Ltd (A Macmillan Company), Chennai, India
Printed in Great Britain

07 08 09 10   10 9 8 7 6 5 4 3 2 1

To Prolet, Marina and Marin

CONTENTS

PREFACE

For a long time, the conventional reliability analyses have been oriented towards selecting the more reliable system and preoccupied with maximising the reliability of systems. On the basis of counterexamples, we demonstrate that selecting the more reliable system does not necessarily mean selecting the system with the smaller losses from failures! As a result, reliability analyses should necessarily be risk-based, linked with the losses from failures.

Accordingly, a theoretical framework, models and algorithms are presented which form the foundations of the risk-based reliability analysis – a reliability analysis linked with the losses from failures. An underlying theme in the book is the basic principle for a risk-based design: *the larger the cost of failure associated with a component, the larger its minimum necessary reliability level*. Even identical components should be designed to different reliability levels if their failures are associated with different losses.

According to a classical definition, the risk of failure is a product of the probability of failure and the cost given failure. This risk measure however, cannot describe the risk of losses exceeding a maximum acceptable limit. As an alternative, an aggregated risk measure based on the cumulative distribution of the potential losses is introduced and the theoretical framework for risk analysis based on the concept *potential losses* is developed. This risk measure incorporates the uncertainty associated with the exposure to losses and the uncertainty in the consequences given the exposure. Historical data related to the loss given failure can only be used to determine the distribution of the conditional losses (given that failure has occurred). Building the distribution of the potential losses however, requires also an estimate of the probability of failure.

Equations related to the probability density distribution, the cumulative distribution and the variance of the potential loss from multiple failure modes have been derived. An upper bound of the variance of the potential loss has been derived in the case where the probability of failure is unknown.

The expected potential loss and its variance however, are still not sufficient to measure the uncertainty associated with the risk and it is demonstrated that using solely the variance to assess uncertainty can be misleading. The cumulative distribution of the potential loss and the maximum potential loss at a specified confidence level have been proposed as an alternative.

Traditionally, the losses from failures have been 'accounted for' by the average production availability (the ratio of the actual production and the maximum production capacity). As demonstrated in Chapter 1 by using a simple counterexample, although the availability level does reflect the cost of lost production which is proportional to the system's downtime, it does not account for the component of the losses which depends on the time of failure occurrence. Because of this, two systems with the same production availabilities can be characterised by very different losses from failures. Furthermore, the average production availability does not reveal the variation of the actual availability and from it, the variation of the lost production.

The combined variation of the losses from failures caused by variations of the actual availability, the number of interventions and the number of failed components can be captured by the cumulative distribution of the potential losses from failures.

For repairable systems with complex topology, the distribution of the potential losses can be revealed by simulating the behaviour of the systems during their life cycle. For this purpose, discrete-event simulators are proposed in Chapter 7, capable of tracking the potential losses for systems with complex topology, containing a large number of components. The proposed algorithms are demonstrated by comparing the losses from failures and the net present values of competing design topologies based on a single-channel and a dual-channel control. To the risk manager, these algorithms provide a pair of spectacles through which the actual operational risk associated with production systems can be seen and subsequently managed effectively. The simulators are based on new, efficient algorithms for reliability analysis of systems comprising thousands of components.

This book also addresses the topic related to a risk-based reliability allocation of complex systems involving a large number of components. A reliability allocation which maximises the net present value for a system consisting of blocks logically arranged in series, is achieved by determining for each block individually, the reliabilities of the components which minimise the sum of the capital cost, operation cost and the losses from failures. By using the principles, algorithms and techniques developed in the book, engineers–designers will be better equipped for developing designs associated with small risks of failure and small total costs.

A net present value cash-flow model has also been proposed, which has significant advantages to traditional cash-flow models based on the expected value of the expenditure. Unlike these models, the proposed model has the capability to reveal the variation of the net present value due to different number of failures occurring during a specified time interval (e.g. 1 year). The model also permits tracking the impact of the actual pattern of failure occurrences and the time dependence of the losses from failures. By using the model, the net present values associated with alternative design solutions can be compared and the solution with the largest net present value selected.

The second half of the book features generic principles and techniques for reducing the risk of failure. These have been classified into three major categories: *preventive* (reducing the likelihood of failure), *protective* (reducing the consequences from failure) and *dual techniques and principles* (reducing both, the likelihood of failure and the consequences). For obvious reasons, preventive (proactive) measures for risk reduction received more emphasis in comparison to protective (reactive) measures. By systematising various techniques and principles for reducing the risk of failure, my intention was to assist engineers in their effort to guarantee an optimal risk-based design early in the conceptual design stage. To the best of my knowledge, such risk-reduction principles and methods have been systematised and presented for the first time. Many of them, for example: *avoiding clustering of events*, *deliberately introducing weak links*, *reducing sensitivity*, *introducing changes with opposite sign*, etc., to the best of my knowledge, are discussed in the reliability and risk literature for the first time.

Significant space has been allocated to component reliability. In Chapter 13, it is shown that increasing the resistance against overstress failures is about selecting parameter values which minimise the integral giving the probability of failure caused by the interaction of the upper tail of the load distribution and the lower tail of the strength distribution. In the last chapter of the book, several applications are discussed of a powerful equation which constitutes the core of a new theory of locally initiated component failure by flaws whose number is a random variable. Among the applications are: (i) determining the probability of overstress failure of loaded components with complex shape, containing flaws; (ii) optimising the design of a component by minimising its vulnerability to an overstress failure; (iii) selecting the type of loading characterised by the smallest probability of overstress failure.

This book has been written with the intention to fill two gaps in the reliability and risk literature: introducing *the risk-based reliability analysis* as a powerful alternative to the traditional reliability analysis and discussing *generic principles for reducing technical risk*. I hope that the principles, models and algorithms presented in this book will help to fill these gaps and make the book useful to reliability and risk-analysts, researchers, consultants, students and practising engineers.

In conclusion, I acknowledge the editing and production staff at Elsevier for their excellent work and in particular the help of Ms Kristi Green, Dr J. Agbenyega and Mr. M. Prabakaran.

Thanks also go to many colleagues from universities and the industry for their useful suggestions and comments.

Finally, I acknowledge the help and support from my family during the preparation of the manuscript.

M.T. Todinov
Cranfield University

1

RISK-BASED RELIABILITY ANALYSIS: A POWERFUL ALTERNATIVE TO THE TRADITIONAL RELIABILITY ANALYSIS

Since removing a failure mode at the design stage is considerably cheaper compared to removing it at the manufacturing stage or during service, it is important that reliability is integrated early into the design of complex systems. Accordingly, for a long time, the conventional reliability analysis has been oriented towards maximising the reliability of a system.

In order to achieve their goal however, designers must also be able to reveal the losses from failures. This creates the possibility to quickly filter out inappropriate design solutions associated with large losses from failures and select solutions associated with minimum losses.

More importantly, on the basis of a simple counterexample, we can demonstrate that selecting the more reliable system does not necessarily mean selecting the system with the smaller losses from failures.

Indeed, consider two very simple systems consisting of only two components, logically arranged in series (Fig. 1.1).

For the first system (Fig. 1.1(a)), suppose that component $A1$ fails on average once a year ($f_{A1} = 1$) and its failure is associated with $C_{A1} = 2000$ units of losses, while component $B1$ fails on average $f_{B1} = 9$ times a year and

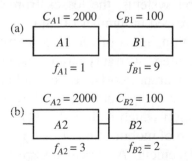

Figure 1.1 Systems composed of two components only, demonstrating that the more reliable system is not necessarily associated with the smaller losses from failures.

1

its failure is associated with $C_{B1} = 100$ units of losses. Suppose now that for an alternative system consisting of the same type of components A and B (Fig. 1.1(b)), the losses associated with failure of the separate components are the same but the failure frequencies are different. Component $A2$ is now characterised by $f_{A2} = 3$ average number of failures per year and component $B2$ is characterised by $f_{B2} = 2$ average number of failures per year. Clearly, the second system is more reliable than the first system because it is characterised by 5 average number of failures per year as opposed to the first system, characterised by 10 average number of failures per year. Since the first system fails whenever either component $A1$ or component $B1$ fails, the expected (average) losses from failures $\overline{L}_1$ for the system are

$$\overline{L}_1 = f_{A1}C_{A1} + f_{B1}C_{B1} = 1 \times 2000 + 9 \times 100 = 2900 \qquad (1.1)$$

while for the second system, the expected losses from failures $\overline{L}_2$ are

$$\overline{L}_2 = f_{A2}C_{A2} + f_{B2}C_{B2} = 3 \times 2000 + 2 \times 100 = 6200 \qquad (1.2)$$

As can be verified, the more reliable system (the second system) is associated with the larger losses from failures!

This simple example shows that a selection of a system solely based on its reliability can be misleading, even if all components in the system are characterised by constant failure rates and are arranged in series. In case of system failures associated with the same cost, a system with larger reliability does mean a system with smaller losses from failures. In the common case of system failures associated with different costs however, a system with larger reliability does not necessarily mean a system with smaller losses from failures.

For many production systems, the losses from failures are extremely high. They can be expressed in number of fatalities, lost production time, volume of lost production, mass of released harmful chemicals into the environment, lost customers, warranty payments, costs of mobilisation of emergency resources, insurance costs, etc. For oil and gas production systems for example, major components of the losses from failures are the amount of lost production which is directly related to the amount of lost production time, the cost of mobilisation of resources and intervention, and the cost of repair/replacement. A critical failure in a deep-water oil and gas production system, in particular, entails long downtimes and extremely high costs of lost production and intervention for repair. Furthermore, such failures can have disastrous environmental and health consequences. The need to measure all types of operational risk is crucial to revealing the magnitude of the existing risk and implementing appropriate risk management procedures. Even if the individual critical failures are associated with relatively small losses, in the long run, particularly if such failures occur with high frequency, the amount of the total accumulated loss can be very large. As a result, reliability analyses related to production systems must necessarily be based on the risk of failure.

Traditionally, the losses from failures have been 'accounted for' by the average production availability (the ratio of the actual production and the maximum production capacity). As we shall demonstrate, two systems with the same production availability can be characterised by very different losses from failures.

The next counterexample clarifies this point. Figure 1.2 features two identical systems with the same operating life cycle of T years, which experience exactly one failure each, associated with the same downtime for repair t_d. For the first system however, the failure occurs towards the end of life in year $k1$ (Fig. 1.2(a)) while for the second system the failure occurs at the beginning of life, in year $k2$ (Fig. 1.2(b)). By assuming an uniform production profile which does not vary during the life cycle of the systems, both systems will be characterised by the same availability $A1 = A2 = (T - t_d)/T$. The component of the cost of lost production C which is proportional to the downtime will also be the same for both systems.

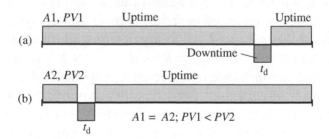

Figure 1.2 Two systems with the same availability and different present values of the losses from failure ($PV1 < PV2$).

Because of the different time at which the failures occur however, the financial impacts of the lost production will be different for the two systems. Indeed, $PV_1 = C/(1 + r)^{k1}$ is the present value of the lost production for the first system and $PV_2 = C/(1 + r)^{k2}$ is the present value of the lost production for the second system. For example, substituting in these formulae a discount rate $r = 7.5\%$, $k1 = 25$ and $k2 = 2$, yields

$$\frac{PV_2}{PV_1} = \frac{(1+r)^{k1}}{(1+r)^{k2}} \approx 5.28$$

Because of the different time of failure occurrence, despite that the availabilities $A1$ and $A2$ are the same for the two systems, the second system (Fig. 1.2(b)) is characterised by more than 5 times bigger losses compared to the first system (Fig. 1.2(a)).

Thus, although the availability does reflect the cost of lost production which is proportional to the system downtime, it does not account for the dependence on the time of failure occurrence. Furthermore, availability does not account for the cost of intervention, which for deep-water oil and gas production, for example, can be significant in relation to the cost of lost production. Relying solely on the availability level does not reveal the real losses from failures.

Apart from estimating the losses from failures, engineers also need to specify reliability requirements regarding components and blocks in the designed systems. None of the popular reliability allocation strategies however, such as the ARINC or the AGREE methods (Ebeling, 1997), are capable of allocating reliability requirements for the components which deliver the minimum losses from failures for the system. Most of the existing methods focus on manipulating the hazard rates of the components so that a particular target system hazard rate is attained. Thus, for a system with components arranged in series, with a required system reliability level R_{sys}, reliabilities $R_i = (R_{\text{sys}})^{1/M}$ are determined for each component. In this way, the requirement $R_{\text{sys}} = R_1 \times R_2 \times \cdots \times R_M$ is indeed fulfilled but an important circumstance is neglected. Component failures are usually associated with *different* losses. Consequently, components associated with large losses from failures should be designed to higher reliability levels.

Since 1977, there have been also a significant number of articles and books (Tillman et al., 1985; Xu et al., 1990; Kuo and Prasad, 2000; Elegbede et al., 2003; Wattanapongsakorn and Levitan, 2004) related to reliability optimisation involving costs. Most of the methods described in these sources, however, are either related to maximising the reliability of a system given an overall budget (a maximum total cost of resources towards the reliability maximisation) or minimising the total cost of resources necessary to achieve a specified level of system reliability. For embedded systems, Wattanapongsakorn and Levitan (2004), for example, presented models for maximising reliability while meeting cost constraints and also for minimising system cost under various reliability constraints. The reliability optimisation involving cost minimisation in Elegbede et al. (2003), for example, was also restricted to maximising the reliability at a minimum cost of the components building the system. These models are not models for risk-based reliability allocation because they do not incorporate the losses associated with system failure. Instead, it is expected that once reliability is maximised, the losses from failures will automatically be minimised, which, as we demonstrated earlier is not necessarily true. Maximising the reliability of a system *does not necessarily guarantee smaller losses from failures*. This conclusion, which does not conform to the current understanding and practice, shows that the risk-based reliability analysis requires a new generation of models and algorithms based on the losses from system failures.

There exists also work related to reliability optimisation based on fuzzy techniques, dealing with the cost of the system and the costs of the separate components (Ravi et al., 2000). The optimal redundancy allocation, however, is again oriented towards maximising the system reliability by minimising the system cost, not minimising the losses associated with system failures.

In Pham (2003), for a parallel system consisting of n components, the optimal sub-system size was determined that minimises the average system cost. The average system cost included the cost of the components and the cost of system failure. For parallel–series systems, the optimal sub-system size was determined that maximises the average system profit.

Optimum reliability minimising the sum of the cost of failure and the cost of reliability has been discussed by Hecht (2004) who acknowledged that the total user cost has a minimum and the failure probability at which the minimum is reached represents the optimum reliability in economic terms.

Often, alternative design solutions are compared and one of them selected. As we demonstrate later, a sound scientific basis for such a selection is the distribution of the potential losses from failures associated with the competing solutions which requires reliability analysis based on the losses from failures.

A fundamental, scientifically sound criterion for assessing competing production architectures is their net present value, after estimating the income stream (inflow) and expenditure stream (outflow) (Wright, 1973; Mepham, 1980; Vose, 2000; Arnold, 2005). The correct estimation of the losses from failures is at the heart of a correct determination of the expenditure stream. All critical failures associated with losses, such as lost production, intervention and repair costs must be tracked throughout the design life of the system and their financial impact assessed. The interaction between the different components of the losses from failures, however, is not well understood.

Currently, sound theoretical models for risk-based reliability analysis involving the losses from failures are difficult to find and this was the major reason which prompted writing this book.

Until recently, one of the main obstacles to developing the theoretical basis of the risk-based reliability analysis, as an alternative to the traditional reliability analysis, was the absence of appropriate models related to the losses from failures from multiple failure modes and the uncertainties associated with the probabilities with which these failure modes are activated. In order to fill this gap, models based on potential losses from failures from multiple mutually exclusive failure modes have been developed by the author (Todinov, 2003, 2006b). The losses were modelled as distribution mixtures and equations related to their cumulative distribution, their variance and its exact upper bound were derived. For systems characterised by a constant hazard rate, a model for determining the optimum hazard rate of the system at which the minimum of the total cost is attained was proposed in Todinov (2004a). Recently, models and algorithms have been developed for determining the expected losses from failures for non-repairable and repairable systems whose components are logically arranged in series and for systems with complex topology (Todinov, 2004c, 2006b, c). These developments form the core of the book.

2

BASIC RELIABILITY CONCEPTS AND CONVENTIONS USED FOR DETERMINING THE LOSSES FROM FAILURES

2.1 RELIABILITY AND FAILURE

Reliability, according to a definition in IEC, 50 (191) (1991) is *'the ability of an entity to perform a required function under given conditions for a given time interval'* which is usually measured by the *probability* of a failure-free operation during a specified time interval $(0, t)$.

A system is said to have a failure, *if the service it delivers to the user deviates from compliance with the specified system function.* Failure has also been defined as 'non-conformance to a defined performance criterion' (Smith, 2001). The failure mode is *the way a system or component fails to function as intended.* It is the effect by which failure is observed. The physical processes leading to a particular failure mode will be referred to as *failure mechanism.* It is important to understand that the same failure mode (e.g. fracture of a component) can be associated with different failure mechanisms. Thus, the fracture of a component could be brittle, ductile or fatigue fracture, fracture due to stress corrosion cracking, etc. In turn, brittle fracture itself can be *cleavage fracture* occurring at low temperature and triggered by cracking of a second-phase particle, *intergranular fracture* caused by segregation of impurities towards the grain boundaries, etc. In each particular case, the failure mechanism is different. The *failure promoting factors* underlie failure mechanisms. Common examples are: high stresses, too high or too low temperature, high humidity, vibration, friction, contamination, cyclic loading, radiation, dust and chemically aggressive atmosphere. *Failure causes* are the particular circumstances that have led to failure. Common failure causes are: errors during design, manufacturing, quality control and assembly, maintenance errors, human errors, etc.

If T is the time to failure, let $F(t)$ be the probability $P(T \le t)$ that the time to failure T will not be greater than a specified time t. *The time to failure distribution $F(t)$ is a fundamental characteristic of components and systems.* Determining the distribution of the time to failure helps to calculate the probability of surviving any specified time interval. In other words, the distribution of the time to failure is a key to determining the reliability of components and systems. Differentiating the cumulative distribution of the time to failure $F(t)$ yields the probability density of the time to failure.

$$f(t) = \frac{\mathrm{d}F(t)}{\mathrm{d}t}\mathrm{d}t \qquad (2.1)$$

$f(t)\,\mathrm{d}t$ gives the probability that failure will occur in the infinitesimal interval $(t, t+\mathrm{d}t)$.

The probability $P(T > t)$ that the time to failure T will be greater than a specified time t is determined from

$$P(T > t) = 1 - P(T \le t)$$

where $R(t) \equiv P(T > t)$ is referred to as *reliability (survival) function*. This is a monotonic non-increasing function, always unity at the start of life $(R(0) = 1, R(\infty) = 0)$. It is linked with the cumulative distribution function of the time to failure $F(t)$ by $R(t) = 1 - F(t)$ (reliability $= 1 -$ probability of failure).

Failures of repairable systems are broadly divided into several basic categories.

- *Critical failures* are present if the system fails to deliver one or more specified functions; for example, at least one of the production units stops production. Critical failures usually require immediate corrective action (e.g. intervention for repair or replacement) in order to return the system into operating condition. Each critical failure is associated with losses due to the cost of intervention, repair and the cost of lost production.

  For the simple production system in Fig. 2.1, an example of a critical failure is the failure of the power block (PB), mechanical device (MD) or failures of both control modules (CM1 and CM2). In all these cases, the mechanical device stops functioning.

- *Non-critical failures* are present if failure of a component does not affect the system's function. Thus, in the simple system from Fig. 2.1, failure of the control module CM1 will not cause the mechanical device to stop

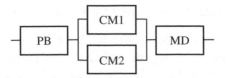

Figure 2.1 A simple system consisting of a power block (PB), two control modules (CM1 and CM2) and a mechanical device (MD).

operation because the redundant control module CM2 will still maintain control over the mechanical device and the system will be operational.
- *Degraded failures* are present if one or more functions are compromised but still delivered.
- *Incipient failures* will occur if a fault is present and will develop into a critical or degraded failure if corrective action is not taken.

For the purposes of illustrating the basic concepts of the risk-based reliability analyses presented in this book, only critical and non-critical failures will be considered.

2.1.1 Mean Time to Failure

A popular reliability measure is the mean time to failure (MTTF) which is *the average time to the first failure*. It can be obtained from the mean of the probability density of the time to failure $f(t)$:

$$\text{MTTF} = \int_0^\infty t\,f(t)\,\mathrm{d}t \tag{2.2}$$

If $R(t)$ is the reliability function, the integral in equation (2.2) becomes $\text{MTTF} = -\int_0^\infty t\,\mathrm{d}R(t)$ which, after integrating by parts (Grosh, 1989), gives

$$\text{MTTF} = \int_0^\infty R(t)\,\mathrm{d}t \tag{2.3}$$

For a constant hazard rate $\lambda = \text{constant}$

$$\text{MTTF} = \theta = \int_0^\infty \exp(-\lambda t)\,\mathrm{d}t = \frac{1}{\lambda} \tag{2.4}$$

In this case, the MTTF is the reciprocal of the hazard rate.

Equation (2.4) is valid *only* for failures characterised by a constant hazard rate. In this case, the probability that failure will occur earlier than the MTTF is approximately 63%. Indeed, $P(T \le \text{MTTF}) = 1 - \exp(-\lambda \text{MTTF}) = 1 - \exp(-1) \approx 0.63$.

2.1.2 Minimum Failure-Free Operating Period

Reliability can be interpreted as the probability of surviving a specified minimum failure-free operating period (MFFOP) without a critical failure.

Within the specified MFFOP, there may be failures of redundant components which do not cause a system failure (Fig. 2.2). The idea behind the MFFOP is to guarantee with high probability no critical failures (associated with losses). Guaranteeing an MFFOP of specified time length with high probability means guaranteeing a large reliability associated with this time interval. This statement however *does not apply* to the MTTF reliability measure. It is true that if the distribution of the time to failure is the negative exponential distribution a large MTTF means large reliability. Not always however, the distribution of the time to failure can be modelled by the negative exponential distribution. In Todinov (2005a), it was demonstrated that for a distribution of the time to failure different from the negative exponential distribution, the MTTF reliability measure *can be misleading*. The component with the larger MTTF is not necessarily the component associated with the larger probability of surviving a specified time interval! This is because a large MTTF can be obtained by aggregating times to failure characterising increased failure frequency at the start of life and very low failure frequency later in life. Indeed, if a reliable work is required from the component during the first 2 years and the component is selected solely on the basis of its MTTF (Fig. 2.3), component '1', characterised by a smaller reliability during the first 2 years will be selected!

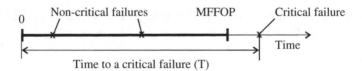

Figure 2.2 Guaranteeing with high probability an MFFOP of specified length, free from critical failures.

2.2 HAZARD RATE AND TIME TO FAILURE DISTRIBUTION

The hazard rate is a notion defined for non-repairable components and systems. The focus is on the time to the first and only failure. In this sense, the hazard rate can also be applied to repairable systems if the focus is on *the first system failure only*.

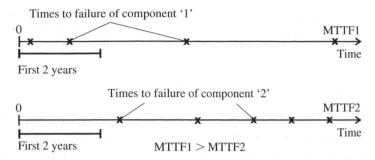

Figure 2.3 Two components of the same type with different MTTF

Suppose that the probability of failure in the small time interval $(t, t + \Delta t)$, depends on the age t of the component/system and is given by $h(t)\Delta t$, where $h(t)$ will be referred to as *hazard rate*. The cumulative distribution of the time to failure $F(t)$ for the component/system is

$$F(t) = 1 - \exp\left(-\int_0^t h(v)\,dv\right) \qquad (2.5)$$

where v is a dummy integration variable. The integral $H(t) = \int_0^t h(v)\,dv$ in equation (2.5) is known as *cumulative hazard rate*. By using the cumulative hazard rate, the cumulative distribution of the time to failure can be presented as $F(t) = 1 - \exp(-H(t))$.

If age t has no significant effect on the probability of failure, then $h(t)\Delta t = \text{constant}$ irrespective of t. This is possible only if the hazard function is constant $(h(t) = \lambda = \text{constant})$ Substituting λ in equation (2.5) results in

$$F(t) = 1 - \exp(-\lambda t) \qquad (2.6)$$

for the distribution of the time to failure, which is the negative exponential distribution. One of the reasons for the fundamental importance of this distribution is its applicability to life distributions of components whose conditional probability of failure within a specified time interval is practically independent of their age. This condition is approximately fulfilled for components which practically do not degrade or wear out with time (e.g. certain electrical components, protected structures, static mechanical components, etc.).

The hazard rate can also be presented as a function of the probability density $f(t)$ of the time to failure (Barlow and Proschan, 1975):

$$h(t) = \frac{f(t)}{R(t)} \tag{2.7}$$

Since $f(t) = -R'(t)$, where $R(t) = 1 - F(t)$ is the reliability function, equation (2.7) can also be presented as $R'(t)/R(t) = -h(t)$. Integrating both sides with the initial condition $R(0) = 1$ yields expression (2.5) which relates the time to failure distribution with the hazard rate function.

There exists a fundamental difference between the failure density $f(t)$ and the hazard rate $h(t)$. Consider an initial population of N_0 components. The proportion $\Delta n / N_0$ of components from the initial number N_0 that fail within the time interval $(t, t + \Delta t)$ is given by $f(t)\Delta t$. In short, the failure density $f(t) = \Delta n / (N_0 \Delta t)$ gives *the percentage of the initial number of items that fail per unit interval.*

Conversely, the proportion $\Delta n / N(t)$ of items which have survived time t and which will fail in the time interval $(t, t + \Delta t)$ is given by $h(t)\Delta t$. In short, the hazard rate $h(t) = \Delta n / (N(t)\Delta t)$ is *the proportion of items in service that fail per unit interval.*

One significant advantage of the distribution of the time to failure is that it is a universal descriptor, applicable to non-repairable and repairable systems alike. The hazard rate, on the other hand, is applicable only to non-repairable components and systems (Ascher and Feingold, 1984).

2.3   HOMOGENEOUS POISSON PROCESS AND ITS LINK WITH THE NEGATIVE EXPONENTIAL DISTRIBUTION

The homogeneous Poisson process is an important model for random events. It is characterised by the following features:

- The numbers of occurrences of a particular event (e.g. failure) in non-overlapping time intervals are statistically independent.
- The probability of occurrence in time intervals of the same length is the same and depends only on the length of the interval, not on its location.
- The probability of more than one occurrence in a vanishingly small time interval is negligible.

A homogeneous Poisson process is characterised by a constant intensity ($\lambda =$ constant) and the mean number of occurrences in the interval $(0, t)$ is λt. If a homogeneous Poisson process with intensity λ is present, the probability density distribution $f(x)$ of the number of occurrences in the time interval $(0, t)$ is described by the Poisson distribution:

$$f(x) \equiv P(X = x) = \frac{(\lambda t)^x \exp(-\lambda t)}{x!} \qquad x = 0, 1, 2, \ldots \qquad (2.8)$$

The function $f(x)$ gives the probability of exactly x occurrences in the time interval $(0, t)$. To determine the probability of r or fewer occurrences in the finite time interval $(0, t)$, the cumulative Poisson distribution:

$$F(r) \equiv P(X \le r) = \sum_{x=0}^{r} \frac{(\lambda t)^x}{x!} \exp(-\lambda t) \qquad (2.9)$$

is used.

From equation (2.9), the probability that the time to the first occurrence will be larger than a specified time t can be obtained directly, by setting $x = 0$ (zero number of occurrences):

$$P(\text{no occurrences}) = \exp(-\lambda t)$$

Consequently, the probability that the time T to the first occurrence will be smaller than t is given by $P(T \le t) = 1 - \exp(-\lambda t)$ which is the negative exponential distribution (2.6). If the times between the occurrences are exponentially distributed, the number of occurrences follows a homogeneous Poisson process and *vice versa*. This important link between the negative exponential distribution and the homogeneous Poisson process will be illustrated by the following example.

Suppose that a component/system is characterised by an exponential life distribution $F(t) = 1 - \exp(-\lambda t)$. After each failure at times t_i, a replacement/repair is initiated which brings the component/system to *as good as new* condition. Under these assumptions, the successive failures of the component/system at times $t_1, t_2, \ldots$ in the finite time interval with length a (Fig. 2.4) follow a homogeneous Poisson process with intensity λ. The mean number of failures in the interval $(0, a)$ is given by λa.

Suppose that load applications (shocks) which exceed the strength of our component/system follow a homogeneous Poisson process with intensity λ. The reliability R associated with a finite time interval with length t is then equal to the probability $R = \exp(-\lambda t)$ that there will be no load

Figure 2.4 Times of successive failures in a finite time interval with length *a*.

application within the specified time interval t. Consequently, the cumulative distribution of the time to failure is given by the negative exponential distribution (2.6). This application of the negative exponential distribution for modelling the time to failure of components and systems which fail whenever a random load exceeds the strength confirms the importance of this distribution. Furthermore, the exponential distribution is an approximate limit failure law for complex systems containing a large number of components which fail independently and whose failures cause a system failure (Drenick, 1960).

The negative exponential distribution, however, is inappropriate to model times to failure caused by damage accumulation (corrosion, fatigue, fracture toughness degradation, wear) if its rate cannot be neglected.

2.4 WEIBULL MODEL FOR THE DISTRIBUTION OF THE TIME TO FAILURE

A widely used model for the times to failure is the Weibull model (Weibull, 1951), with a cumulative distribution function

$$F(t) = 1 - \exp[-(t/\eta)^m] \tag{2.10}$$

In equation (2.10), $F(t)$ is the probability that failure will occur before time t, η is the characteristic lifetime and m is a shape parameter. If $m = 1$, the Weibull distribution transforms into the negative exponential distribution (2.6) with parameter $\lambda = 1/\eta$.

Differentiating equation (2.10) with respect to t gives the probability density function of the Weibull distribution:

$$f(t) = \frac{m}{\eta}\left(\frac{t}{\eta}\right)^{m-1} \exp\left[-\left(\frac{t}{\eta}\right)^m\right] \tag{2.11}$$

Since the hazard rate is defined by equation (2.7), where $f(t)$ is given by equation (2.11) and the reliability function is $R(t) = \exp[-(t/\eta)^m]$, the

Weibull hazard rate function becomes

$$h(t) = \left(\frac{m}{\eta}\right)\left(\frac{t}{\eta}\right)^{m-1} \tag{2.12}$$

As can be verified from equation (2.12), for $m < 1$, the hazard rate is decreasing; for $m > 1$, it is increasing and $m = 1$ corresponds to a constant hazard rate.

For a shape parameter $m = 1$, the Weibull distribution transforms into the negative exponential distribution and describes the useful life region of the bathtub curve (Fig. 2.5), where the probability of failure within a specified time interval practically does not depend on age. For components, for which early-life failures have been eliminated and preventive maintenance has been conducted to replace worn parts before they fail, the hazard rate tends to remain constant (Villemeur, 1992). A value of the shape parameter smaller than one indicates infant mortality failures while a value greater than one indicates wearout failures.

2.5 RELIABILITY BATHTUB CURVE FOR NON-REPAIRABLE COMPONENTS/SYSTEMS

The hazard rate of non-repairable components and systems follows a curve with bathtub shape (Fig. 2.5), characterised by three distinct regions. The first region, referred to as *early-life failure region or infant mortality region*, comprises the start of life and is characterised by initially high hazard rate which decreases with time. Most of the failures in the infant mortality region are quality related overstress failures caused by inherent defects due to poor design, manufacturing and assembly. Since most substandard components fail during the infant mortality period and the experience of the personnel

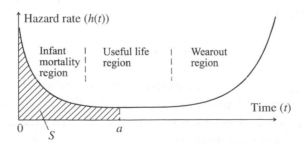

Figure 2.5 Reliability bathtub curve.

operating the equipment increases with time, the initially high hazard rate gradually decreases.

The second region of the bathtub curve, referred to as *useful life region*, is characterised by approximately constant hazard rate. The negative exponential distribution is the model of the times to failure in this region. Failures in this region are not due to age, wearout or degradation.

The third region, referred to as *wearout region*, is characterised by an increasing with age hazard rate due to accumulated wear and degradation of properties (e.g. wear, erosion, corrosion, fatigue, creep).

As we demonstrate in Chapter 6, if the times to failure follow a non-homogeneous Poisson process, the area S of the hatched region beneath the hazard rate curve within the time interval $(0, a)$ gives the expected number of failures in this time interval.

2.6   PRODUCTION AVAILABILITY

The average production availability A_P is defined as the ratio

$$A_P = \frac{\text{Average actual production}}{\text{Maximum production capacity}} \qquad (2.13)$$

Assuming that the average actual production is proportional to the average actual production time, the production availability can also be defined in terms of production time:

$$A_P = \frac{\text{Average actual production time}}{\text{Maximum possible production time}} \qquad (2.14)$$

Considering that the average actual production time is a function of the average lost production time, for n production units, the production availability can also be defined as

$$A_P = 1 - \frac{L_d}{M_d} \qquad (2.15)$$

where $L_d = \sum_{i=1}^{n} l_{d,i}$ is the expected value of the total lost production time during a specified time interval (e.g. the life cycle of the system); $l_{d,i}$ is the expected lost production time for the i-th production unit during the specified time interval; $M_d = \sum_{i=1}^{n} m_{d,i}$ is the maximum possible production time during the specified time interval; $m_{d,i}$ is the maximum possible production time for the i-th production unit during the specified time interval.

2.7 TIME TO FAILURE DISTRIBUTION OF A SERIES ARRANGEMENT, COMPOSED OF COMPONENTS WITH CONSTANT HAZARD RATES

A typical system with components logically arranged in series is obtained when assessing the probability of a leak to the environment from an interface. With regard to the failure mode 'leak to the environment', all interfaces in a system are logically arranged in series despite their actual physical arrangement. Indeed, a leak to the environment is present if at least one of the interfaces leaks. Another typical example of a series arrangement is present where a system contains a number of components and failure of any of them causes system failure.

The reliability of a system composed of n independently working components logically arranged in series is $R = R_1 \times R_2 \times \cdots \times R_n$ where $R_1, R_2, \ldots, R_n$ are the reliabilities of the components. In the special case where the components are characterised by constant hazard rates: $\lambda_1, \lambda_2, \ldots, \lambda_n$, the reliabilities of the components are $R_1 = \exp(-\lambda_1 t), \ldots, R_n = \exp(-\lambda_n t)$. Consequently, the time to failure distribution of a series arrangement becomes the negative exponential distribution

$$F(t) = 1 - \exp[-(\lambda_1 + \lambda_2 + \cdots + \lambda_n)t] \qquad (2.16)$$

The failure times follow a homogeneous Poisson process with intensity $\lambda = \sum_{i=1}^{n} \lambda_i$. This additive property is the basis for the widely used *parts count method* for predicting system reliability (Bazovsky, 1961; MIL-HDBK-217F, 1991). The method however is suitable only for systems including independently working components, logically arranged in series, characterised by constant hazard rates.

If the components are not logically arranged in series, the system's rate of occurrence of failures $\lambda = \sum_{i=1}^{n} \lambda_i$ calculated on the basis of the parts count method is an upper bound of the real rate of occurrence of failures. One downside of this approach is that the reliability predictions are too conservative. Another downside is that the constant hazard rate assumption is not valid for components whose failure is caused by damage accumulation or by any other type of deterioration.

2.8 REDUNDANCY

As production systems become more complex, their analysis becomes increasingly difficult. Complexity increases the risks of both random component failures and *design-related* failures. Incorporating redundancy in the systems is particularly effective where random failures predominate. *Redundancy* is a technique whereby one or more components in a system are replicated in order to increase reliability (Blischke and Murthy, 2000). Since a design fault would usually be common to redundant components, design-related failures cannot be reduced in the same way. While for an *active redundancy* no switching is required to make the alternative component available, *passive (standby) redundancy* requires a switching operation to make the redundant component available.

2.8.1 Active Redundancy

Active redundancy is present if all redundant components are in operation and share the load with the main unit from the time the system is put in operation. It must be pointed out that the fact that the components are logically arranged in parallel does not necessarily mean that they are connected in parallel physically. A typical example is a two-engine aircraft capable of flying on one engine only. The converse is also true. Components connected physically in parallel, may not necessarily be logically arranged in parallel. Parallel pipelines transporting toxic chemicals are a good example. Accident/failure associated with a release of toxic substance occurs whenever at least one of the pipelines looses containment.

2.8.1.1 *Full Active Redundancy*: Full active redundancy is present where the assembly is operational if at least one of the units is operational. Suppose that the reliabilities of the separate components are $r_1, \ldots, r_n$. Since the system fails only when all of the components fail, in the special case where the component failures are statistically independent, the probability of failure of the system (Fig. 2.6) is

$$p_f = (1 - r_1) \times \cdots \times (1 - r_n)$$

and its reliability is

$$R = 1 - (1 - r_1) \times (1 - r_2) \times \cdots \times (1 - r_n) \qquad (2.17)$$

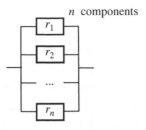

Figure 2.6 A system with n components logically arranged in parallel.

The reliability of a full active redundancy is greater than the reliability of its most reliable component. Suppose that the components have reliabilities greater than zero ($r_i > 0$). Since $1 - r_i < 1$ for all $i = 1, 2, \ldots, n$, with increasing the number of components, the probability of failure p_f tends to zero as a product of large number of terms smaller than unity and the system reliability R in equation (2.17) can be made arbitrarily close to unity.

2.8.1.2 *Partial Active Redundancy (k-out-of-n Redundancy)*: An active redundant system of n units which works if and only if at least k out of the n units work is called a *partially redundant system* or *k-out-of-n* system. Suppose that all units in the system are identical and characterised by the same reliability r. The probability of obtaining a number of successes greater than or equal to k is

$$P(X \geq k) = \sum_{x=k}^{n} \frac{n!}{x!(n-x)!} r^x (1-r)^{n-x} \qquad (2.18)$$

Equation (2.18) is in fact the sum of the probabilities of the following mutually exclusive events: (i) exactly k units are operating, (ii) exactly $k+1$ units are operating,..., exactly n units are operating. The probability that exactly x units will be operating is given by the binomial distribution $P(X = x) = [n!/(x!(n-x)!)]r^x(1-r)^{n-x}$.

The assumption that the failures of components are statistically independent is not always true. Usually the failure of one component causes redistribution of the load on the rest of the working components and, as a result, the observed reliability is lower than the reliability predicted by assuming statistically independent components. Another limitation is that the incremental gain in reliability with the addition of replicate components decreases significantly beyond a certain point (Blischke and Murthy, 2000).

2.8.1.3 *Consecutive k-out-of-n Redundancy* There are systems that are not considered failed until at least k consecutive components have failed. An example of a consecutive k out of n system can be given with a telecommunication system consisting of relay stations. A signal emitted from the first station is received by k other consecutive stations, but not by the $k + 1$-st. A signal emitted from the second station is received by k other consecutive stations and so on until the signal reaches the last station. Failure occurs only if k consecutively numbered stations fail to emit a signal (Chiang and Niu, 1981).

An algorithm determining the probability of failure of a consecutive *k-out-of-n* system is presented in Appendix 2.1.

Using the algorithm, the probability of having four consecutive failures out of 30 components has been determined, given that the probability of failure of a single component is $p = 0.3$. The probability of 0.147 determined by using the algorithm in Appendix 2.1 has been confirmed by a direct Monte Carlo simulation.

Another related application of interest is present if a supply system can handle only a limited number of k demands in consecutive discrete time intervals (e.g. seconds, minutes, hours, days) without overloading and failure. The total number of such intervals is n. After k consecutive demands, a single discrete time interval without a demand is necessary for the system to recover and be ready for another series of consecutive demands. Each discrete time interval (e.g. an hour or a day) is characterised by a constant probability that there will be a demand during its duration. The probability of a reliable operation is then equal to the probability that there will be no more than k consecutive demands during all n time intervals.

2.8.2 Passive (Standby) Redundancy

In cases of passive redundancy, the redundant components do not share any load with the operating component. The redundant components are put in use one at a time after failure of the currently operating component and the remaining components are kept in reserve. If the operating component fails, one of the components on standby is put into use through switching (Fig. 2.7).

Standby components do not operate until they are sequentially switched in. Thus, the second standby component does not operate until the first component fails. Then the second standby component is automatically switched in by the switch S. The system does not fail until the n-th standby component

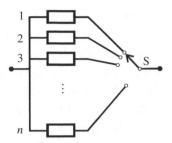

Figure 2.7 A passive (standby) redundancy.

fails. In contrast to an active redundant system based on n components operating in parallel, the components in the standby system operate one after
another. The switch used to disconnect a failed component and connect a
working component can be either perfect (100% reliable) or imperfect (less
than 100% reliable).

A *cold standby redundancy* (Fig. 2.7) is composed of standby components
which have zero hazard rates ($\lambda_{cold} = 0$) in standby mode and non-zero
hazard rates when they are in operation. This distinguishes the cold standby
redundancy from the *hot standby redundancy* where a component in standby
has the same hazard rate as in operation ($\lambda_{hot} = \lambda_{op}$) and from the warm
standby system where a component in standby mode is characterised by a
smaller but non-zero hazard rate compared to the component in operation
($0 < \lambda_{warm} < \lambda_{op}$). Components in warm standby tend to deteriorate and fail
before they are put in use. They are partially energised as opposed to being
fully energised in hot standby and not energised in cold standby. This is
reflected by their hazard rate which is between zero and the hazard rate of
a hot standby component. For cold standby components, an assumption is
often made that the standby component practically does not deteriorate until
it is switched in. The component is considered *as good as new* irrespective of
the length of time it has spent on standby. In some practical applications, this
assumption may not hold, especially where some form of environmentally
induced deterioration (e.g. corrosion, stress corrosion cracking, hydrogen
embrittlement) is involved.

The switch can have two basic failure modes, depending on whether the
system operates or fails given that the currently switched in component is
operating. For the first failure mode, the system fails immediately if the
switch fails. In other words, the switch fails open. The switch can also fail
closed. If this failure mode is activated, the system does not fail immediately.
It continues its operation until the currently working standby component

fails and there is a need for a changeover. A good example illustrating these failure modes is the ordinary multi-pole electrical switch. The first failure mode is present if a mechanical failure destroys the conduction of current through the switch (the switch fails open). The second failure mode is present if, for example, the switch is jammed in one position (e.g. welded) and is incapable to make a changeover. In this case, the switch fails closed. This type of switch failure will be illustrated by a section from a hydraulic liquid supply system from two lines (Fig. 2.8). The hydraulic fluid in each line passes through the filters F1 and F2, the control valves CV1 and CV2 and flows into the *shuttle valve* SV. The shuttle valve always connects the line with the higher pressure with the output line. The idea is to provide hydraulic fluid in the output line even if pressure in one of the input lines drops because of failure.

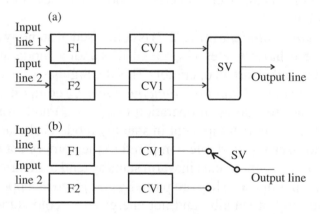

Figure 2.8 Functional (a) and reliability block diagram (b) of a shuttle valve.

The system in Fig. 2.8(a) is essentially a warm standby system where the shuttle valve can be modelled as imperfect switch. Suppose that while the first hydraulic supply line is switched in and is in operation, the shuttle valve is jammed with debris and can no longer perform switching of the second input line. The hydraulic supply however will still be present as long as the current input line is operating. The system will fail when, for example, a failure in the current input line causes the pressure to drop. Then, it would be impossible to switch in the second input line because of jamming of the valve.

Standby systems are not limited to systems where all standby units are identical. For example, an electrical device can have a hydraulic backup device.

2.9 BUILDING RELIABILITY NETWORKS

Let us start with an example where the same message is transmitted from two identical sources $a1$ and $a2$ to two identical receivers $b1$ and $b2$ (Fig. 2.9). The message can be transmitted directly from $a1$ to $b1$ or from $a2$ to $b2$, but cannot be sent directly from $a1$ to $b2$ or from $a2$ to $b1$. Instead, the message must be sent first to the transmitter t, as shown in Fig. 2.9.

The logical arrangement of the components in this system can be represented by the *reliability network* in Fig. 2.10.

The operation logic of this system can be modelled by a set of nodes (the filled circles in Fig. 2.10, numbered from 1 to 4) and components ($a1$, $a2$, $b1$, $b2$ and t) connecting them. The system works (the message is transmitted) only if there exists a path through working components between the start node '1' and the end node '4' (Fig. 2.10).

Reliability networks can be modelled conveniently by graphs. The nodes are the *vertices* and the components that connect the nodes are the *edges* of the graph. Each component is connected to exactly two nodes. If nodes i and j are connected with an edge (component), the nodes are said to be *adjacent*.

Now, suppose that in order to increase the reliability of the system in Fig. 2.9, a redundancy has been included for each source and receiver. The corresponding reliability network is given in Fig. 2.11.

Thus, between any two adjacent nodes, there may be more than one component and the corresponding edges in the graph are called *parallel*

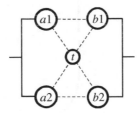

Figure 2.9 A functional diagram of the message transmitting system.

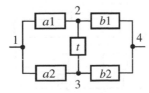

Figure 2.10 A reliability network of the message transmitting system in Fig. 2.9 with no redundancies for the sources and the receivers. The nodes have been marked by black circles.

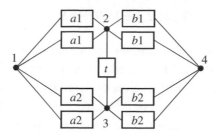

Figure 2.11 A reliability network of the message transmitting system with redundancies for the sources and the receivers. The nodes have been marked by black circles.

edges. Since each edge of the graph (component) is associated with a non-ordered pair of vertices (nodes), the graph is *undirected.*

For production systems based on multiple production units, to each production unit corresponds a single node (see nodes 12–19 in Fig. 7.1). These nodes will subsequently be referred to as 'production nodes'.

Often, reliability networks show no resemblance to the initial system. Such is for example the system in Fig. 2.12. It represents a reservoir from which working fluid is distributed through two pipelines. On each pipeline, there are two valves operated by actuators. The actuators are in turn controlled by a control module (CM). The whole system is powered by a power

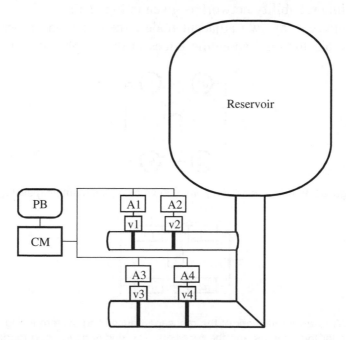

Figure 2.12 A functional diagram of a reservoir distributing working liquid into two pipelines.

Figure 2.13 Reliability block diagram related to the delivery of working fluid through both pipelines.

block (PB). One of the functions of the system in Fig. 2.12 is to open all valves on command in order to allow the process fluid to flow through both pipelines. A system failure is present if at least one of the valves remains closed on command. This means that the flow will be blocked and fluid will not be delivered through one or both pipelines.

A system failure is present if at least one of the valves remains closed on command. This means that the flow will be blocked and fluid will not be delivered through one or both pipelines. As can be verified, a system failure is present if at least one of the devices: the power block (PB), the control module (CM), any of the actuators or any of the valves fails to operate. Consequently, with respect to delivering working fluid in both pipelines, all components are logically arranged in series (Fig. 2.13).

Now let us explore the other basic function of the system in Fig. 2.12 – 'to contain the fluid in the reservoir'. Failure to contain the fluid in the reservoir occurs only if both of the valves in one or both pipelines fail to close. The reliability network corresponding to containing the fluid in the reservoir is represented in Fig. 2.14.

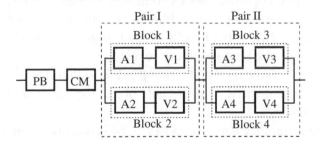

Figure 2.14 Reliability network, related to containing the fluid in the reservoir.

It was built by using the top-down approach. The system is divided into several large blocks, logically arranged in a particular manner. Next, each block is further developed into several smaller blocks. These blocks are in turn developed and so on, until the desired level of indenture is achieved for all of the blocks.

Failure to isolate the fluid is considered at the highest indenture level: at the level of pairs of valve blocks. (A valve block consists of a valve and an actuator.) On each pipeline, there is one such pair of valve blocks

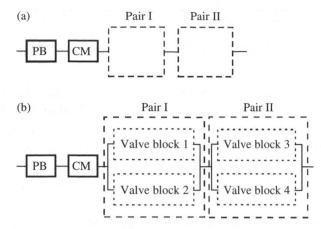

Figure 2.15 (a) First stage and (b) second stage of detailing the reliability network from Fig. 2.13.

(Fig. 2.12). Since failure to contain the fluid in the reservoir occurs if either the first pair of valve blocks denoted by Pair I in Fig. 2.14 or the second pair of valve blocks denoted by Pair II fails to contain the fluid, the pairs of valve blocks are arranged logically in series (Fig. 2.15(a)).

Failure to isolate the fluid also occurs if the power block (PB) or the control module (CM) fails. Consequently, they are arranged logically in series with the pairs of valve blocks. Within each pair of valve blocks, only the operation of one of the valve blocks is sufficient to isolate the fluid. Consequently, the two valve blocks in each pair are logically arranged in parallel and this is the next indenture level.

A valve block fails to isolate the fluid if either the actuator or the valve fails. Consequently, within each block the valve and the actuator are arranged in series.

Finally, the reliability network in Fig. 2.14 is obtained and this is the last level of indenture. Deeper levels of indenture are also possible, for example, within the actuator and the valves.

2.10 TYPE OF COMPONENTS IN A RELIABILITY NETWORK

Some of the basic type of components which are often present in a reliability network are listed below. The first, most common type of component will be referred to as type A component. It is connected to the network by two nodes (nodes '1' and '2' in Fig. 2.16). This is a single active component connected all the time to nodes 1 and 2 and is described by its cumulative distribution

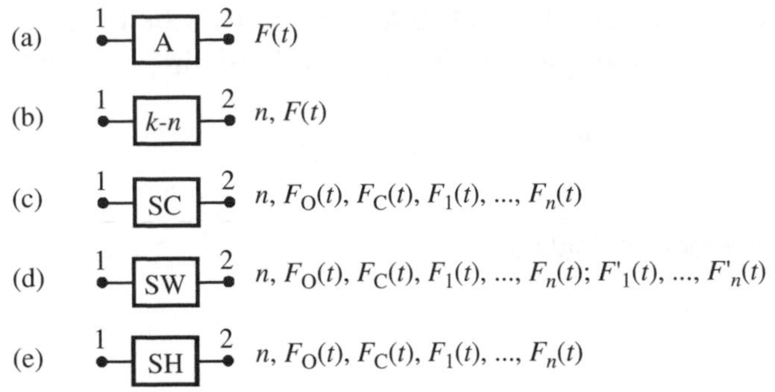

(a) $\overset{1}{\bullet}\!-\!\boxed{\text{A}}\!-\!\overset{2}{\bullet}$ $F(t)$

(b) $\overset{1}{\bullet}\!-\!\boxed{k\text{-}n}\!-\!\overset{2}{\bullet}$ $n, F(t)$

(c) $\overset{1}{\bullet}\!-\!\boxed{\text{SC}}\!-\!\overset{2}{\bullet}$ $n, F_O(t), F_C(t), F_1(t), ..., F_n(t)$

(d) $\overset{1}{\bullet}\!-\!\boxed{\text{SW}}\!-\!\overset{2}{\bullet}$ $n, F_O(t), F_C(t), F_1(t), ..., F_n(t); F'_1(t), ..., F'_n(t)$

(e) $\overset{1}{\bullet}\!-\!\boxed{\text{SH}}\!-\!\overset{2}{\bullet}$ $n, F_O(t), F_C(t), F_1(t), ..., F_n(t)$

Figure 2.16 Basic type of components building a reliability network.

of the time to failure. The second type of component is a *k-out-of-n* component which consists of n identical components working in parallel. The *k-n* component works only if at least k out of the n components work. Its description involves specifying the cumulative distribution of the time to failure $F(t)$ of a single component and the total number of components n.

The SC component in Fig. 2.16(c) is a cold standby component. Its description requires specifying the cumulative distribution of the time to failure $F_O(t)$ and $F_C(t)$ for the basic failure modes of the switch (failed open and failed closed), and the number n of switched in standby components characterised by n cumulative distributions of the times to failure. The SW component in Fig. 2.16(d) is a warm standby component. While the life of a cold standby component starts to be consumed only after the component has been switched in, for a warm standby component, its life starts to be consumed immediately after installing it (this also applies to the hot standby component SH in Fig. 2.16(e)). The warm standby component requires as input data the times to failure distributions $F_i(t)$ of the n standby components in a state of operation and their time to failure distributions $F'_i(t)$ in a state of standby. For the hot standby components, only a single set of times to failure is required, because there is no difference between the distributions of the times to failure for the components in operation and the components in standby.

2.11   PSEUDO-CODE CONVENTIONS USED IN THE ALGORITHMS FOR RISK-BASED RELIABILITY ANALYSIS

In describing the algorithms presented in the next Chapters, a number of conventions are used. Thus, the statements in braces {*Statement 1*; *Statement 2*; *Statement 3*;...} separated by semicolons are executed as a single block. In the conditional statement below, the block of statements in the braces is executed only if the specified condition is true:

If (Condition) **then** {*Statement 1*; . . . ;*Statement n*;}

The construct:
For $i = 1$ **to** Number_of_trials **do**
   {
    ...
   }

is a loop with a control variable i, accepting successive values from one to the total number of trials (Number_of_trials).

The loop executes the block of statements in the braces Number_of_trials number of times. If a statement **break** is encountered in the body of the loop, the execution of statements continues with the next statement immediately after the loop (*Statement n+1* in the next example) thereby skipping all statements between the statement '**break**' and the end of the loop:

For $i = 1$ **to** Number_of_trials **do**
   {
    Statement 1;
    ...
    break;
    ...
    Statement *n-1*;
    Statement *n*;
   }

    Statement *n+1*;

The construct:
While (Condition) **do** {*Statement 1*;...;*Statement n*;}

is a loop which executes the block of statements repeatedly as long as the specified condition is true. If the variable Condition is false before entering the loop, the block of statements is not executed at all. A similar construct is the loop

repeat
   Statement 1;
   ...
   Statement n;
until (Condition);

which repeats the execution of all statements between **repeat** and **until** in its body until the specified condition becomes true.

A *procedure* is a self-contained section designed to perform a certain task. The procedure is called by including its name ('*proc*' in the next example) in other parts of the algorithm.

procedure *proc*()
{
  Statement 1;
  ...
  Statement n;
}

A *function* is also a self-contained section which returns value and which is called by including its name in other parts of the algorithm. Before returning to the point of the function call, a particular value *p* is assigned to the function name ('*fn*' in the next example) with the statement **return**.

function *fn*()
{
  Statement 1;
  ...
  Statement n;
return *p*;
}

Text in italic between the symbols '/*' and '*/' or after the symbol '//' is comments.

APPENDIX 2.1

Determining the probability of failure of *k-out-of-n* consecutive system.

If p is the probability of failure of a single component, the probability of failure of a k-out-of-n consecutive system can be determined from the following probabilistic argument.

The probability of failure $F(n)$ of k consecutive components out of n components is equal to the sum $F(n) = S(n - k + 1) = s_1 + s_2 + \cdots + s_{n-k+1}$ of the probabilities that the series of k consecutive failures will start at the first, at the second, . . ., or at the $n - k + 1$-st component. For $n - k + 1 < i \leq n$, the probabilities s_i are zero ($s_i = 0$) because a consecutive series of k failed components cannot fit into a space containing fewer than k components. The probability $s_1 = p^k$ because all of the first k outcomes must be failures in order for the consecutive failure sequence to start from the first component; $s_2 = (1-p)p^k$ because in order for the consecutive sequence to start from the second component, the first outcome must not be failure; $s_3 = (1-p)p^k P_0(3-2)$ because in order for the consecutive sequence to start from the third component, the second outcome must not be a failure and there must not be a consecutive sequence of k failures within the first $3-2$ components (the probability of this event has been denoted by $P_0(3-2)$); $s_k = (1-p)p^k P_0(k-2)$ because in order for the consecutive sequence to start from the k-th component, the k-1-st outcome must not be a failure and there must not be a consecutive sequence of k failures within the first k-2 components, the probability of which is denoted by $P_0(k-2)$. If $i \le k-1$ then $P_0(1) = 1, \ldots, P_0(k-1) = 1$ is fulfilled for the probabilities $P_0(i)$ because there can be no sequence of k consecutive failures within a space of fewer than k components. If $i \ge k$, $P_0(i) = 1 - F(i)$, where $F(i)$ is the probability that there will be a sequence of k consecutive failures out of i components.

Since $F(i) = S(i-k+1) = s_1 + s_2 + \cdots + s_{i-k+1}$, the algorithm for determining the probability $F(n)$ reduces to determining the sum $F(n) = S(n-k+1)$. The first $k+1$ partial sums S are initialised as follows:

$$S[1] = p^k, \quad \text{for } i = 2, k+1 \quad S[i] = S[i-1] + (1-p)p^k \qquad (2.19)$$

Partial sums $S[i]$ greater than $k+2$ are determined as follows:

$$S[i] = S[i-1] + (1-p)p^k S[i-2-k+1] \qquad (2.20)$$

Since $S[i-2-k+1]$ has already been determined, it is simply substituted in (2.19) to determine $S[i]$. The process of determining $S[i]$ continues until i becomes equal to $n-k+1$. Then $F(n)$ is simply equal to $S[n-k+1]$.

3

METHODS FOR ANALYSIS OF COMPLEX RELIABILITY NETWORKS

3.1 NETWORK REDUCTION METHOD FOR RELIABILITY ANALYSIS OF COMPLEX SYSTEMS AND ITS LIMITATIONS

In the reliability literature, there exist a number of methods for system reliability analysis oriented mainly towards systems with simple topology. Such are for example the method of *network reduction* and the *event-tree method* (Billinton and Allan, 1992). The essence of the *network reduction method* for example is reducing the entire system to a single equivalent element, by systematically combining appropriate series and parallel branches of the reliability network. At the end, the reliability of the remaining equivalent element equals the reliability of the initial system.

Suppose that all components in a system are specified with their most fundamental characteristic – the cumulative distribution of the time to failure. For a system containing M components, $F_i(t)$ $i = 1, \ldots, M$ distributions are specified. For any specified time interval with length a, the reliabilities $R_i = 1 - F_i(a)$ of the separate components related to the specified time interval can be determined. These reliabilities are subsequently used to calculate the reliability of the system associated with the time interval with length a.

Given the reliabilities of the separate components, the reliability network from Fig. 2.14, for example, with components logically arranged in series and parallel, can be reduced in complexity to a simple series system, in stages, as shown in Fig. 3.1.

In the first stage, the components in series with reliabilities R_3, R_4, R_5, R_6, R_7, R_8, R_9 and R_{10} are reduced to four components with reliabilities $R_{34} = R_3 R_4$, $R_{56} = R_5 R_6$, $R_{78} = R_7 R_8$ and $R_{910} = R_9 R_{10}$ (Fig. 3.1(b)).

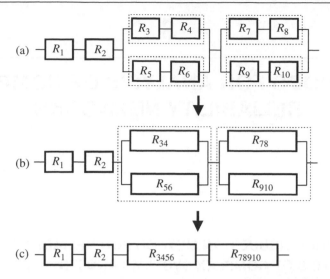

(a)

(b)

(c)

Figure 3.1 Determining the reliability of a system including components logically arranged in series and parallel.

Next, the components in parallel, with reliabilities R_{34} and R_{56} are reduced to a component with reliability $R_{3456} = 1 - (1 - R_{34})(1 - R_{56})$ and the components in parallel, with reliabilities R_{78} and R_{910} are reduced to a component with reliability $R_{78910} = 1 - (1 - R_{78})(1 - R_{910})$ (Fig. 3.1(c)).

Next, the remaining components are reduced to a single component with reliability $R_{12345678910} = R_1 \times R_2 \times R_{3456} \times R_{78910}$.

The main drawback of the network reduction method is its limited application area – only to networks with relatively simple (usually series–parallel) topology. Systems like the one in Fig. 3.3, for example, cannot be handled by this method. Many production systems however, such as the dual-control production system in Fig. 7.2, do not have a simple series–parallel topology and cannot be handled by this method. The decomposition method described next, avoids this limitation.

3.2 DECOMPOSITION METHOD FOR RELIABILITY ANALYSIS OF SYSTEMS WITH COMPLEX TOPOLOGY AND ITS LIMITATIONS

The decomposition method is based on conditioning a complex system on the state of a key component K_1. As can be verified from the Venn diagram in Fig. 3.2, the event S denoting the system's success (system is working) can be presented as the union of two mutually exclusive events: (i) $K_1 \cap S$

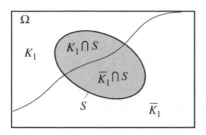

Figure 3.2 The event S (system is working) is the union of two mutually exclusive events: $K_1 \cap S$ and $\overline{K}_1 \cap S$.

the key component will not fail and the system will be working and (ii) $\overline{K}_1 \cap S$ the key component will fail and the system will be working.

The idea of this method is to decompose the initial system into two systems $K_1 \cap S$ and $\overline{K}_1 \cap S$ with simpler topology. According to the total probability theorem (discussed in all introductory books on probability and statistics), the probability $P(S)$ of event S that the system will be working, can be presented as a sum of the probabilities of the two mutually exclusive events:

$$P(S) = P(S \cap K_1) + P(S \cap \overline{K}_1) \tag{3.1}$$

Since $P(S \cap K_1) = P(S|K_1)P(K_1)$ and $P(S \cap \overline{K}_1) = P(S|\overline{K}_1)P(\overline{K}_1)$, finally:

$$P(S) = P(S|K_1)P(K_1) + P(S|\overline{K}_1)P(\overline{K}_1) \tag{3.2}$$

In equation (3.2), $P(S|K_1)$ is the probability that the system will not fail given that the key component will not fail and $P(S|\overline{K}_1)$ is the probability that the system will not fail given that the key component will fail. $P(K_1)$ and $P(\overline{K}_1)$ are the probabilities that the key component will not fail and will fail, correspondingly.

If the probability of success for any of the simpler systems is difficult to calculate, another decomposition can be made by selecting another key component K_2 and so on, until trivial systems are obtained whose reliability can be evaluated easily.

Consider the topologically complex system S in Fig. 3.3 which consists of six identical components with reliabilities r. The system works if a path between the start node '1' and the end node '4' exists.

This system is not trivial but it can be simplified if a key element K_1 is selected. The probability of system success $P(S)$ (the reliability of the system) can be determined from equation (3.2). Since the probability

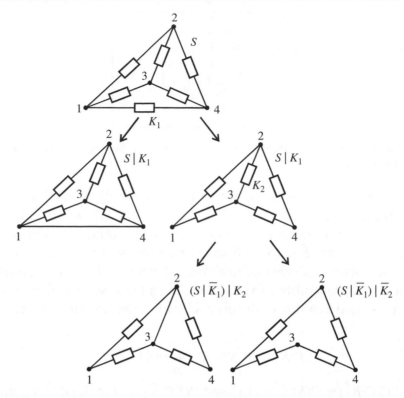

Figure 3.3 Determining the reliability of a topologically complex system by using the decomposition method.

$P(S|K_1)=1$, $P(K_1)=r$ and $P(\overline{K}_1)=1-r$, the reliability of the system becomes

$$P(S) = r + P(S|\overline{K}_1) \times (1-r) \tag{3.3}$$

One of the obtained systems from the decomposition, with reliability $P(S|\overline{K}_1)$, is not trivial. By selecting another key component K_2 however, it can be decomposed into two trivial systems whose reliabilities are easily determined (Fig. 3.3).

Consequently

$$P(S|\overline{K}_1) = P[(S|\overline{K}_1)|K_2] \times P(K_2) + P[(S|\overline{K}_1)|\overline{K}_2] \times P(\overline{K}_2) \tag{3.4}$$

Since $P[(S|\overline{K}_1)|K_2]=[1-(1-r)^2]^2$, $P[(S|\overline{K}_1)|\overline{K}_2]=1-(1-r^2)^2$, $P(K_2)=r$ and $P(\overline{K}_2)=1-r$, substituting in equation (3.4) yields

$$P(S|\overline{K}_1) = 2r^2 + 2r^3 - 5r^4 + 2r^5$$

Substituting this expression in equation (3.2) finally yields

$$P(S) = r + 2r^2 - 7r^4 + 7r^5 - 2r^6 \tag{3.5}$$

This method, however, also has significant limitations. It is not suitable for large systems. Each selection of a key component splits a large system into two systems each of which is in turn split into two new systems and so on. For a large number n of components in the initial system, the number of product systems generated from the selection of the key components quickly increases and becomes unmanageable.

3.3 METHODS FOR SYSTEM RELIABILITY ANALYSIS BASED ON MINIMUM PATH SETS AND CUT SETS AND THEIR LIMITATIONS

One of the most important methods for determining the reliability of complex networks are based on minimal paths and cut sets (Hoyland and Rausand, 1994).

A path is *a set of components which, when working, connect the start node with the end node through working components thereby guaranteeing that the system is in working state*. A minimal path is *a path from which no component can be removed without disconnecting the link it creates between the start and the end node*. Consequently, minimal paths are free of loops. In other words, in each minimal path a particular node may appear only once.

A cut set is *a set of components which, when failed, disconnect the start node from the end node and the system is in a failed state*. A minimal cut set is *a cut set for which no component can be returned in working state without creating a path between the start and the end node, thereby returning the system into a working state*.

The network in Fig. 3.4 illustrates the use of minimal paths and cut sets for estimating the system reliability. The system will function if at least a single minimal path through working components exists, connecting the start node '1' and the end node '4'. In other words, the system will function if at least one of the minimal paths (A, C) or (B, D) or (A, E, D) or (B, E, C) exists. The system's success (the system is working) is the compound event

$$S = (A \cap C) \cup (B \cap D) \cup (A \cap E \cap D) \cup (B \cap E \cap C)$$

where for example the intersection $A \cap C$ means that both components A and C are working. As can be verified, the compound event 'system is working' is the union of all minimal paths.

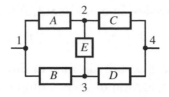

Figure 3.4 A bridge-type reliability network.

Assume that the separate components work independently from one another and let us denote the reliabilities of the components simply by a, b, c, d and e. The union $\bigcup_{i=1}^{n} A_i = A_1 \cup A_2 \cup \cdots \cup A_n$ of n events is the event which contains all outcomes x belonging to *at least one* of the events A_i (Fig. 3.5):

$$\bigcup_{i=1}^{n} A_i = A_1 \cup A_2 \cup \cdots \cup A_n = \{x | x \in A_1 \text{ or } x \in A_2 \text{ or } \cdots x \in A_n\}$$

The probability of the union is determined from:

$$P\left(\bigcup_{i=1}^{n} A_i\right) = \sum_{i=1}^{n} P(A_i) - \sum \sum_{i<j} P(A_i \cap A_j) +$$

$$\sum \sum_{i<j<k} \sum P(A_i \cap A_j \cap A_k) - \cdots + (-1)^{n+1} P(A_1 \cap A_2 \cap \cdots \cap A_n)$$

$$(3.6)$$

This expression is also known as *the inclusion–exclusion expansion*.

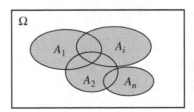

Figure 3.5 A union of n non-disjoint events.

According to the inclusion–exclusion expansion formula, the probabilities of all single events are first added. However, this means that the probability of the intersections of any pair of events has been added twice and should therefore be subtracted. Next, the probabilities of all intersections of two events are subtracted from the previous result. During these operations however, the contribution of the intersection of any three events has been added 3 times by adding the probabilities of the single events and subsequently subtracted 3 times by subtracting the probabilities of the twofold intersections. Consequently, the probabilities of all threefold intersections must be added. In short, terms with odd number of intersections are added while terms with even number of intersections are subtracted.

For n statistically independent components, the probability that at least one component will be working can be obtained by subtracting from unity the probability of the complementary event: none of the components will be working. As a result:

$$P\left(\bigcup_{i=1}^{n} A_i\right) = 1 - [1 - P(A_1)] \times [1 - P(A_2)] \times \cdots \times [1 - P(A_n)]$$

which, after expanding leads to equation (3.6).

Applying the inclusion–exclusion expansion and the *idempotent* law $A \cap A = A$ related to events, results in

$$R = P(S) = ac + bd + aed + bec - abcd - aced - abec - abde - bdec + 2abcde$$

Assuming equal probabilities of failure $a = b = c = d = e = r$ for all components results in

$$R = 2r^2 + 2r^3 - 5r^4 + 2r^5 \qquad (3.7)$$

for the reliability of the system.

Alternatively, minimal cut sets can be used to determine the reliability of the network. The minimal cut sets are AB, CD, AED and BEC. The system will fail if at least a single cut set is present. Therefore, the system failure is described by the compound event

$$F = (\overline{A} \cap \overline{B}) \cup (\overline{C} \cap \overline{D}) \cup (\overline{A} \cap \overline{E} \cap \overline{D}) \cup (\overline{B} \cap \overline{E} \cap \overline{C})$$

where $\overline{A} \cap \overline{B}$, for example, means that both A and B are in a failed state. As can be verified, this compound event is the union of all minimal cut sets in the reliability network.

Suppose that the separate components work independently from one another and let us denote the probabilities of failure of the components simply by $\overline{a}, \overline{b}, \overline{c}, \overline{d}$ and $\overline{e}$.

The expression regarding the probability of an union of statistically independent events can then be determined from:

$$P(F) = \overline{a}\overline{b} + \overline{c}\overline{d} + \overline{a}\,\overline{e}\overline{d} + \overline{b}\overline{e}\,\overline{c} - \overline{a}\overline{b}\overline{c}\overline{d} - \overline{a}\overline{b}\,\overline{e}\overline{d} - \overline{a}\overline{b}\overline{e}\,\overline{c} - \overline{a}\overline{c}\,\overline{e}\overline{d} - \overline{b}\overline{c}\,\overline{e}\overline{d} + 2\overline{a}\overline{b}\overline{c}\,\overline{d}\,\overline{e}$$

If the probabilities of failure of all components are the same $(\overline{a} = \overline{b} = \overline{c} = \overline{d} = \overline{e} = \overline{r})$, the probability of failure p_f becomes

$$p_f \equiv P(F) = 2\overline{r}^2 + 2\overline{r}^3 - 5\overline{r}^4 + 2\overline{r}^5$$

from which, the reliability of the system is determined from

$$R = 1 - [2\overline{r}^2 + 2\overline{r}^3 - 5\overline{r}^4 + 2\overline{r}^5] \tag{3.8}$$

Since between the probability of failure $\overline{r}$ of a component and the reliability r, the relationship $\overline{r} = 1 - r$ holds, substituting it in equation (3.8) yields equation (3.7).

3.4 MONTE CARLO SIMULATION ALGORITHMS FOR SYSTEM RELIABILITY ANALYSIS BASED ON TESTING MINIMAL PATHS OR MINIMAL CUT SETS

Methods for system reliability analysis based on all listed minimal paths or cut sets can be created by using dynamic arrays where all paths or cut sets are stored. Suppose that a dynamic array $A[i]$, contains the i-th minimal path or cut set. $A[i][0]$ holds the number of components in the i-th minimal path or cut set; $A[i][j]$ holds the index of the j-th component from the i-th minimal path or cut set.

3.4.1 A System Reliability Analysis Algorithm Based on Testing Minimal Paths

Suppose that the number of components is stored in the variable *Number_of_components* and the number of minimal paths is stored in the variable *Number_of_paths*. All minimal paths are stored in dynamic arrays $A[i]$ with number equal to the number of paths, $i = 1, 2, \ldots$, *Number_of_paths*.

The elements of the array *Failed[]* contain the state of the components in the network. If the component with index j is in a failed state, the corresponding element from the array *Failed* is unity (*Failed[j]=1*) otherwise, the corresponding element is zero (*Failed[j]=0*).

A main feature of the algorithm given below are the two nested loops with control variables k and m. The loop with control variable k goes through all minimal paths in the system while the loop with control variable m goes through all components in each minimal path. If in the loop m, it is discovered that a component from the examined path has failed then this path is marked 'blocked' and the loop m is exited immediately with a statement **break**. After it has been established that the path is blocked, there is no need for further checks. If the loop m has been passed without executing a statement **break** (which means that a path through working components connecting the start and the end node does exist), the outer loop with index k is exited immediately with a statement **break**. Indeed, after it has been established that the system is working (a minimal path from the start to the end node exists), there is no need for further checks. Each time it has been established that a path exists, the success counter $s\_counter$ is incremented. Reliability is estimated by dividing the content of the $s\_counter$ to the number of simulation trials.

Algorithm 3.1

```
function generate_time_to_failure( i )
{
  // Returns a random time to failure for the i-th component.
}
s_counter = 0;
a = Number of years; /* Specifies the finite time interval regarding which the
                        reliability is calculated */
For t = 1 to Number_of_trials do
  {
  /* Generate the times to failure of all components */
  For i=1 to Number_of_components do
        {
        time_to_failure = generate_time_to_failure( i );
        if (time_to_failure > a) then  Failed[i]=0;
        else  Failed[i]=1;
        }
  For  k=1 to  Number_of_paths do
  {
    path=1;
    For  m=1 to  A[k][0] do   {
                              tmp=A[k][m];
                              if(Failed[tmp]=1) {path=0; break; }
                              }
```

      if (path=1) **then** {s_counter = s_counter+1; **break;**}
  }
}
Reliability = s_counter / Number_of_trials.

3.4.2   A System Reliability Analysis Algorithm Based on Testing Minimal Cut Sets

A computer-based method for system reliability analysis based on testing minimal cut sets can be created by using dynamic arrays where all minimal cut sets are stored. The number of minimal cut sets is stored in the variable *Number_of_cut_sets*. All data structures are similar to the ones from the previous algorithm. The only difference is that now, the dynamic arrays $A[i], i = 1, 2, \ldots$, *Number_of_cut_sets* contain minimal cut sets instead of minimal paths.

Similar to the previous algorithm, the main feature of the algorithm based on cut sets are again two nested loops with control variables k and m. The loop with control variable k goes through all minimal cut sets of the system while the loop with control variable m goes through all components in each cut set. If inside loop m, it is discovered that a component from the cut set is working, then this is no longer a cut set, and there is no need to check other components from the same cut set. Hence, the loop m is exited immediately with the statement **break**.

After exiting the loop m, a check is performed whether the state of the variable path has remained *path = 0*. If this is the case the system is certainly in a failed state and no more cut sets need to be checked. Consequently, the failure counter *f_counter* is incremented and the k-loop is exited immediately with a statement **break**.

Reliability is obtained by subtracting from unity the ratio of the content of the *f_counter* and the number of simulation trials.

Algorithm 3.2

f_counter = 0;
a = Number of years; /* *Specifies the finite time interval for which the*
                              reliability is calculated */
For t = 1 to Number_of_trials **do**
   {
/* *Generate the times to failure of all components* */
   For i=1 to Number_of_components **do**
      { time_to_failure = **generate_time_to_failure(** i **);**

```
        if(time_to_failure > a) then  Failed[i]=0;
        else  Failed[i]=1;
  For  k=1 to  Number_of_cut_sets do
  {
   path=0;
   For  m=1 to  A[k][0] do  {
                            tmp=A[k][m];
                            if(Failed[tmp]=0) {path=1; break;}
                            }
    if (path=0) then { f_counter = f_counter+1; break;}
   }
}
```

Reliability = 1- f_counter / Number_of_trials.

3.5   DRAWBACKS OF THE METHODS FOR SYSTEM RELIABILITY ANALYSIS BASED ON MINIMUM PATH SETS AND MINIMUM CUT SETS

The main drawback of methods based on minimal paths and minimal cut sets is that the number of minimal paths or cut sets increases quickly with increasing the size of the system. For large systems, the increase of the number of paths and cut sets leads to a combinatorial explosion, which can be demonstrated by using the simple network in Fig. 3.6.

For the parallel–series network in Fig. 3.6, a minimal cut set is present if a single component fails in each of the N branches composed of M components arranged in series. Since each branch can fail in M different ways, the number of ways in which all parallel branches can fail is M^N. Even for a system containing only two components in a branch ($M = 2$), for $N = 50$ branches in parallel we already have 2^{50} different cut sets, which is

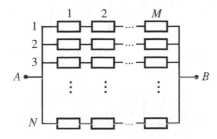

Figure 3.6 A simple network with M^N minimal cut sets.

a very large number. Not only the manipulation of such a large number of cut sets is impossible, but even their storage is a problem.

3.6 SYSTEM RELIABILITY ANALYSIS BASED ON FINDING PATHS THROUGH WORKING COMPONENTS IN RELIABILITY NETWORKS

3.6.1 Presenting the Topology of a Reliability Network by an Adjacency Matrix

A graph can be represented by drawing it. If a computer is used for analysing the graph however, a more formal representation is needed. A graph (reliability network) can be presented by an adjacency matrix. In order to represent the reliability network in Fig. 3.4 for example, a square matrix is constructed with number of rows and columns equal to the number of nodes. The rows and the columns are labelled with the ordered nodes. An element a_{ij} of the matrix is an integer number greater than zero if the row and the column vertices (nodes) are adjacent and zero otherwise. An integer number greater than zero specifies the number of edges between adjacent nodes i and j. The adjacent matrix is symmetric about the main diagonal ($a_{ij} = a_{ji}$) because if vertices i, j are adjacent, vertices j, i are also adjacent, with the same number $k > 0$ of parallel edges between them ($a_{ij} = a_{ji} = k > 0$). Alternatively, if vertices i, j are not adjacent, vertices j, i are not adjacent either ($a_{ij} = a_{ji} = 0$). For convenience, with respect to the algorithmic implementation, all diagonal elements are set to zero: $a_{ii} = 0$. While space can be saved by storing only half of this symmetric matrix, the algorithms are in fact, simpler with a full adjacency matrix. The adjacency matrix representation of the reliability network in Fig. 3.4 is

$$A \equiv \begin{Bmatrix} 0 & 1 & 1 & 0 \\ 1 & 0 & 1 & 1 \\ 1 & 1 & 0 & 1 \\ 0 & 1 & 1 & 0 \end{Bmatrix}$$

while the corresponding representation of the reliability network in Fig. 3.7 is

$$A \equiv \begin{Bmatrix} 0 & 2 & 2 & 0 \\ 2 & 0 & 1 & 2 \\ 2 & 1 & 0 & 2 \\ 0 & 2 & 2 & 0 \end{Bmatrix}$$

$$A \equiv \begin{Bmatrix} 0 & 2 & 2 & 0 \\ 2 & 0 & 1 & 2 \\ 2 & 1 & 0 & 2 \\ 0 & 2 & 2 & 0 \end{Bmatrix}$$

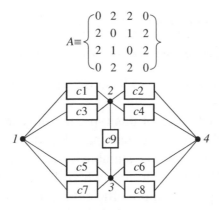

Figure 3.7 Reliability network with parallel edges.

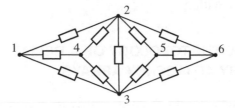

Figure 3.8 A dense reliability network.

The adjacency matrix representation is suitable for dense networks, characterised by a significant number of components. Such is for example the network in Fig. 3.8, whose adjacency matrix is

$$A \equiv \begin{Bmatrix} 0 & 1 & 1 & 1 & 0 & 0 \\ 1 & 0 & 1 & 1 & 1 & 1 \\ 1 & 1 & 0 & 1 & 1 & 1 \\ 1 & 1 & 1 & 0 & 0 & 0 \\ 0 & 1 & 1 & 0 & 0 & 1 \\ 0 & 1 & 1 & 0 & 1 & 0 \end{Bmatrix}$$

Sparse networks result in sparse adjacent matrices with too many zeros. Such is for example the adjacency matrix

$$A \equiv \begin{Bmatrix} 0 & 1 & 0 & 0 & 0 \\ 1 & 0 & 1 & 0 & 0 \\ 0 & 1 & 0 & 1 & 0 \\ 0 & 0 & 1 & 0 & 1 \\ 0 & 0 & 0 & 1 & 0 \end{Bmatrix}$$

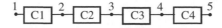

Figure 3.9 A system with components logically arranged in series.

for the simple network in Fig. 3.9 including four components logically arranged in series.

For all reliability networks of this type, the only non-zero elements are the elements from the two diagonals parallel to the main diagonal. The adjacency matrix representation is suitable for very dense networks, because the matrix requires V^2 bits of storage, where V is the number of nodes.

If the network is sparse, a more efficient representation is by adjacency lists. This representation is also suitable in the cases of dense reliability networks. In the adjacency list representation, for each node a list of all adjacent nodes is provided.

3.7 PRESENTING THE TOPOLOGY OF A RELIABILITY NETWORK BY ADJACENCY ARRAYS

The reliability network in Fig. 3.4 can also be represented by adjacency dynamic arrays which are essentially adjacency lists. The topology of the network in Fig. 3.4 is fully represented by an array *Node[4]* of four pointers (memory addresses) which correspond to the four nodes in the network. For each pointer *Node[i]*, an exact amount of memory is reserved to accommodate the indices (names) of all adjacent nodes to the i-th node. In this way, the topology of the network is described by four dynamic arrays where *Node[i]*, $i = 1, 4$ is the address of the i-th dynamic array. *Node[i][0]* is reserved for the number of nodes adjacent to the i-th node. Thus *Node[2][0] = 3*, because node 2 has exactly three adjacent nodes. The indices of the actual nodes adjacent to the i-th node are stored sequentially in *Node[i][1], Node[i][2],*

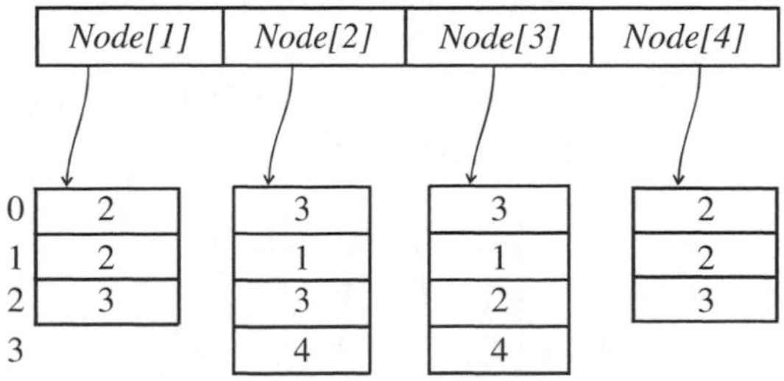

Node[1]	Node[2]	Node[3]	Node[4]

0	2	3	3	2
1	2	1	1	2
2	3	3	2	3
3		4	4	

Figure 3.10 Presentation of the topology of the reliability network from Fig. 2.10 by dynamic arrays (adjacency lists).

etc. For the second node for example, $Node[2][1] = 1$; $Node[2][2] = 3$ and $Node[2][3] = 4$, because the nodes with indices '1', '3' and '4' are adjacent to node '2'.

As can be verified, the dynamic array representation is very suitable for sparse networks. If a single node requires 2 bytes memory, the space required by the dynamic arrays is $2(V + 2E)$, where V is the number of nodes and E is the number of edges. For a sparse network, this storage space is considerably smaller than the space of $2 \times V^2$ bytes required by the adjacency matrix representation. This is because storage space is reserved only for the adjacent nodes and there are no zero elements in the arrays. Thus, for a system containing 100 components in series, the required storage space by the dynamic arrays representation is $2(101 + 2 \times 100) = 602$ bytes while the adjacency matrix representation requires $2 \times 100^2 = 20,000$ bytes.

In order to keep track of the parallel edges in the network, another set of dynamic arrays called *Link[4]* are created, whose elements correspond one-to-one to the elements of *Node[4]* dynamic array. Instead of indices of adjacent nodes, however, *Link[4]* arrays contain the number of parallel edges between adjacent nodes.

Link[1][1], for example, gives the number of parallel edges between node 1 and its neighbour, whose index is listed first in *Node[1]* dynamic array (the index is kept in *Node[1][1]*).

The structure of the *Link* arrays for the network in Fig. 3.7 is given in Fig. 3.11.

The *Node* arrays together with the link arrays describe fully the entire topology of the network. Linked lists (Horowitz and Sahni, 1997) can also be used to present the topology of a reliability network. However, the main features of the linked lists, such as the possibility to insert and delete an element without the need to move the rest of the elements are not used here, because there is no need for such operations. As a result, the proposed presentation based on adjacency arrays is sufficient, simpler and more efficient compared to a presentation based on linked lists.

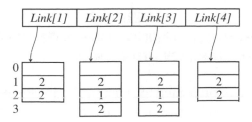

Figure 3.11 Structure of the *Link* array for the network in Fig. 3.7.

3.8   UPDATING THE ADJACENCY MATRIX AND THE ADJACENCY ARRAYS AFTER A COMPONENT FAILURE

The adjacency matrix or the adjacency dynamic arrays keep track of the system topology which changes dynamically, with the failures of the separate components. When components fail, some of the links between the nodes of the reliability network disappear and the adjacency matrix or the dynamic arrays needs to be updated. In case of a presentation based on adjacency matrix, a copy of the original adjacency matrix is kept in the memory. Each component is identified by its index and the two nodes to which it is connected. Suppose that component c_m is connected to nodes i and j.

Failure of component c_m is indicated by subtracting unity from elements a_{ij} and a_{ji} in the adjacency matrix. This reflects the circumstance that due to failure of component c_m, one of the links between nodes i and j disappears.

Similar updating is performed if the reliability network is represented by adjacency arrays. For this purpose, two specially designed arrays named *IJ-link* and *JI-link* with length equal to the number of components in the network are created. In order to update efficiently the *Link* arrays upon failure of components, for each component, the two addresses of the entries into the *Link[]* dynamic arrays are stored under the component index in arrays *IJ-link* and *JI-link*, respectively. This is done to avoid unnecessary searching through the *Node* and *Link* dynamic arrays, each time when a component fails.

The mechanism of this updating will be illustrated by an example. Let a component with index x be connected to nodes i and j, respectively. Then *IJ-link[x]* and *JI-link[x]* contain the addresses of *Link[i][j]* and *Link[j][i]*, related to the number of edges between the i-th and the j-th node to which the component with index x is connected. In case of failure of component x, two values in the *Link* arrays are immediately updated: $Link[i][j] = Link[i][j] - 1$ and $Link[j][i] = Link[j][i] - 1$.

3.9   AN ALGORITHM FOR DETERMINING THE EXISTENCE OF A PATH THROUGH WORKING COMPONENTS IN COMPLEX RELIABILITY NETWORKS

Reliability of a system presented by a network can conveniently be defined as *the probability of existence of a path through working components, from the start to the end node, at the end of a specified time interval.*

Consequently, central to the system reliability analysis is the algorithm for determining the existence of a path through working components in a reliability network. An algorithm in pseudo-code, for determining the existence of a path in any network is presented below.

Algorithm 3.3

function **path**()
{ /* *Returns '1' if a path between the first and the end node exists and '0' otherwise* */
stack[];   /* *An empty stack* */
sp;   /* *Stack pointer* */
Num_nodes; /* *Number of nodes* */
marked[];   /* *An array containing information about which nodes have been visited* */

/* *Mark all nodes as 'not visited'* */;
For i=1 **to** Num_nodes **do** marked[i]=0;

sp=1; stack[sp]=1; /**Add the first node into the stack* */
While (sp > 0) **do** /* *while the stack is not empty do the following statements* */
{

   r_node = stack[sp]; /* *Take a node from the top of the stack* */
   marked[r_node]=1; /* *Mark the node as 'visited'* */

   sp=sp-1; /* *Remove the visited node from the stack* */

/* *Find all unmarked nodes adjacent to the removed node* r_node*/
   For i=1 **to** Num_nodes **do**
     if (marked[i]=0) **then** /* *if node 'i' is not marked* */
        if (*node i is adjacent to* r_node) **then**
                {
                   if (*node 'i' is the end node*)
                   then return 1; /* *a path has been found* */
                   else {sp=sp+1; stack[sp]=i;} /* *put the i-th node into the stack* */
                }
}
   return 0; /* *a path has not been found between the start and the end node* */
}

The function **path()** checks whether there exists a path through working components from the start to the end node. It works as follows. A stack is created first where initially, only the start node resides. Then, until there exists at least a single node in the stack, the node from the top of the stack is removed and marked as 'visited'. A check is then conducted whether the end node is among the adjacent non-marked (non-visited) nodes of the removed node r_node. If the end node is among them, the function *path()* returns immediately true ('1'). Otherwise, all non-marked adjacent nodes of the removed node are stored in the stack. If node i is adjacent to the removed node r_node, this will be indicated by a greater than zero element A[r_node][i] in the adjacency matrix. The algorithm then continues with removing another node from the top of the stack. In this way, the algorithm first traverses the network in depth and if a path to the end node is not found, a non-visited node is pulled from the top of the stack and an alternative path is explored. If the stack is empty and the end node still has not been reached, no path exists between the start and the end node.

Suppose that n is the total number of nodes in the system. As can be verified, if adjacency matrix is used to represent the reliability network, for each node, n checks are performed to find its neighbours. This results in algorithm complexity $O(n^2)$ which guarantees a good calculation speed for systems with relatively small number of nodes. With increasing the number of nodes however, the computation time increases polynomially. If lists of the neighbours of each node are kept however, it will be no longer necessary to search for neighbours and the computational efficiency can be increased substantially.

3.10   AN EFFICIENT ALGORITHM FOR DETERMINING THE EXISTENCE OF A PATH IN A COMPLEX RELIABILITY NETWORK REPRESENTED BY ADJACENCY ARRAYS

An algorithm based on adjacency arrays is a more efficient alternative to an algorithm based on adjacency matrix representation. Furthermore, parallel edges between pairs of adjacent nodes are represented easily by using adjacency arrays. The work of the algorithm for finding a path in a reliability network represented by adjacency arrays is given below.

Algorithm 3.4

function **path()**
{ /* Returns '1' if a path between the first and the end node exists and '0' otherwise */
stack[];   /* An empty stack */
sp;   /* Stack pointer */
Num_nodes;   /* Number of nodes */
marked[];   /* An array containing information about which nodes have been visited */

```
/* Mark all nodes as 'not visited' */;
For i=1 to Num_nodes do marked[i]=0;

sp=1; stack[sp]=1; /*Add the first node into the stack */
While (sp > 0) do /* while the stack is not empty do the following statements */
{
   r_node = stack[sp]; /* Take a node from the top of the stack */
   marked[r_node]=1; /* Mark the node as 'visited' */

   sp=sp-1; /* Remove the visited node from the stack */

/* Find all unmarked nodes adjacent to the removed node  r_node */
   For i=1 to  Node[r_node][0] do /* Go through the nodes adjacent to  r_node */
     {node_index = Node[r_node][i];
      if(marked[node_index]=0 and  Links[r_node][i]>0) then
                {
                  if (node_index i is the end node)
                  then return 1; /* a path has been found */
                  else {sp=sp+1; stack[sp]=node_index;} /* put the
                  i-th node into the stack */
                }
     }
}
   return 0; /* there is no path between the start and the end node */
}
```

Until there exists at least a single node in the stack, the node from the top of the stack is removed and marked as 'visited'. The difference with the previous algorithm is in the way adjacent nodes are determined. The statement node_index = Node[r_node][i] retrieves the index of the node which is adjacent to node r_node, removed from the top of the stack.

By using the comparison Links[r_node][i] > 0, another check is performed in the *Links* dynamic arrays whether there are still parallel edges remaining between r_node and the current adjacent node. Links[r_node][i]=0 indicates that all parallel edges (components) between adjacent nodes r_node,i have failed. The algorithm then continues with removing another node from the top of the stack until the stack is empty or the end node has been reached. If the stack is empty and the end node still has not been reached, no path exists between the start and the end node.

Now let us assess the computational efficiency of this algorithm. Suppose that n is the total number of nodes in the system. Denoting by b_1, $b_2, \ldots, b_n$ the number of neighbours of each node i, at most $b_1 + b_2 - 1 + \cdots + b_{n-1} - 1 = (n-1)\bar{b} - (n-2)$ nodes (where $\bar{b} = \frac{1}{n-1}\sum_{i=1}^{n-1} b_i$ is the average number of adjacent nodes for all nodes except the last node) will be processed (loaded and checked in the stack) before all nodes have been marked as visited or a path has been found. After the first node, for each subsequent loading of non-marked neighbours in the stack, at least one neighbour has already been marked and removed. As can be verified, the complexity of this algorithm is $O(n \times \bar{b})$. Suppose that the average number of adjacent nodes does not exceed a particular quantity k. If the size of the system is increased, no matter how complicated the system topology becomes, if the average number of neighbours never exceeds k, the computation time is proportional to the number of nodes and the algorithm's performance is not worse than the performance of an algorithm with linear complexity $O(kn)$. A comparison can now be made between an algorithm of polynomial complexity $O(n^2)$ and an algorithm with linear complexity $O(8n)$ where, for example, the average number of adjacent nodes is 8. Suppose that a simulation based on the polynomial complexity algorithm, for a network with 10 nodes requires 1s computational time. The same simulation, for a similar system containing 10,000 nodes would require approximately $10,000^2/10^2 = 1,000,000$s which is approximately 11.57 days. Now suppose that a simulation based on the linear complexity algorithm of a system with 10 nodes requires 1s. The same simulation for a system with 10,000 nodes would now require $8 \times 10,000/(8 \times 10) = 1000$s which is less than 17 minutes! Note that the relative increase in the computation time for the algorithm with complexity $O(kn)$ is not affected by the value of constant k.

3.11 AN EFFICIENT ALGORITHM FOR DETERMINING THE EXISTENCE OF k OUT OF n PATHS IN COMPLEX RELIABILITY NETWORKS CONTAINING MULTIPLE END NODES

In many cases, the network consists of a large number of end nodes, each for example corresponding to a production component (see Fig. 7.1). The system is considered in operating state if at least k out of n end components are operating. This is equivalent to the existence of paths to at least k end nodes.

Many production systems based on n production components are examples of n out of n systems (see Chapter 7). A critical failure is present if at least one of the production components has stopped production. A key part of the reliability analysis algorithm for these systems is determining after each failure of a component whether there are paths through working components to each production node in the network. If a path to at least one of the production nodes does not exist, a critical failure is registered and repair is initiated. The existence of paths through working components to all production nodes can be determined from the following algorithm.

Algorithm 3.5

function **paths()**
{ /* *Returns '1' if there are paths to all production nodes and '0' otherwise* */
stack[];   /* *An empty stack* */
sp;   /* *Stack pointer* */
Num_nodes; /* *Number of nodes* */
num_prod_nodes; /* *Number of production nodes.* */

cmp_index = Num_nodes - num_prod_nodes + 1; /* *All production nodes have
                indices greater than or equal to Num_nodes - num_prod_nodes + 1* */

marked[];   /* *An array containing information about which nodes have been visited* */
/* *Mark all nodes as 'not visited'* */;
For i=1 **to** Num_nodes **do** marked[i]=0;

sp=1; stack[sp]=1; /**Add the first node into the stack* */
s=0; /**Accumulates the number of visited production nodes* */

While (sp > 0) **do** /* *while the stack is not empty do the following statements* */
{
 r_node = stack[sp]; /* *Take a node from the top of the stack* */
 marked[r_node]=1; /* *Mark the node as 'visited'* */

 sp=sp-1; /* *Remove the visited node from the stack* */

/* *Find all unmarked nodes adjacent to the removed node* r_node */
  For i=1 **to** Node[r_node][0] **do** /* *Go through the nodes adjacent to* r_node */
   { node_index = Node[r_node][i];
    if(marked[node_index]=0 **and** Links[r_node][i]>0) **then**
        {
        if(node_index>=cmp_index) /* *check if node node_index is one
            of the production nodes*/*)
            then s=s+1; /* *if one of the production nodes
                    then increment the number of visited nodes* */

```
                    if (s = num_prod_nodes) then return 1  //there are paths to all
                                                                       prod. nodes
                    else  {sp=sp+1; stack[sp]=node_index;}  /*
                    put the i-th node into the stack */
                    }

        }
}
     return 0;   //A critical failure. Paths have not been found to all production nodes
}
```

A characteristic feature of this algorithm is that a stack is first created, where initially, only the start node resides. Then, until there exists at least a single node in the stack, the node from the top of the stack is removed and marked as 'visited'. For each adjacent node of the removed node, a check is performed whether it is a production node. Since all production nodes have indices greater than or equal to cmp_index = Num_nodes - num_prod_units + 1, the check consists simply of comparing the node index with the value stored in cmp_index. If the node index is equal or greater than cmp_index, the node is a production node and the value of counter 's' is incremented by one. Subsequently, the value of counter 's' is checked and if it is equal to the number of production nodes (all production nodes have been visited) the algorithm returns '1'. This means that paths have been found to all production nodes in the reliability network. If the content of counter 's' is smaller than the number of production nodes, the current node is pushed into the stack and the process continues.

If all nodes in the network have been visited and no paths to all production components have been found, the algorithm returns '0'.

This algorithm can also be modified to be applicable in cases where the absence of a critical failure requires not all of the production units to be working but at least a minimum critical number n_crt_nodes units out of num_prod_nodes. In this case, the statement:

if (s = num_prod_nodes) **then return 1;**   /* *no critical failure*/

should be modified to

if (s = n_crt_nodes) **then return 1;**   /* *no critical failure* */

The function ***paths()*** then has to be called with a parameter n_crt_nodes.

3.12 AN ALGORITHM FOR DETERMINING THE RELIABILITY OF A COMPLEX RELIABILITY NETWORK

The algorithm works as follows. A counter with name s_counter is initialised first, where the number of trials resulting in existence of a path from the start to the end node will be accumulated. Next, a Monte Carlo simulation loop with control variable k is entered. For each Monte Carlo simulation trial, a fresh copy of the *Link* dynamic arrays is made, which is updated each time when failure occurs within the specified time interval with length a. Due to the two additional arrays *IJ_link* and *JI_link* containing the entry addresses in the *Link* dynamic arrays which need to be updated upon failure of a component, the *Link* arrays are updated very efficiently in case of component failure. After updating the corresponding parts of the *Link* arrays upon component failure, by calling the function *paths()*, a check is performed whether there exist paths from the start to all production nodes. If such paths exist, the s_counter is incremented. For the new Monte Carlo simulation trial, the content of the *Link* dynamic arrays is restored by making a copy from the original arrays, corresponding to the case where no components have failed.

Algorithm 3.6

procedure create_and_initialise_the_dynamic_arrays()
{ /* *creates and initialises the dynamic arrays by reading the input*
     data from a text file
}
function paths() { /* *determines the existence of paths from the*
                  start to all of the end nodes. Returns '1' *if such paths exist* */ }
procedure   create_a_copy_of_the_Link_dynamic_arrays()
                  { /* *creates a copy of the Link[] arrays* */ }
function generate_time_to_failure(k)
                  { /* *generates the time to failure of the k-th component* */ }
function generate_time_to_failure(k) { /* *Returns the time to failure for the component*
                  with index 'k' */ }
s_counter = 0;
a = Number_of_years; /* *Specifies the finite time interval for which*
                  reliability is calculated */
For k = 1 to Number_of_trials **do**
   {
   create_a_copy_of_the_Link_dynamic_arrays();

   /* *Generate the times to failure of all components* */

```
For i=1 to Number_of_components do
    {
        time_to_failure = generate_time_to_failure( i );
        if(time_to_failure < a) then
        Update the copy of the Link-arrays by using the IJ- and JI-Link arrays;
    }
    if (paths()=1) s_counter = s_counter + 1;
}
Reliability = s_counter / Number_of_trials;
```

The efficiency of the described algorithm can be improved significantly if a reduction of the number of nodes is implemented. Thus, a branch containing a large number of components logically arranged in series is reduced to a single equivalent component which fails whenever any of the components in the original branch fails. This process of reduction of the number of nodes has been illustrated in Fig. 3.12 where branches containing hundreds of components in series have been replaced by single equivalent components e_1, e_2, e_3 and e_4. As a result, the initial system in Fig. 3.12(a) containing hundreds of nodes has been reduced to the system in Fig. 3.12(b) containing only three nodes!

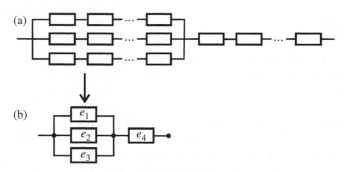

Figure 3.12 Reducing the number of nodes in a system, by substituting branches containing components in series with single equivalent components.

3.13 APPLICATIONS: RELIABILITY ANALYSIS OF COMPLEX RELIABILITY NETWORKS INCLUDING A LARGE NUMBER OF COMPONENTS

The algorithm for system reliability analysis described in the previous section will be demonstrated on two test networks of type 'full square lattice' and 'quasi-complete graph'. These systems were selected because: (i) both

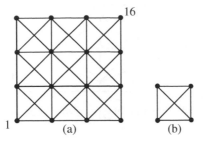

Figure 3.13 (a) A system of type 'full square lattice', and (b) elementary building cell of the system.

are dense and complex; (ii) both can be scaled, that is similar systems of larger size can be obtained easily; and (iii) both networks contain a large number of paths and loops which will test the capability of the developed algorithms. The nodes in the networks are denoted by filled circles; the 'components' are the edges connecting them. Only edges can fail, not nodes. Let us postulate the node with the lowest index to be the start node and the node with the largest index to be the end node Fig. 3.13. Reliability is then defined as *the probability of existence of a path through working edges, from the start to the end node, at the end of the specified time interval.*

3.13.1 A network of Type 'Full Square Lattice'

The elementary building blocks of the system of type *full square lattice* in Fig. 3.13 are the cells in Fig. 3.13(b). The smallest system similar to the one in Fig. 3.13(a) involves only a single cell (Fig. 3.13(b)).

We will refer to this system as 'full square lattice of order 1'. Other systems of type full square lattice of different order $(1, 2, \ldots, k)$ can be built by a simple translation of the elementary building cell in the plane. The order of a system obtained in this way is equal to the number of elementary cells on each side of the square lattice. As can be verified, a lattice of order k has $(k + 1)^2$ nodes and

$$E = 2k(2k + 1) \qquad (3.9)$$

number of edges.

3.13.2 A Network of Type 'Almost-Complete Graph'

In a 'complete graph' every node is connected to every other node. A system for which every node is connected to every other node, but there is a missing

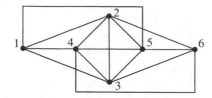

Figure 3.14 A system of type 'almost-complete graph'.

connection between the start and end node we will refer to as *almost-complete graph*.

Such a system, with $N = 6$ nodes (of order 6) is given in Fig. 3.14.

The adjacency matrix representation of these systems is simple: $a_{ij} = 1$, if $i <> j$ and i, j are not the start and the end nodes. The initialisation of the adjacency matrix $\{a_{ij}\}$ for a system with N nodes can therefore be done by using the next fragment:

```
For  i = 1 to N do
For  j = 1 to N do
   {If  (i <> j) then  a[i,j]=1; else  a[i,j] = 0;}

a[1,N] = 0; a[N,1] = 0;
```

In this way, systems of arbitrary size (order) can be generated easily. As can be verified, the number of edges E in a system with k nodes is given by

$$E = 0.5\,k(k-1) - 1 \tag{3.10}$$

3.13.3 Reliability of a Network of Type 'Full Square Lattice'

All components (edges) connecting the nodes of the lattice are assumed to be identical, characterised by a constant hazard rate $\lambda = 0.5$ year^{-1} and working independently from one another. Here are the probabilities of surviving 2 years (the reliability associated with 2 years operation; Table 3.1):

As can be verified from these results, for the selected hazard rate $\lambda = 0.5$ year^{-1}, with increasing the size of the system of type 'full square lattice' the reliability decreases monotonically. With increasing the size of the system, the rate of the decrease diminishes significantly.

Table 3.1 Reliability associated with 2 years operation as a function of the size (order) of the system of type 'full square lattice' (hazard rate $\lambda = 0.5$ year^{-1}).

Size of the system (order of the square lattice)	Reliability
1	0.553
2	0.433
3	0.395
4	0.379

3.13.4 Reliability of a Network of Type 'Almost-Complete Graph'

To test the performance of the system reliability algorithm on systems of type 'dense graph' such as the system in Fig. 3.14, systems of different order were generated. The probability that the systems will survive 2 years of continuous operation was determined under the assumption that all components (edges) connecting the nodes are identical, characterised by a constant hazard rate $\lambda = 1.5$ year^{-1} and working independently from one another.

Again, reliability is defined as *the probability of existence of a path through working edges, from the start to the end node, at the end of the specified time interval.* Here are the probabilities of surviving 2 years (the reliability associated with 2 years operation; Table 3.2).

The system reliability estimates have been obtained on the basis of 100,000 Monte Carlo simulations. Even for the very large number of

Table 3.2 Reliability associated with 2 years operation as a function of the size (order) of the system of type 'quasi-complete graph'.

Size of the system (number of nodes)	Reliability	Calculation time (s)
6	0.011	0.05
15	0.060	0.30
25	0.196	0.84
35	0.457	1.67
45	0.699	2.78
55	0.836	4.18
65	0.908	5.83
75	0.948	7.75

components corresponding to 75 nodes (the number of components is $E = 0.5 \times 75 \times (75 - 1) - 1 = 2774$), the computational time remains in the range of few seconds.

With increasing the size of the system of type 'almost-complete graph', reliability increases monotonically. With increasing the size of the system however, the rate of the reliability increase diminishes significantly.

4

PROBABILISTIC RISK ASSESSMENT AND RISK MANAGEMENT

The purpose of risk analysis is to provide support in making correct management decisions. By evaluating the risk associated with a set of decision alternatives, the risk analysis helps to identify the alternative which maximises the expected utility for the stakeholders by complying with a set of specified criteria and constraints. According to a classical definition (Henley and Kumamoto, 1981; Vose, 2000), the risk of failure K is defined as:

$$K = p_f C \qquad (4.1)$$

where p_f is the probability of failure and C is the cost given failure. To an operator of production equipment for example, the cost given failure C may include several components: cost of lost production, cost of cleaning up polluted environment, medical costs, insurance costs, legal costs, costs of mobilisation of emergency resources, cost of loss of business due to loss of reputation and low customer confidence, etc. The cost of failure to the manufacturer of production equipment may include: warranty payment if the equipment fails before the agreed warranty time, loss of sales, penalty payments, compensation and legal costs. Most of the losses from engineering failures can be classified in several major categories:

- *Loss of life or damage to health.*
- *Losses associated with damage to the environment and the community infrastructure.*
- *Financial losses* including loss of production, loss of capital assets, loss of sales, cost of intervention and repair, compensation payments, penalty payments, legal costs, reduction in benefits, losses due to change of laws, product liability, cost overruns, inflation, capital costs changes, exchange rate changes, etc.

- *Loss of reputation* including loss of market share, loss of customers, loss of contracts, impact on share value, loss of confidence in the business, etc.

Depending on the category, the losses can be expressed in monetary units, number of fatalities, lost time, volume of lost production, volume of pollutants released into the environment, number of lost customers, amount of lost sales, etc. Often losses from failures are expressed in monetary units and are referred to as cost of failure.

4.1   TRADITIONAL ENGINEERING RISK ASSESSMENT

The theoretical justification of equation (4.1) can be made on the basis of the following thought experiment. Suppose that a particular non-repairable equipment is put in operation for a length of time a. If the equipment fails before the specified time a, its failure is associated with a constant loss C, which combines the cost of intervention, the cost of replacement and the cost of lost production. Next, another identical piece of non-repairable equipment is put in operation for the same time a. Suppose that the experiment uses N identical pieces of equipment, N_f of which fail before time a. Since only failure before time a is associated with losses, the total loss generated by failures during N trials is $N_f \times C$. The average (expected) loss is then $(N_f \times C)/N$. If the number of trials N is sufficiently large, $p_f = \lim_{N \to \infty} (N_f/N)$ approximates the probability of failure of the equipment before time a. According to the empirical definition of probability, the probability of failure is a limit of the ratio of failure occurrences from a large number of trials. Usually, a relatively small number of trials N gives a sufficiently accurate estimate of the true probability of failure $p_f \approx N_f/N$. As a result, equation (4.1) describes the average (expected) loss from failure before time a. This is one of the reasons why, in the engineering context, risk is often treated as *expected loss from failure*.

The risk is not synonymous with the magnitude of the loss, irrespective of whether this magnitude is constant or variable. Without exposure to a loss-generating factor the loss will never materialise. Despite the large number of definitions and interpretations of risk (e.g. risk as uncertainty about the performance of a system (Aven, 2003) or risk as a measure of how variable the losses are (Crouhy et al., 2006)), the risk seems to incorporate two basic

Probability of failure (*pf*)

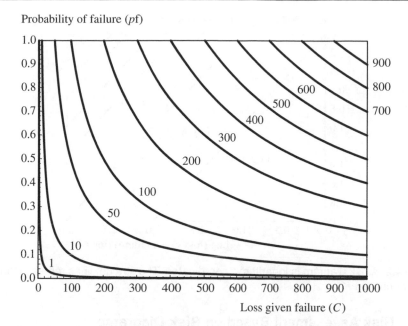

Loss given failure (*C*)

Figure 4.1 Representation of the 3D diagram related to the risk of failure.

elements: (i) the uncertainty of an exposure to a loss-generating factor (hazard, circumstances, failure) and (ii) the magnitude of the exposure.

The risk of failure can be presented by a 3D diagram (Fig. 4.1).

The *C*-axis of the diagram represents the loss given failure while the p_f-axis represents the probability of failure. The *K*-axis which is perpendicular to the $p_f - C$ plane represents the risk of failure, *K*. If this 3D diagram is sectioned by planes perpendicular to the risk axis (the *K*-axis), the projections of the lines of intersection on the $p_f - C$ plane represent a combination of probability of failure and loss given failure whose product results in a constant level of risk. This diagram is shown in Fig. 4.1 where the solid lines represent risk levels *K* equal to $1, 10, 50, \ldots, 900$ units. The risk diagram can also be represented in logarithmic coordinates. Taking logarithms from both sides of equation (4.1) results in

$$\log K = \log p_f + \log C \qquad (4.2)$$

which, in coordinates $\log p_f$ versus $\log C$ is an equation of a straight line with negative slope equal to -1. The levels of constant risk $\log K$ are then parallel straight lines with negative slopes equal to -1, representing in logarithmic coordinates the levels of constant risk from Fig. 4.1.

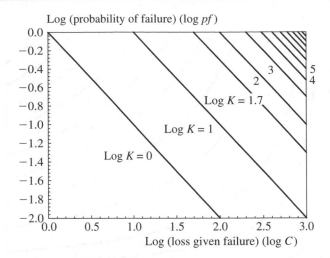

Figure 4.2 A risk diagram in logarithmic coordinates. The straight lines represent levels of constant risk log K.

4.1.1 Risk Assessment Based on Risk Diagrams

Assessing the risk associated with a single failure scenario starts with assessing its likelihood and consequences. Each combination of values for the probability of failure and the loss given failure defines a point in the risk diagram (Fig. 4.3).

The area corresponding to a high-risk occupies the upper right corner of the diagram while the area corresponding to a low-risk occupies the lower left corner. The area between these two zones corresponds to medium risk.

There exists also a qualitative approach, which is subjective and less precise but requires less effort. The likelihood of failure is not quantified

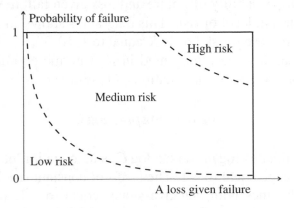

Figure 4.3 A three-regions risk diagram.

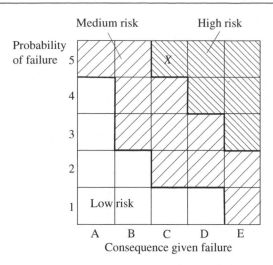

Figure 4.4 A qualitative risk matrix.

but ranked in classes, for example 1: very low; 2: low; 3: medium; 4: high and 5: very high (Fig. 4.4). Similarly, the consequences from failure are also ranked in classes A, B, C, D and E, ranging for example from very low to very high. The traditional approach to risk assessment is based on these three-regions diagrams (Vose, 2000).

The next technique is also known as *semi-quantitative risk analysis* (Vose, 2000). A set of scores $p_1, \ldots, p_5$ are assigned to the ranked probabilities of failure and another set of scores $c_A, \ldots, c_E$ to the ranked consequences given failure. The product of the scores expressing the likelihood of failure and the consequence from failure gives the risk score. The risk scores measure the risk magnitude and can be used to rank risks associated with identified failure scenarios. Subsequently, the risks are segregated according to their magnitude. Suppose that the scores measuring the likelihood of failure are $p_1 = 1, \ldots, p_5 = 5$ and the scores measuring the consequences given failure are $c_A = 1, \ldots, c_E = 5$. In Fig. 4.4, squares with scores greater than 14 define the high-risk region, squares with scores smaller than 5 define the low-risk region while squares with scores between 5 and 14 define the medium-risk region.

For a single failure scenario, the risk-assessment procedure works as follows. The likelihood and the consequences associated with the failure scenario are ranked by using scores and the product of these scores gives the risk score. If the risk score defines a square in the low-risk region, the risk is so low that it is considered negligible and no response is required. If the square representing the risk is in the high-risk region, the risk is considered intolerable. For example, a failure scenario with likelihood score 5 and consequence score 3 produces a risk score of 15 and defines a square marked by 'x' in the high-risk region (Fig. 4.4). Risk reduction measures are required to exit this region.

Risk represented by a square in the intermediate region, requires risk reduction measures to reduce it to a level which is as low as reasonably practicable (ALARP). Cost-benefit analysis can be used to verify that the benefits from a risk reduction will outweigh the cost.

A similar procedure based on assessing the risk scores associated with all individual risks, ranking them according to their scores and determining for each individual risk whether response is required, recommended or not necessary has been described, for example, in Heldman (2005).

4.1.2   Drawbacks of a Risk Assessment Based on Assessing Individual Risks

The traditional approach to risk assessment based on a risk matrix is suitable in cases where the system failure is caused by a single failure scenario. In the very common case where the system can fail due to multiple failure scenarios, the traditional approach reveals a major weakness. Often, each individual risk corresponding to the separate failure scenarios is in the low-risk region (therefore acceptable) which creates a dangerous perception of safety. In many cases, however, the aggregated risk from all failure scenarios cannot be tolerated. Despite that all individual risks may have low scores and for none of them a response plan is required, the total aggregated risk may not be acceptable. Indeed, for M mutually exclusive failure scenarios for example, the aggregated risk is determined from (Todinov, 2004c):

$$K = p_1 C_1 + \cdots + p_M C_M \qquad (4.3)$$

where K is the total risk, p_i is the likelihood of the i-th failure scenario and C_i is the expected loss, associated with it. Although each individual risk $K_i = p_i C_i$ may be tolerable, the aggregated risk K may not be acceptable. This can be illustrated by a simple example involving failure of a system from two mutually exclusive failure scenarios characterised by probabilities of occurrence $p_1 = 0.1$ and $p_2 = 0.2$, and associated with expected losses $C_1 = \$20,000$ and $C_2 = \$10,000$, correspondingly. If the maximum tolerable risk is \$2100, both individual risks will be acceptable ($K_1 = p_1 C_1 = \$2000 < \$2100$; $K_2 = p_2 C_2 = \$2000 < \$2100$) but the total aggregated risk $K = K_1 + K_2 = p_1 C_1 + p_2 C_2 = \$4000 > \$2100$ will be nearly twice the maximum tolerable limit.

This simple example shows that reducing each individual risk below the maximum tolerable level does not necessarily reduce the aggregated risk. A large aggregated risk from multiple failure scenarios, each characterised by risk below the tolerable level can be just as damaging as a large risk resulting from a single failure scenario.

4.1.3 Assessing the Aggregated Risk Associated with Multiple, Mutually Exclusive Failure Scenarios

In order to avoid the outlined drawbacks, for multiple, mutually exclusive failure scenarios, the risks associated with them should be assessed and accumulated into a total risk. The total risk should subsequently be assessed by comparing it to risk acceptability criteria, similar to the risk associated with a single failure scenario. Accordingly, assessing the total risk from mutually exclusive failure scenarios includes:

1. Identifying all potential hazards and failure scenarios.
2. Estimating the probability of occurrence of each failure scenario.
3. Estimating the consequences (losses) from each failure scenario *given* its occurrence.
4. Estimating the risk associated with each failure scenario.
5. Estimating the total risk by accumulating the risks associated with the separate failure scenarios.
6. Comparing the estimated total risk with risk acceptability criteria.

4.2 A RISK ACCEPTABILITY CRITERION BASED ON A SPECIFIED MAXIMUM TOLERABLE RISK LEVEL

Let us present the risk equation (4.1) as:

$$p_f = \frac{K}{C} \tag{4.4}$$

If K_{max} is the maximum acceptable risk of failure and $p_{f\,max}$ is the corresponding maximum acceptable probability of failure, equation (4.4) can also be presented as:

$$p_{f\,max} = \frac{K_{max}}{C} \tag{4.5}$$

For a specified loss given failure C, from equations (4.4) and (4.5) it follows that limiting the risk of failure K below K_{max} is equivalent to limiting the probability of failure p_f below the maximum acceptable level $p_{f\,max}$ ($p_f \leq p_{f\,max}$). This leads to the cost-of-failure concept for setting reliability requirements limiting the risk of failure proposed in (Todinov, 2003):

$$p_f \leq \frac{K_{max}}{C} \tag{4.6}$$

Whenever $p_f \leq p_{f\,max} = K_{max}/C$ is fulfilled, the risk of failure K is limited below K_{max} ($K \leq K_{max}$). Denoting the ratio $r_{max} = K_{max}/C$ ($0 \leq r_{max} \leq 1$) as a *maximum acceptable fraction of the cost of failure*, the cost-of-failure concept for setting reliability requirements which limit the risk of failure can also be presented as:

$$p_f \leq r_{max} \tag{4.7}$$

In words, equation (4.7) states that the probability of failure should be smaller than the maximum acceptable fraction of the cost given failure. If for example only 40% of the cost of failure can be tolerated, the probability of failure should not be greater than 40%. The ratio r_{max} can also be interpreted as the maximum fraction of the cost of failure which the owner of the risk is prepared to accept. Using equation (4.7), reliability requirements can be specified without the need to know the absolute value of the cost given failure.

Components associated with large losses from failure should be designed to a higher reliability level. Indeed, let C denote the cost given failure. Suppose that the system consists of a single component only. If the maximum tolerable risk is K_{max}, the maximum tolerable probability of failure $p_{f\,max}$ is given by equation (4.5).

Since $R_{min} = 1 - p_{f\,max}$ is the minimum reliability of the component required to keep the risk of failure at least equal to the maximum tolerable risk K_{max}, we obtain:

$$R_{min} = 1 - \frac{K_{max}}{C} \tag{4.8}$$

Equation (4.8) essentially states that in order to keep the risk of failure below the maximum tolerable level K_{max}, *a component whose failure is associated with large losses should be more reliable compared to a component whose failure is associated with smaller losses* (Todinov, 2006c). This is the principle of the risk-based design which applies even to identical components in systems with hierarchy. The higher the component in the hierarchy, the more production units will be affected by its failure, the larger the required minimum reliability level of this component should be.

An important application of relationship (4.6) can be obtained immediately for a non-repairable system, whose failure is associated with constant cost C and which is characterised by a constant hazard rate λ. Such is, for example, the system composed of n components logically arranged in series and characterised by constant hazard rates $\lambda_1, \lambda_2, \ldots, \lambda_n$. Because the system fails whenever any of its components fails, the system's hazard rate λ is a sum of the hazard rates of its components $\lambda = \lambda_1 + \lambda_2 + \cdots + \lambda_n$. The probability of failure before time a of such a system is given by $p_f = 1 - \exp(-\lambda a)$. For a maximum acceptable risk K_{max} of failure (related to a finite time interval with length a) and a constant hazard rate λ, inequality (4.6) becomes $p_f = 1 - \exp(-\lambda a) \le K_{max}/C$, from which

$$\lambda^* = -\left(\frac{1}{a}\right) \ln\left[1 - \frac{K_{max}}{C}\right] \tag{4.9}$$

is obtained for the upper bound of the system hazard rate λ^* which still guarantees that the risk of failure does not exceed K_{max} (Todinov 2003). In other words, whenever the system hazard rate λ lies within the as-determined bound ($\lambda \le \lambda^*$), the risk K of failure before time a remains within the maximum acceptable level ($K \le K_{max}$). According to equation (4.9), identical components, whose failures are associated with different losses are characterised by different upper bounds of their hazard rates. The component whose failure is associated with the largest losses has the smallest upper bound of the hazard rate. Consider now an example of setting reliability requirements by using equation (4.9).

Suppose that a failure-free service for a time interval of length at least a is required from an electrical component. A premature failure of the component entails a loss of expensive unit and the associated cost C is significant. Consequently, the designer wants to limit the expected loss from failure per electrical component below K_{max}. What should be the maximum possible hazard rate characterising the component so that the risk of failure still remains within the specified level K_{max}?

Equation (4.9) provides a solution. The hazard rate envelope $\lambda^* = -(1/a)\ln[1 - r_{max}]$, where $r_{max} = K_{max}/C$, guarantees that the risk of failure will be smaller than the specified maximum acceptable level. Substituting for example $a = 2$ years and $r_{max} = K_{max}/C = 0.1$, yields a hazard rate envelope $\lambda^* \approx 0.05$ year^{-1}. Hence, an electrical component with hazard rate smaller than 0.05 year^{-1} limits the risk of failure before 2 years below 10% of the cost of failure.

4.3 RISK OF FAILURE IN CASE OF A TIME-DEPENDENT COST OF FAILURE

4.3.1 Risk of Failure for a Time-Dependent Cost of Failure and Multiple Failure Modes

The cost of failure $C(t)$ may also be specified as a discrete function accepting constant values $C_1, C_2, \ldots, C_N$ in N sub-intervals (years).

The loss from failure occurring at the end of the design life of N years is smaller than the loss from failure occurring at the start of life. This is true even if the cost of failure is expressed by the same amount C. Indeed, because of the time value of money, expenditure C on intervention and repair following failure, has a present value

$$C_{PV} = \frac{C}{(1+r)^i} \tag{4.10}$$

where i is the year in which the failure occurs and r is the risk-free discount rate with which the time value of money is calculated. The present value C_{PV} of the loss for $C = \$2000$ and $r = 6\%$, for $i = 1, 2, \ldots, 20$ years has been presented in Fig. 4.5.

As can be verified from the graph, the earlier failure occurs, the greater its present value, the greater its financial impact. The cost of failure may diminish with time because of other reasons, for example, the production capacity may decline. This is exactly the case in subsea oil and gas production where the recoverable reserves with oil and gas significantly diminish with time. As a result, the cost of lost production and therefore, the cost of failure, also diminish significantly.

Consider now a component/non-repairable system characterised by a non-constant hazard rate $h(t)$ during the finite time interval of N years. The hazard rate $h(t)$ has been approximated by constant hazard rates $h_1, \ldots, h_N$ in the separate years. The risk of premature failure before the end of N years is given by

$$K = \sum_{i=1}^{N} C_i [\exp(-H_{i-1}) - \exp(-H_i)] \tag{4.11}$$

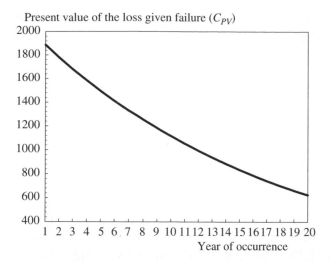

Figure 4.5 A graph showing the present value of the loss from failure with cost $2000 depending on its time of occurrence.

where $H_{i-1} = \sum_{k=1}^{i-1} h_k$ ($H_0 = 0$) and $H_i = \sum_{k=1}^{i} h_k$ are the cumulative hazard functions at the beginning and at the end of the i-th year. The probability that the premature failure will occur during the i-th year is $\exp(-H_{i-1}) - \exp(-H_i)$ which is the basis of equation (4.11). If the hazard rate is constant, $h(t) \equiv \lambda = \text{constant}$ [years^{-1}] and equation (4.11) transforms into

$$K = \sum_{i=1}^{N} C_i[\exp(-(i-1)\lambda) - \exp(-i\lambda)] \qquad (4.12)$$

which is the risk of premature failure before the end of the N-th year. If the cost of system failure C is constant, considering that C_i varies because of the time value of money, equation (4.12) becomes

$$K = C \sum_{i=1}^{N} \frac{[\exp(-(i-1)\lambda) - \exp(-i\lambda)]}{(1+r)^i} \qquad (4.13)$$

An important application of equations (4.11) and (4.13) is for determining whether the failure rates characterising a particular operating equipment are sufficient to guarantee risk of failure below maximum acceptable levels.

For a topologically complex system comprising M components, the general equation

$$K_{\max} - \sum_{i=1}^{N} C_i p_i(\lambda) = 0 \qquad (4.14)$$

implicitly specifies a domain **D** for the hazard rate which limits the risk of failure below the maximum acceptable level $K_{\max}$. Whenever the hazard rate λ belongs to the domain **D**, the risk of premature failure K in the finite time interval of N years is within the maximum acceptable level $K_{\max}$. In equation (4.14), N is the number of years, $\lambda \equiv \{\lambda_1, \lambda_2, \ldots, \lambda_M\}$ is the hazard rate vector, $p_i(\lambda)$ is the probability of system failure in the i-th year, C_i is the expected value of the cost of failure in the i-th year. (For the sake of simplicity, it is assumed that the cost of failure C_i in the i-th year does not depend on which component has caused the system failure.) For the practically important special case of M components logically arranged in series, $\lambda = \lambda_1 + \lambda_2 + \cdots + \lambda_M$ and $p_i(\lambda) = \exp(-(i-1)\lambda) - \exp(-i\lambda)$. Equation (4.14) transforms into

$$K_{\max} - \sum_{i=1}^{M} C_i \frac{\left[\exp(-(i-1)\lambda) - \exp(-i\lambda)\right]}{(1+r)^i} = 0 \qquad (4.15)$$

The hazard rate envelope λ^* which limits the risk of premature failure below the maximum acceptable level $K_{\max}$ can then be obtained as a numerical solution of equation (4.15) with respect to λ. Whenever the system hazard rate is in the specified envelope ($\lambda \leq \lambda^*$), the risk of failure will be within the maximum acceptable limit $K_{\max}$ ($K \leq K_{\max}$).

4.3.2   Risk of Failure for a Time-Dependent Cost of Failure and Multiple Failure Modes

Suppose that the component or system fails due to M mutually exclusive failure modes or due to failure of any of M components logically arranged in series, characterised by constant hazard rates $\lambda_1, \lambda_2, \ldots, \lambda_M$. The expected costs given failure $C_{k|f}$ of the components/failure modes are time dependent. Suppose that the main component of the cost given failure $C_{k|f}$ associated with each component/failure mode is the cost of the consequences of system failure. Consider the risk of system failure associated with N years. Since the times to failure are described by the negative exponential distribution, the probability of failure in the i-th year is $\exp(-(i-1)\lambda) - \exp(-i\lambda)$. According to the derivations in Chapter 5, the cost given failure, associated with M failure modes/components characterised by constant hazard rates $\lambda_k (k=1,\ldots,M)$ is $\sum_{k=1}^{M}(\lambda_k/\lambda)\overline{C}_{k|f}$, where $\lambda=\lambda_1+\ldots+\lambda_M$. Considering also the time value of money, the present value of the cost of failure in the i-th year calculated at an opportunity cost of capital (discount rate) r is $[1/(1+r)^i]\sum_{k=1}^{M}(\lambda_k/\lambda)\overline{C}_{k|f}$. Applying the total probability theorem yields

$$K = \sum_{i=1}^{N}\left\{[\exp(-(i-1)\lambda) - \exp(-i\lambda)] \times \frac{1}{(1+r)^i}\sum_{k=1}^{M}\frac{\lambda_k}{\lambda}\overline{C}_{k|f}\right\} \quad (4.16)$$

for the risk of failure of a system with components logically arranged in series.

Equation (4.16) has been verified by a Monte Carlo simulation.

Thus, for a system comprising $M=3$ components with hazard rates $\lambda_1 = 0.09$ years^{-1}, $\lambda_2 = 0.14$ years^{-1}, $\lambda_3 = 0.22$ years^{-1}, and costs given failure $\overline{C}_{1|f} = \$1200$, $\overline{C}_{2|f} = \$3400$ and $\overline{C}_{3|f} = \$4800$, the theoretical relationship (4.16) yields $K = \$3126$ for the risk of failure. The same result is obtained from a Monte Carlo simulation based on 10 million trials.

During the calculations, the length of the time interval was $N=17$ years while the discount rate r was taken to be 6% ($r=0.06$).

4.3.3 Risk of Failure Associated with Multiple Threats

Consider now the case where multiple threats to a particular target arrive randomly during a time interval $(0,t)$. The threats follow a non-homogeneous Poisson process with intensity $\lambda(t)$. Given that a threat has arrived, whether the target will be damaged or not, depends on its strength and degree of protection. Similar to the qualitative risk matrix discussed earlier, a matrix can be built in coordinates probability that a threat will arrive – probability of damaging the target given that a threat has arrived. If the probability of damaging the target given that a threat has arrived is p, the probability that the target will be damaged by a threat during the time interval $(0,t)$ is

$$p_f = 1 - \exp\left[-p\int_0^t \lambda(t)\mathrm{d}t\right] \quad (4.17)$$

For threat arrivals following a homogeneous Poisson process with density $\lambda(t) \equiv \lambda = \text{const.}$, the probability of damaging the target becomes

$$p_f = 1 - \exp(-\lambda p t) \tag{4.18}$$

Given that the cost of damage is C, the risk of failure is the product of p_f and C.

4.4   RISK-ASSESSMENT TOOLS

Risk assessment involves a number of well-documented procedures and tools. Thus, hazard and operability studies (HAZOP) and preliminary hazard analysis (PHA) are widely employed in the industry for identifying possible hazards and their effect (Sundararajan, 1991; Sutton, 1992), especially at the conceptual design stage.

Risk is closely associated with *hazards*: *anything with a potential for causing harm*. In broader sense, hazards are phenomena with the potential to adversely affect targets. A large amount of fuel stored in a fuel depot is an example of a major safety hazard. Chemicals, nuclear wastes, suspended heavy objects, black ice on the road, pressurised gas, etc. are all examples of hazards. The same hazard can have different targets. Thus, the blast wave and the heath radiation from an explosion of a fuel tank affects buildings and people, while the toxic fumes affect people and the environment.

Gathering a group of stakeholders and experts followed by brainstorming is a common technique for identifying hazards and failure scenarios which could inflict losses. Structural thinking, analysis of past failures and lessons learned, going through various scenarios, are other useful techniques which are frequently used. In case of large uncertainty regarding the system under consideration, a number of assumptions are often made about events and processes which need to be carefully tested. A question should always be asked what could happen if any of these assumptions does not hold. A number of components/stages critical to the success of the system/process need to be carefully analysed in order to identify how could they fail thereby causing a system/process failure.

The HAZOP study involves a team of individuals with different backgrounds and expertise. By bringing various expertise together, through a collective brainstorming effort, the purpose is to perform a thorough review of the system or process and identify as many hazards/failure scenarios as possible. The HAZOP study also involves screening for the causes of accidents and failures: human errors, hazardous actions, particular sequences of events, external events. It also involves identifying combinations of hazards, events and latent faults which lead to a failure scenario. The PHA usually looks at the big picture of the system. It consists of (Sundararajan, 1991):

(i) Examining all of the available information about the system's layout, process flow, environment and operating conditions.
(ii) Hazard identification.
(iii) Identification of barriers.
(iv) Assessing the impact of the hazards on the system.

Using standard lists of hazardous sources also helps to identify risks. Hazardous sources cause failure scenarios under certain triggering conditions, events or chain of events. Such sources can be particular actions, flammable materials, highly reactive chemical compounds, toxic substances, explosive materials, materials with very high or very low temperature, high voltage, high-intensity electromagnetic fields and high-intensity electromagnetic radiation (microwaves, ultraviolet (UV), infrared, γ-radiation), high-intensity sound and vibration, fast moving or falling objects, etc.

Hazardous sources are also objects with significant amount of potential or kinetic energy: gases under pressure, large masses of water with significant potential energy, loaded elastic components or springs, suspended heavy objects, parts rotating or moving at a high speed.

Hazards may not pose danger under normal conditions. A *triggering condition*, a particular action, event or a chain of events are often necessary for developing a failure scenario. Examples of triggering conditions are: human errors, latent faults and flaws, component failure, overloading, impact, material degradation, leak, power failure, sparks, excessive temperature, etc.

A failure scenario may result from a single triggering condition. Such is the case where a falling object penetrates a pipeline and causes loss of containment and a release of toxic substance. In order for an accident to occur, usually a combination of hazards, faults, actions or chain of events is required.

Such is, for example, the combination of a material defect and a design fault leaving an ignition source (e.g. a motor with spark brushes) close to a metal reservoir containing flammable material. If the material defect in the reservoir causes pitting corrosion, a puncture of the reservoir by a corrosion pit will cause a release of flammable substance which, if ignited, could cause fire.

Finally, the impact of the identified hazards on the system is assessed. A hazard may cause damage ranging from 'insignificant' to 'catastrophic'. The PHA is usually followed by more detailed analyses which focus on the failure modes of the system and its components. In order to identify possible failure modes, design analysis methods such as FMEA (failure mode and effect analysis) (MIL-STD-1629A, 1977) and its extension FMECA (failure modes, effects and criticality analysis) including criticality analysis can be used (Andrews and Moss, 2002). These ensure that as many as possible potential failure modes are identified and their effect on the system performance assessed. The objective is also to identify critical areas where design modifications can reduce the consequences from failure.

After identifying the failure scenarios and their impacts (consequences), the next important step is assessing their likelihood. A number of techniques, such as *reliability networks, fault trees, event trees, load-strength interference and various simulation and analytical techniques* are currently available for assessing the likelihood of failure. A comprehensive overview of various load-strength interference techniques, Monte Carlo simulation and analytical techniques for assessing the likelihood of failure is given in Todinov (2005a).

Reliability networks, fault trees and event trees used for assessing the likelihood of failure can be illustrated on simple circuits preventing a liquid from reaching dangerously low or high level in a tank (Fig. 4.6). The circuit includes two low-level switches LS1 and LS2, two high-level switches HS1 and HS2, a mechanical pump (MP) and a control module (CM) which operates the pump according to the information received from the switches about the water level. All devices operate independently from one another.

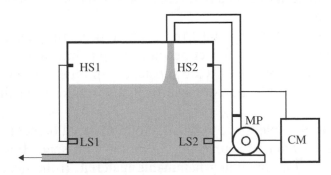

Figure 4.6 A simple system which controls filling of a tank with liquid.

The pump is required to start pumping liquid into the tank if the control module (CM) receives a signal from low-level switches LS1 or LS2 that the liquid in the tank has reached dangerously low level. The event 'no liquid supply when the liquid level is dangerously low' occurs when the first switch LS1 fails to send a signal to CM (event A) and the second switch LS2 fails too (event B) or the control module (CM) fails to switch on the pump (event C) or the mechanical pump (MP) fails to operate (event D). Suppose that the probabilities of these undesirable events are $P(A) = a$, $P(B) = b$, $P(C) = c$ and $P(D) = d$, correspondingly. The probability of the undesirable event 'no liquid supply in case of a dangerously low liquid level' can be obtained by building the reliability network in Fig. 4.7, illustrating the logical arrangement of the components. In case of a dangerously low liquid level, the pump will start operating if and only if a path through working components can be found from the start node '1' to the end node '4' (in Fig. 4.7 the nodes have been marked by filled circles).

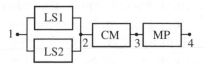

Figure 4.7 A reliability network of the circuit preventing dangerously low liquid level. The circuit consists of two low-level switches (LS1, LS2), a control module (CM) and a mechanical pump (MP).

The probability of a liquid supply on demand is then equal to the probability of existence of a path through working components. Such a path exists if and only if, in case of a low liquid level at least one of the level switches is working, the CM is working and the MP is working. The probabilities of these events are as follows: $1 - ab$ (the probability that at least one of the level switches will be working on demand), $1 - c$ (the probability that the CM will be working on demand) and $1 - d$ (the probability that the MP will be working on demand). The probability of a liquid supply on demand is then $P(S_1) = (1 - ab)(1 - c)(1 - d)$ from which, the probability of failure to supply liquid on demand in case of a dangerously low liquid level is $P(F_1) = 1 - P(S_1)$. After simplifying, $P(F_1)$ becomes

$$P(F_1) = ab + c + d - abc - abd - cd + abcd \qquad (4.19)$$

Fault trees are constructed by starting from a top undesired event (Sundararajan, 1991). The undesired event could be an accident or system failure. A logic gate is then used to combine events causing the top event.

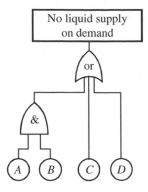

Figure 4.8 A fault tree of the simple system for liquid supply on demand.

For each of the causal events, logic gates are in turn used to combine events causing them and so on. The process continues until a set of basic events are reached which have no causal events. These terminate the branches of the fault tree. The probability of the top event is then determined as a function of the probabilities of these basic events. A number of fault trees can be constructed for each system. Each undesired top event will have a different fault tree. For the system in Fig. 4.6, the fault tree is presented in Fig. 4.8.

It includes one AND gate marked by '&' and one OR gate marked by 'or'. The basic events A, B, C and D denote failure of the first switch, failure of the second switch, failure of the control module and failure of the pump, correspondingly. Since all basic events are statistically independent, the top failure event F_1 (no liquid supply on demand) can be expressed as an union of minimal *cut sets*. A cut set in a fault tree is a collection of basic events such that if they all occur, this will cause the top event to occur too. A *minimal cut set* is a cut set such that if any basic event is removed from it, the top event will not necessarily occur if all remaining events in the cut set occur. Determining the minimal cut sets of a fault tree containing AND and OR gates involves two basic steps:

   (i) Boolean expressions are first created by substituting the AND gates with an operation logical AND ('·') and the OR gates by an operation logical OR ('+').
  (ii) Using the laws of Boolean algebra, the Boolean expressions are expanded, and subsequently reduced to a sum of products form and the redundancies in the expressions are removed. The sum of products Boolean expression corresponding to the fault tree in Fig. 4.8 is

$$F_1 = AB + C + D \tag{4.20}$$

The minimal cut sets are (A, B), C and D. The probability of the top event is determined by using the *inclusion–exclusion expansion* formula (see Chapter 2) for obtaining the probability of an union of statistically independent events:

$$P(F_1) = P(AB \cup C \cup D)$$
$$= P(AB) + P(C) + P(D) - P(ABC) - P(ABD)$$
$$- P(CD) + P(ABCD) \qquad (4.21)$$

Substituting the probabilities $P(A) = a$, $P(B) = b$, $P(C) = c$ and $P(D) = d$ in equation (4.21) yields expression (4.19).

Event trees are widely used for safety-oriented systems or standby systems where the chronological order in which events occur is essential. They are particularly useful for modelling accidents caused by a chain of events. Building the event tree starts from an initiating event. In the case of events characterised by two states only, the event tree will be a binary tree. In this case, depending on whether the next event from the chain occurs or not, the main branch splits into two branches. Each of these splits into two new branches depending on whether the third event occurs or not. This process continues until all events from the chain have been considered. For a chain of n events, there will be 2^n possible final states. A unique path will correspond to each final state. Paths which obviously do not lead to the undesirable event may not be developed. The probability of a particular state is equal to the probability of the path leading to this state. This probability is determined as a product of the probabilities of the branches composing the path. The probability of the undesirable event is the sum of the probabilities of all paths (outcomes) which lead to this event. The event tree of the liquid supply system is presented in Fig. 4.9.

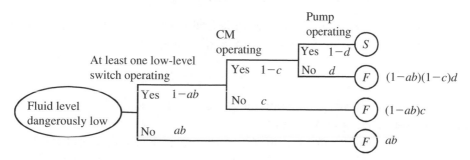

Figure 4.9 An event tree of the system for liquid supply on demand.

Three of the obtained mutually exclusive states are failure states (no liquid supply in case of dangerously low liquid level) marked by '*F*' and one state is a 'success state' marked by '*S*'. The probability of each failure path (failure state) is obtained by multiplying the probabilities of the branches composing it. The total probability of the event *no liquid supply on demand* is a sum of the probabilities of the failure states. In this case:

$$P(F_1) = (1 - ab)(1 - c)d + (1 - ab)c + ab \tag{4.22}$$

which, after simplifying, yields expression (4.19).

Another set of level switches (switches HS1 and HS2 in Fig. 4.6) has also been installed in a separate circuit whose function is to switch off the pump if the liquid level becomes dangerously high. Failure to switch off the pump, in case of dangerously high liquid level occurs if both switches fail to send a signal to the CM or the CM fails itself. In the reliability network of the switch off circuit (Fig. 4.10) the block marked by MP is missing. Clearly, if the power supply to the pump is cut off by the control module (CM), the pump will certainly switch off irrespective of whether it is in working state or not.

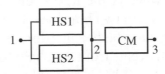

Figure 4.10 A reliability network of the circuit controlling switching the pump off if the level of the supplied liquid reaches a dangerously high level.

Suppose that the probabilities of the switches not sending a signal if the liquid level is dangerously high are k and m, respectively, and the probability that the control module (CM) will fail to switch off the pump is p. The probability of a successful operation on demand is

$$P(S_2) = P[(HS1 \cup HS2) \cap CM] = (1 - km)(1 - p)$$

from which, the probability of the undesirable event 'failure to switch off the pump in case of dangerously high liquid level' is

$$P(F_2) = 1 - P(S_2) = km + p - kmp \tag{4.23}$$

After the risks have been evaluated, a decision needs to be made whether the total risk is acceptable or not.

Suppose that the cost of failure C_l given the undesirable event 'no liquid supply on demand' is much lower compared to the cost of failure C_h given the undesirable event 'failure to switch off the pump on demand'. This is indeed the case if the liquid is flammable and failure to switch off the pump leads to an overflow and formation of flammable vapours which could ignite and cause fire.

Suppose also that the probabilities of failure of the control module and the pump are negligible and the probability of failure of the circuits is dominated by the probabilities of failures of the switches. For identical low-level switches LS1 and LS2, characterised by probabilities of failure $a = b = l$, the risk due to failure of both switches becomes $K_l = l^2 C_l$. Similarly, for identical high-level switches characterised by probabilities of failure $k = m = h$, the risk due to failure of both switches becomes $K_h = h^2 C_h$. Given that no control over the consequences from failure exists, the risk can be reduced by reducing the probabilities of failure l^2 and h^2 of the switches. This can be done either by improving the reliability of the switches (reducing the probabilities of failure l and h characterising the switches) or by introducing redundant switches working in parallel.

Including a sufficiently large number n of redundant switches ($n > 2$) can reduce significantly the probabilities of failure from l^2 and h^2 to l^n and h^n.

The next example illustrates the principle of the risk-based design introduced earlier: *the larger the loss from failure the larger the reliability of the component.*

In order to maintain the same maximum acceptable risk level K_{max}, the high-level switches HS1 and HS2 whose failure is associated with a large loss, must be designed to a higher reliability level compared to the low-level switches LS1 and LS2. Indeed, for the same risk level K_{max} and n redundant switches, $K_{max} = l^n C_l$, $K_{max} = h^n C_h$ and after dividing the two equations

$$\frac{h}{l} = \left(\frac{C_l}{C_h} \right)^{1/n} \tag{4.24}$$

is obtained. Suppose that the loss associated with failure of the high-level switches is 100 times larger than the loss associated with the low-level switches: $C_h \approx 100 \times C_l$. Then, for a single low-level switch and a high-level switch ($n = 1$), in order to maintain the same risk level K_{max} for both circuits, the probability of failure of the high-level switch must be 100 times smaller than the probability of failure of the low-level switch. For a pair of redundant switches in each circuit ($n = 2$), the probability of failure of the high-level switch must be 10 times smaller than the probability of failure of the low-level switch and so on. Increasing the number of redundant switches reduces the required difference in the reliability levels to which the low-level switches and the high-level switches should be designed.

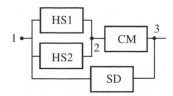

Figure 4.11 A reliability block diagram of the circuit controlling switching the pump off if an extra safety device (SD) is added.

The process of increasing the number of switches, however, increases the reliability of the switch off circuit only to a certain level. An increase of the number of redundant switches beyond this level does not substantially increase the reliability of the switch off circuit because its reliability is limited by the reliability of the control module (CM). An increase of the reliability of the switch off circuit can be achieved by installing an additional safety device (SD) (Fig. 4.11) for interrupting the power supply of the pump in case of an overflow or for draining the extra liquid beyond a certain level. If the probability of failure of this device is $s \ll 1$, the overall probability of failure of the switch off circuit is further reduced to

$$P(F_3) = s \times (km + p - kmp) \tag{4.25}$$

A further decrease of the risk of overflow can be achieved if for example, the pump is switched on only for a limited amount of time after which it is automatically switched off.

Despite all these improvements, there is still some risk of overflow which can be avoided if the design of the tank is altered (e.g. by encapsulation).

4.5 RISK MANAGEMENT

Managing operational risks is at the heart of any management strategy related to production assets. Controlling operational risk depends on measuring it, understanding it, and knowing how to reduce it. Consequently, the process of managing operational risk can be summarised by the following stages:

- Risk assessment and risk prioritising:
    - Identification of possible failure scenarios.
    - For each failure scenario estimating its likelihood and consequences (impact).
    - Prioritising risks according to their magnitude.
    - Estimating the total risk.

- Assessing to what extent risk can be managed and selecting appropriate risk response strategy:
  - Avoiding the risk.
  - Reducing the risk through appropriate risk reduction measures and techniques.
  - Accepting the risk.
  - Transferring the risk partially or fully to another party (e.g. transferring the risk by contracting, through purchasing insurance, warranties, etc.).
  - Spreading the risk (e.g. by a joint venture, alliances, risk apportionment through contracts between several parties, etc.).
- Implementing the selected response strategy, reviewing and maintaining the implemented measures.

Central to the risk management is assessing to what extent risk can be managed and selecting appropriate risk response strategy. Avoiding the risk altogether is the best prevention measure because it eliminates the cause of risk. Thus, the risk of chemical poisoning is avoided if non-toxic substances are used. The cost of risk avoidance is often very small compared to the cost of the consequences should the risk materialises. Just as it is in the all-familiar case, where the extra few minutes to check the traffic route before leaving to the airport and selecting an appropriate alternative route, avoids the cost of missing the flight and its consequences.

The problem with the risk avoidance strategy is that it is not always possible or appropriate for every risk. As a result, various risk reduction measures are implemented.

In cases where the intervention for repair is very difficult or very expensive (e.g. deep-water oil and gas production), preventive approach to risk reduction should be used which consists of reducing the likelihood of failure modes. Preventive measures should be preferred to protective measures wherever possible because while protective measures mitigate the consequences from failure, preventive measures exclude failures altogether or reduce the likelihood of their occurrence.

Protective measures are often preferred in cases where the likelihood of failure is significant and little or no control over the failure occurrence exists. Protective measures are also efficient against low-probability high-impact events.

A basic step of the risk management is the identification of as many as possible failure scenarios, assessing their likelihood and impacts. After the total risk associated with the identified failure scenarios has been estimated, the focus is on making a decision. If the risk is low, the risk is accepted and no further action is taken. Otherwise, the risk must be transferred, spread or reduced.

If the risk can be managed easily by a risk reduction, a large total risk would require selecting and implementing appropriate risk reduction measures.

After assessing the risks corresponding to the separate failure scenarios they are prioritised. The risks $p_i C_i$ associated with the separate failure scenarios are ranked in order of magnitude. A Pareto chart can then be built on the basis of this ranking and from the chart, the failure scenarios accountable for most of the total risk are identified (Fig. 4.12). Risk reduction efforts are then concentrated on the few failure scenarios accountable for most of the total risk.

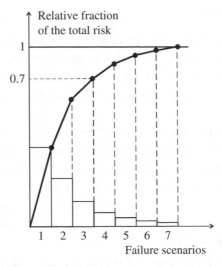

Figure 4.12 Ranking the risks of failure associated with the identified failure scenarios.

Appropriate risk reduction measures are identified which will reduce the risks associated with these few failure scenarios. Next, new failure scenarios are identified, and the total risk is estimated and assessed again. This iterative process continues until the risk-assessment procedure indicates that the total risk is acceptable. Consequently, the process of risk reduction can be described by the block diagram in Fig. 4.13, the main feature of which is the iterative loop related to selecting appropriate risk reduction measures.

Deciding upon and selecting particular risk reduction measures may not necessarily reduce the total risk. Indeed, a common situation during the design of complex systems exists when design modifications to eliminate a particular failure mode often create another failure mode. In order to reduce the possibility of introducing new failure modes, each time after

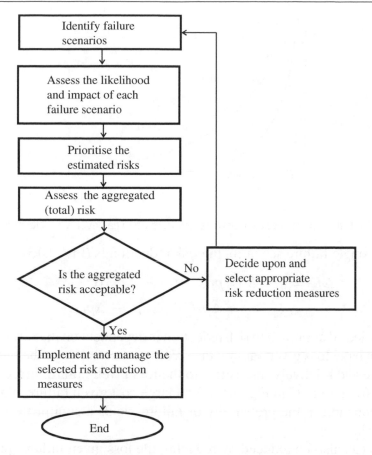

Figure 4.13 A block diagram of risk management through risk reduction.

deciding upon and selecting appropriate risk reduction measures (e.g. a design modification), possible failure scenarios are identified and assessed again. Furthermore, risks are often interrelated. Decreasing the risk of a particular failure scenario may increase the risk of other failure scenarios. Thus, building a tourist attraction on a remote place with sunny weather reduces the risk of reduced number of customers due to bad weather but simultaneously increases the risk of reduced number of customers due to higher transportation expenses (Pickford, 2001). The only protection against interrelated risks is *integrated risk management* which includes assessment of all individual risks and the total risk after deciding upon each risk reduction measure.

Risk can be reduced from a level K to a lower level K' either by reducing the loss given failure or by reducing the probability of failure or by reducing both (point A in Fig. 4.14).

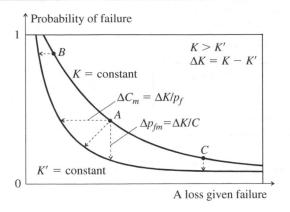

Figure 4.14 Different ways of reducing the risk from an initial level K to a level K' ($K > K'$).

For a single failure scenario, the risk reduction is (Fig. 4.15)

$$\Delta K = K - K' = p_f C - (p_f - \Delta p_f)(C - \Delta C)$$
$$= \Delta p_f C + \Delta C p_f - \Delta p_f \Delta C \qquad (4.26)$$

The selected approach to risk reduction is dependent on the *risk profile*. In case of a large loss given failure, the risk is very sensitive to the probability of failure and relatively insensitive to the loss given failure. Indeed, as can be seen from point C in Fig. 4.14, for a large cost given failure, a relatively small reduction in the probability of failure yields a significant reduction of the risk.

Risks can also be reduced by reducing the loss given failure (point B in Fig. 4.14). A risk reduction of magnitude ΔK can be achieved solely by reducing the probability of failure by $\Delta p_{fm} = \Delta K / C$. Conversely, the same risk reduction ΔK can also be achieved solely by reducing the loss given failure by $\Delta C_m = \Delta K / p_f$ (Fig. 4.14, point A). The same risk reduction ΔK can be achieved at various combinations of the probability of failure Δp_f and the losses from failure ΔC which vary in the intervals $0 \leq \Delta p_f \leq \Delta p_{fm}$ and $0 \leq \Delta C \leq \Delta C_m$. The decision regarding which type of risk reduction should be preferred depends also on the cost of investment to achieve the reduction. In other words, the values Δp_f and ΔC should be selected in such a way that the risk reduction ΔK is achieved at minimal cost.

From equation (4.26), for M mutually exclusive failure scenarios, the expression

$$\Delta K = K - K' = \sum_{i=1}^{M} \Delta p_{fi} C_i + \sum_{i=1}^{M} \Delta C_i p_{fi} - \sum_{i=1}^{M} \Delta p_{fi} \Delta C_i \qquad (4.27)$$

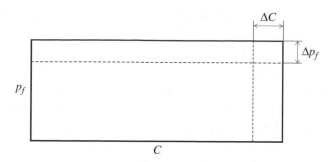

Figure 4.15 A risk reduction from level $K = p_f C$ to a level $K' = (p_f - \Delta p_f)(C - \Delta C)$.

is obtained for the risk reduction, where Δp_{fi} and ΔC_i vary in the intervals $0 \le \Delta p_{fi} \le \Delta K / C_{im}$ and $0 \le \Delta C_i \le \Delta K / p_{fim}$. Again, the values Δp_{fi} and ΔC_i should correspond to a risk reduction ΔK achieved at a minimal cost.

An optimum balance of the expenditure Q towards a risk reduction and the total risk of failure K must be achieved wherever possible. Too little investment towards risk reduction results in too large risk of failure. Too large investment towards risk reduction means unnecessary costs which cannot be outweighed by the risk reduction. The right balance is achieved at the optimum level of expenditure Q^* which minimises the total cost $G = Q + K$ (Fig. 4.16).

Consider for example a problem from risk-based inspection related to determining the optimum number of independent inspections which minimises the sum of the cost of inspections and the risk of fatigue failure.

Failure is caused by a particular defect and is associated with expected cost with magnitude C. Let p denote the probability that such a defect will reside in the high-stress region of the component and will certainly cause

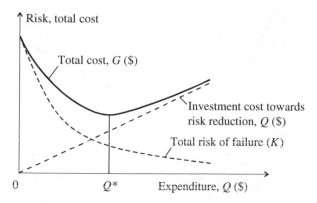

Figure 4.16 Total cost as a function of the expenditure towards risk reduction. The optimal expenditure towards risk reduction Q^* corresponds to the minimum total cost.

fatigue failure if it goes unnoticed during inspection. Each independent inspection is associated with probability q that the defect will be identified given that it is present in the inspected region. Consequently, given that the defect resides in the high-stress region of the component, the probability of missing it after n independent inspections is $(1 - q)^n$. The probability that the defect will be present in the high-stress region after n inspections is $p(1 - q)^n$ which is the product of the probability that the defect will reside in the high-stress region and the probability that it will be missed by all independent inspections. Suppose also, that the cost of each inspection is Q. The risk of failure after n inspections is then $p(1 - q)^n C$, because $p(1 - q)^n$ is the probability of failure of the component. The risk of failure decreases exponentially with increasing the number of inspections (Fig. 4.17).

The cost of inspection is nQ, and increases linearly with increasing the number of inspections (Fig. 4.17). The objective function $f(n)$ to be minimised is the total cost, which is a sum of these two costs

$$f(n) = p(1 - q)^n C + nQ \tag{4.28}$$

where n can only accept integer non-negative values in the range $n = 0, 1, 2, \ldots, n_{max}$ where n_{max} is an upper limit of the possible number of inspections. The optimal number of inspections n_{opt} minimising the total cost (expenditure) can then be determined easily by using a standard numerical algorithm for non-linear optimisation (Press et al., 1992).

Figure 4.17 represents the function $f(n)$ for the numerical values $p = 0.05$, $q = 0.7$, $C = \$30,000$ and $Q = \$200$.

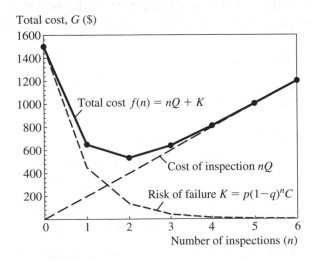

Figure 4.17 Total cost versus number of inspections.

As can be verified from the graph, the minimum of the total cost is attained for $n = 2$ independent inspections. A smaller number of inspections is associated with greater total cost because of the greater risk of fatigue failure. Similarly, a larger number of inspections is also associated with greater total cost because of the excessive cost of inspections which cannot be outweighed by the risk reduction.

4.6 REDUCING THE RISK OF FAILURE BY DESIGNING AND MAINTAINING BARRIERS

The probability of an accident or failure can be reduced by designing and maintaining barriers (Haddon, 1973). Barriers are *physical and engineered systems, procedures, instructions, practices or human actions designed to reduce either the likelihood of failure or the consequences from failure or both.* Engineered systems and human actions often work together to guarantee and maintain a barrier. We will distinguish between *preventive*, *protective* and *dual* barriers. Preventive barriers reduce the likelihood of failure while protective barriers eliminate or reduce the consequences given that failure has occurred. Dual barriers reduce both: the likelihood of failure and the consequences from failure.

Preventive barriers should be preferred because while protective barriers mitigate the consequences from failure, preventive barriers reduce or exclude the possibility of failure altogether. This is particularly relevant in cases where the intervention for repair is very difficult or very expensive (e.g. deep-water oil and gas production).

The reliability of barriers is assessed as a probability that the barrier will perform on demand. Barriers can be active or passive and involve physical and non-physical components (Fig. 4.18).

In order to be activated, *active barriers* require a detection mechanism. Often the detector is directly linked to an actuator as in the case of a detector of toxic fumes and a ventilation system or a fire detector and sprinklers. Active barriers follow the sequence Detect–Diagnose–Activate–Respond. They often involve a combination of hardware, software and human action. A typical example of an active barrier is the quality control of a weld involving an ultrasonic detection device and an operator who performs the diagnosis and responds accordingly. The reliability of an active barrier depends on the reliability of its building blocks: detector, operator, activation and response. In many automatic active barriers, the activation

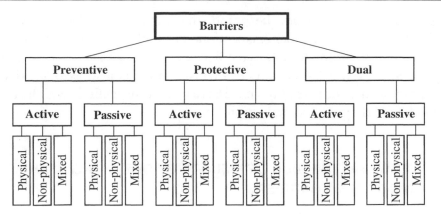

Figure 4.18 A generic classification of barriers against failures.

mechanism is incorporated in the detector. For active barriers involving operators, a decision following a diagnosis is often part of the activation mechanism. Often the diagnosis requires evaluation of several input signals before an action is performed. Hardware actuators can be automatic (e.g. an automatic shutdown actuator) or manual (e.g. a control panel button). The response mechanisms perform the barrier function (e.g. a mechanism for closing a control valve).

Passive barriers do not need a detector and they are constantly ready to perform. Passive barriers such as walls, guards, dykes, fire insulation and minimum separating distances perform their function simply through their design and placing.

Barriers have to be selected, designed, installed and maintained according to the risk assessment of the corresponding hazards. In order to fulfil their role, the strength and performance of the designed barriers should be reviewed, assessed, monitored and maintained constantly during their life cycles. Passive barriers require periodic inspection and maintenance while active barriers may also require adjustment and testing (Hale et al., 2004).

The interdependence between the separate barriers needs to be understood well, since failure of one barrier may automatically compromise other barriers (e.g. failure of the detection system may compromise all subsequent barriers which depend on a successful detection of failure. Barriers often suffer common cause failures. Poor safety culture for example, may compromise most of the non-physical barriers based on strict adherence to safety rules, instructions and practices.

5

POTENTIAL LOSS FROM FAILURE FOR NON-REPAIRABLE COMPONENTS AND SYSTEMS WITH MULTIPLE FAILURE MODES

5.1 DRAWBACKS OF THE EXPECTED LOSS AS A MEASURE OF THE LOSS FROM FAILURES

The risk equation (4.1) only estimates the average value of the potential loss from failure. A decision criterion based on the expected loss would prefer the design solution characterised by the smallest expected potential loss. What is often of primary importance however *is not* the expected (average) loss, but the deviation from the expected loss (the unexpected loss). This is for example the case where a company estimates the probability that its potential loss will exceed a particular critical value after which the company will essentially be insolvent. Despite that the expected loss gives the long-term average of the loss, *there is no guarantee that the loss will revert quickly to such average* (Bessis, 2002). This is particularly true for short time intervals where the variation of the number of failures is significant.

Let us consider a real-life example where a selection needs to be made between two competing identical systems which differ only by the time to repair. A critical failure of the first system is associated with a time for repair which follows a normal distribution. As a consequence, the lost production due to the critical failure also follows a normal distribution. Suppose that this distribution is characterised by mean $\overline{C}_1$ and variance σ_1^2. The second system is associated with a constant time for repair and constant cost of lost production $\overline{C}_2 > \overline{C}_1$. The two systems are characterised by the same probability of failure $p_{1f} = p_{2f} = p_f$. Equation (4.1) yields $K_1 = p_f\overline{C}_1$ for the risk of failure characterising the first system and $K_2 = p_f\overline{C}_2$ for the risk of failure characterising the second system. Clearly, $K_1 < K_2$ because

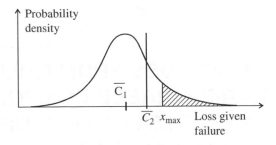

Figure 5.1 Distributions of the loss given failure for two systems.

$\overline{C}_1 < \overline{C}_2$. However, as can be verified from Fig. 5.1, the probability that the loss given failure will exceed a critical maximum acceptable value x_{max} is zero for the system characterised by the larger risk and non-zero the system characterised by the smaller risk.

In other words, *smaller expected loss does not necessarily mean smaller probability that the loss will exceed a particular limit.*

If the expected value of the loss given failure was selected as a utility function, the first system would be selected by a decision criterion based on minimising the expected loss.

Suppose that x_{max} is the maximum amount of reserves available for covering the loss from critical failure. No recovery can be made from a loss exceeding the amount of x_{max} and production cannot be resumed. With respect to whether a recovery from a critical failure can be made, the first system is associated with risk while the second system is not.

In order to make a correct selection of a system minimising the risk of exceeding a maximum acceptable limit by using the statistical decision theory based on maximising the expected utility (Roberts, 1979), the utility function should reflect whether the loss exceeds the critical limit x_{max} or not.

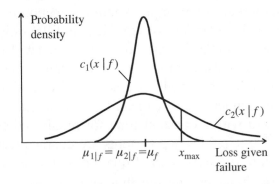

Figure 5.2 The variance of the loss given failure is strongly correlated with the probability that the loss will exceed a specified quantity.

Increasing the variance of the loss given failure increases the risk that the loss will exceed a specified maximum tolerable level. This is illustrated in Fig. 5.2 by the probability density distributions of the loss given failure $c_1(x|f)$ and $c_2(x|f)$ of two systems characterised by different variances $(\sigma_1^2 < \sigma_2^2)$.

A new measure of the loss from failure which avoids the limitations of the traditional risk measure (equation (4.1)) is the *cumulative distribution of the potential loss*.

5.2 POTENTIAL LOSS, CONDITIONAL LOSS AND RISK OF FAILURE

Here, several concepts will be introduced: (i) *potential loss related to a single and only failure* before a specified time a; (ii) *conditional loss given that failure has occurred* and (iii) *potential losses related to multiple failures* in the time interval $(0, a)$, Fig. 5.3.

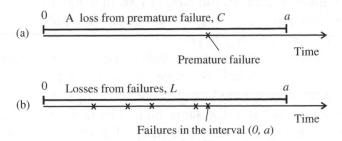

Figure 5.3 (a) A premature failure and (b) multiple failures in the interval (0, a).

The concepts *potential loss* and *conditional loss* apply to both non-repairable and repairable systems while the concept *potential losses* applies only to repairable systems. The quantity *a loss given failure* is a conditional quantity because it is defined *given that failure has occurred*. This is in sharp contrast with the *potential loss* which is *unconditional quantity* and is defined *before failure occurs*. While the conditional distribution of the loss given failure can be used to determine the probability that given failure, the loss will be larger than a specified limit, the distribution of the potential loss combines the probability that there will be failure and the probability that the loss associated with it will be larger than a specified limit. In other words, the measure 'potential loss' incorporates the uncertainty associated with the exposure to losses and the uncertainty associated with the consequences given exposure.

Historical data related to the losses from failures can only be used to determine the distribution of the conditional loss. Building the distribution of the potential losses, however, requires also an estimate of the probability of failure.

Both the conditional loss and the potential loss are random variables. Thus, in the failure event leading to a loss of containment of a reservoir or a pipeline transporting fluids, the conditional loss will depend on how severe is the damage of the container.

Since the potential loss is a random variable, it is characterised by a cumulative distribution function $C(x)$ and a probability density function $c(x)$. The probability density function $c(x)$ gives the probability $c(x)\,dx$ (before failure occurs) that the potential loss X will be in the infinitesimal interval x and $x + dx$ $(P(x < X \leq x + dx) = c(x)\,dx)$.

Accordingly, the conditional loss (the loss *given* failure) is characterised by a cumulative distribution function $C(x|f)$ and the probability density function $c(x|f)$. The conditional probability density function $c(x|f)$ gives the probability $c(x|f)\,dx$ that the loss X will be in the infinitesimal interval x and $x + dx$ given that failure has occurred $(P(x < X \leq x + dx|f) = c(x|f)\,dx)$.

Let S be a non-repairable system composed of M components, logically arranged in series, which fails whenever any of the components fails. It is assumed that the components' failures are mutually exclusive; that is, no two components can fail at the same time. The reasoning below and the derived equations are also valid if instead of a set of components, a set of M mutually exclusive system failure modes are considered; that is, no two failure modes can initiate failure at the same time. Since the system is non-repairable, the losses are associated with the first and only failure of the system. The reasoning below, however, is also valid for a repairable system if the focus is on the loss from the first failure only.

The cumulative distribution function $C(x) \equiv P(X \leq x)$ of the potential loss gives the probability that the potential loss X will not be greater than a specified value x. A loss is present only if failure is present. Consequently, the unconditional probability $C(x) \equiv P(X \leq x)$ that the potential loss X will not be greater than a specified value x is equal to the sum of the probabilities of two mutually exclusive events: (i) failure will not occur and the loss will not be greater than x and (ii) failure will occur and the loss will not be greater than x. The probability of the first compound event is $(1 - p_f) \times H(x)$, where p_f is the probability of failure and $H(x)$ is the conditional probability that the loss will not be greater than x given that no failure has occurred. This conditional probability can be presented by the Heaviside unit step function (Abramowitz and Stegun, 1972) $H(x) = \begin{cases} 1, & x \geq 0 \\ 0, & x < 0 \end{cases}$. The probability of the second compound event is $p_f C(x|f)$ where $C(x|f)$ is the conditional probability that *given* failure, the loss will not be greater than x. Consequently, the probability $C(x)$ that the potential loss X will not be greater than x is given by the distribution mixture:

$$C(x) \equiv P(X \leq x) = (1 - p_f) \times H(x) + p_f \times C(x|f) \qquad (5.1)$$

The difference between a potential and conditional loss is well illustrated by their distributions in Fig. 5.4. A characteristic feature of the cumulative distribution of the potential loss is the concentration of probability mass with magnitude $1 - p_f$ at point A (Fig. 5.4(b)) because there exists a probability $1 - p_f$ that failure will not occur and the potential loss will be zero.

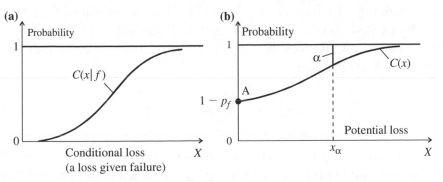

Figure 5.4 (a) A conditional loss (a loss given failure); (b) A potential loss and maximum potential loss x_α at a pre-set level α.

If a level α for the probability of obtaining as-extreme or more extreme loss is specified, a maximum potential loss x_a can be determined which corresponds to the specified level; α is the probability that the potential loss will exceed this maximum specified loss x_α ($\alpha = P(X > x_a)$, Fig. 5.4(b)). Then, the maximum potential losses $x_{\alpha,i}$ characterising different design solutions can be compared.

The maximum potential loss at a pre-set level is a risk measure which specifies the limit, whose probability of exceeding is not greater than the pre-set level.

The maximum potential loss at a pre-set level serves to determine the risk-based capital required to absorb the loss associated with failure. If x_α is the available resource of capital, the pre-set level α is the probability that the actual loss will exceed it thereby triggering insolvency. The expected loss is not sufficient to define the necessary resource of capital because the actual loss varies randomly around it.

Let $C_k(x|f)$ be the conditional cumulative distribution of the loss (the loss *given* failure) characterising the kth failure mode, and $p_{k|f}$ be the conditional probability that *given* failure, the kth failure mode has initiated it first ($\sum_{k=1}^{M} p_{k|f} = 1$). The conditional probability distribution $C(x|f) \equiv P(X \le x|f)$ that the loss X *given* that failure has occurred will not be greater than a specified value x can be presented by the union of the following mutually exclusive and exhaustive events: (i) It is the first failure mode that has initiated the failure and the loss X is not greater than x (the probability of which is $p_{1|f}C_1(x|f)$). (ii) It is the second failure mode that has initiated the failure and the loss X is not greater than x (the probability of which is $p_{2|f}C_2(x|f)$).... The final compound event is the Mth failure mode has initiated the failure and the loss X is not greater than x (the probability of which is $p_{M|f}C_M(x|f)$). The probability of a union of mutually exclusive events equals the sum of the probabilities of the separate events. As a result, the conditional distribution of the loss given failure (the conditional loss) becomes:

$$C(x|f) = \sum_{k=1}^{M} p_{k|f} C_k(x|f) \qquad (5.2)$$

The distribution of the conditional loss $C(x|f)$ is a mixture of the distributions of the conditional losses $C_k(x|f)$ characterising the individual failure modes, scaled by the conditional probabilities $p_{k|f}$ of initiating failure first *given* that failure has occurred ($\sum_{i=1}^{M} p_{k|f} = 1$). Finally, equation (5.1) regarding the cumulative distribution of the potential loss becomes

$$C(x) = (1 - p_f) H(x) + p_f \sum_{k=1}^{M} p_{k|f} C_k(x|f) \qquad (5.3)$$

The product of the probability of failure p_f and the probability $p_{k|f}$ that given failure, the kth failure mode has initiated it is simply equal to the probability p_k that the kth failure mode will initiate failure first ($p_f p_{k|f} = p_k$). Considering this relationship and also the relationship $\sum_{i=1}^{M} p_k = p_f$, equation (5.3) can also be presented as

$$C(x) = (1 - p_f) H(x) + \sum_{k=1}^{M} p_k C_k(x|f) \qquad (5.4)$$

Equations (5.3) and (5.4) are fundamental and give the cumulative distribution of the potential loss associated with mutually exclusive failure modes. Differentiating equation (5.4) with respect to x results in

$$c(x) = (1 - p_f)\,\delta(x) + \sum_{k=1}^{M} p_k c_k(x|f) \tag{5.5}$$

where $c(x) \equiv dC(x)/dx$ is the probability density distribution of the potential loss and $c_k(x|f) \equiv dC_k(x|f)/dx$ are the conditional probability density distributions of the loss *given* that failure has occurred, associated with the separate failure modes/components.

In equation (5.5), $\delta(x)$ is the Dirac's delta function which is the derivative of the Heaviside function $dH(x)/dx$ (Abramowitz and Stegun, 1972). The expected value of the potential loss from failures $\overline{C}$ is obtained by multiplying equation (5.5) by x and integrating it ($\int x\delta(x)dx = 0$):

$$\overline{C} = \int x\,c(x)\,dx = \sum_{k=1}^{M} p_k \int x c_k(x|f)\,dx = \sum_{k=1}^{M} p_k \overline{C}_{k|f} \tag{5.6}$$

where $\overline{C}_{k|f} = \int x\,c_k(x|f)\,dx$ are the expected values of the loss given that failure has occurred, characterising the individual failure modes/components.

For a single failure mode, equation (5.6) transforms into

$$\overline{C} = p_f \overline{C}_f \tag{5.7}$$

which is equivalent to the risk equation (4.1). Clearly, the *risk of failure K in equation (4.1) can be defined as the expected value of the potential loss.*

Equation (5.4) can be used for determining the probability that the potential loss will exceed a specified critical quantity x. This probability is

$$P(X > x) = 1 - C(x) = 1 - (1 - p_f)\,H(x) - \sum_{k=1}^{M} p_k C_k(x|f)$$

which, for $x > 0$, becomes

$$P(X > x) = \sum_{k=1}^{M} p_k[1 - C_k(x|f)] \tag{5.8}$$

Equation (5.8) can also be presented as

$$P(X > x) = p_f \sum_{k=1}^{M} p_{k|f}[1 - C_k(x|f)] \tag{5.9}$$

where p_f is the probability of failure and $p_{k|f}$ is the conditional probability that given failure, it is the kth failure mode which initiated it. The sum $P(X > x|f) = \sum_{k=1}^{M} p_{k|f}[1 - C_k(x|f)]$ can be interpreted as the conditional probability that given failure, the loss will be greater than x.

The probability that the potential loss will exceed a specified quantity is always smaller than the probability that the conditional loss will exceed the specified quantity.

Suppose now that the times to failure characterising M statistically independent failure modes are given by the cumulative distribution functions $F_k(t)$, $k = 1, 2, \ldots, M$, with corresponding probability density functions $f_k(t) = \mathrm{d}F_k(t)/\mathrm{d}t$. The probability that the first failure mode will initiate failure can then be determined by using the following probabilistic argument.

Consider the probability that the first failure mode will initiate failure in the infinitesimal time interval $(t, t + \mathrm{d}t)$. This probability can be expressed as a product $p_1(t) = f_1(t)[1 - F_2(t)] \cdots [1 - F_M(t)] \, \mathrm{d}t$ of the probabilities of the following statistically independent events: (i) the first failure mode will initiate failure in the time interval $(t, t + \mathrm{d}t)$, the probability of which is $f_1(t)\mathrm{d}t$ and (ii) the other failure modes will not initiate failure before time t, the probability of which is given by $[1 - F_2(t)] \times \cdots \times [1 - F_M(t)]$. According to the total probability theorem, the total probability that the first failure mode will initiate failure in the time interval $(0, a)$ is

$$p_1 = \int_0^a f_1(t)[1 - F_2(t)] \cdots [1 - F_M(t)] \, \mathrm{d}t$$

Similarly, for the kth failure mode, this probability is

$$p_k = \int_0^a f_k(t)[1 - F_1(t)] \cdots [1 - F_{k-1}(t)][1 - F_{k+1}(t)] \cdots [1 - F_M(t)] \, \mathrm{d}t \tag{5.10}$$

Substituting these probabilities in equation (5.8) yields the probability that the potential loss from multiple failure modes with known distributions of the time to failure will exceed a critical value x.

For failure modes characterised by constant hazard rates λ_k, $f_k(t) = \lambda_k \exp(-\lambda_k t)$ and $F_k(t) = 1 - \exp(-\lambda_k t)$. Substituting these in equation (5.10) and integrating, results in

$$p_k = \int_0^a \lambda_k \exp\left[-(\lambda_1 + \lambda_2 + \cdots + \lambda_M)t\right] dt$$

$$= \frac{\lambda_k(1 - \exp\left[-(\lambda_1 + \lambda_2 + \cdots + \lambda_M)a\right])}{\lambda_1 + \lambda_2 + \cdots + \lambda_M} \tag{5.11}$$

Thus, for a system with failure modes characterised by constant hazard rates $\lambda_1, \lambda_2, \ldots, \lambda_M$, where M is the number of failure modes, the probability that the potential loss in a specified interval $(0, a)$ will exceed a specified value x is given by equation (5.8) where the probabilities p_k, $k = 1, 2, \ldots, M$ are given by equation (5.11).

For failure modes characterised by constant hazard rates λ_k, the probability of failure before time a is $p_f = 1 - \exp\left[-(\lambda_1 + \lambda_2 + \cdots + \lambda_M)a\right]$ and from $p_k = p_f p_{k|f}$, the relationship

$$p_{k|f} = \frac{\lambda_k}{\lambda_1 + \cdots + \lambda_M}$$

is obtained.

The conditional probability $p_{k|f}$ that given failure in the finite time interval $(0, a)$, it is the kth failure mode that has initiated it is given by

$$p_{k|f} = \frac{1}{p_f} \int_0^a f_k(t)[1 - F_1(t)] \cdots [1 - F_{k-1}(t)]$$

$$\times [1 - F_{k+1}(t)] \cdots [1 - F_M(t)] dt \tag{5.12}$$

where $p_f = 1 - [1 - F_1(a)][1 - F_2(a)] \cdots [1 - F_M(a)]$ is the probability of failure before time a. Now, we can verify that the conditional probabilities $p_{k|f}$ add up to unity. Indeed, the sum of these probabilities can be presented as

$$\sum_{k=1}^M p_{k|f} = -\frac{1}{p_f} \int_0^a d([1 - F_1(t)][1 - F_2(t)] \cdots [1 - F_M(t)])/dt$$

$$= \frac{1}{p_f}\{-[1 - F_1(t)][1 - F_2(t)] \cdots [1 - F_M(t)]\}\,|_0^a = 1$$

because $[1 - F_1(0)][1 - F_2(0)] \cdots [1 - F_M(0)] = 1$.

Following equation (5.6), regarding the expected loss given failure, the expected value of the potential loss (the risk) becomes

$$K = (1 - \exp[-(\lambda_1 + \cdots + \lambda_M)a]) \times \sum_{k=1}^{M} \frac{\lambda_k}{\lambda_1 + \cdots + \lambda_M} \overline{C}_{k|f} \quad (5.13)$$

where the sum

$$\overline{C}_f = \sum_{k=1}^{M} \frac{\lambda_k}{\lambda_1 + \cdots + \lambda_M} \overline{C}_{k|f} \quad (5.14)$$

can be interpreted as the expected conditional loss (given that failure has occurred before time a).

Considering that for a non-repairable system, the probability of failure before time a is $1 - \exp\left(-\int_0^a \lambda(t)dt\right)$, where $\lambda(t)$ is the hazard rate of the system, for the expected value of the potential loss (the risk) before time a, the equation

$$\overline{C} \equiv K = \left(1 - \exp\left[-\int_0^a \lambda(t)dt\right]\right) \times \sum_{i=1}^{M} p_{k|f} \overline{C}_{k|f} \quad (5.15)$$

is obtained.

5.3   VARIANCE OF THE CONDITIONAL LOSS AND THE POTENTIAL LOSS FROM MULTIPLE MUTUALLY EXCLUSIVE FAILURE MODES

5.3.1   Variance of the Conditional Loss

Suppose that a non-repairable system is characterised by M mutually exclusive failure modes and the distributions of the conditional loss related to these failure modes are characterised by expected values $\mu_{1|f}$, $\mu_{2|f}, \ldots, \mu_{M|f}$ and standard deviations $\sigma_{1|f}, \sigma_{2|f}, \ldots, \sigma_{M|f}$. According to the previous discussion, the distribution of the conditional loss characterising the system is the distribution mixture given by equation (5.14).

The variance of the conditional loss σ_f^2 can be determined from the equation regarding a variance of a distribution mixture (Todinov, 2002):

$$\sigma_f^2 = \sum_{k=1}^{M} p_{k|f}[\sigma_{k|f}^2 + (\mu_{k|f} - \mu_f)^2] \quad (5.16)$$

where $\sigma^2_{k|f}$, $k=1,\ldots,M$ are the variances characterising the loss given failure related to the M failure modes, and

$$\mu_f = p_{1|f}\mu_{1|f} + \cdots + p_{M|f}\mu_{M|f} \tag{5.17}$$

is the mean of the loss given failure; $p_{k|f}$ in equations (5.16) and (5.17) are given by equation (5.12). These are the conditional probabilities that given failure, it is the kth failure mode which has initiated it ($\sum_{i=1}^{M} p_{k|f} = 1$).

Equation (5.16) regarding the variance of a distribution mixture can also be presented as (Todinov, 2003)

$$\sigma^2_f = \sum_{k=1}^{M} p_{k|f}\sigma^2_{k|f} + \sum_{i<j} p_{i|f}p_{j|f}(\mu_{i|f} - \mu_{j|f})^2 \tag{5.18}$$

The expansion of $\sum_{i<j} p_{i|f}p_{j|f}(\mu_{i|f} - \mu_{j|f})^2$ has $M(M-1)/2$ number of terms, equal to the number of different pairs (combinations) of indexes among M indexes. Thus, for three failure modes ($M=3$, Fig. 5.5), equation (5.18) becomes

$$\sigma^2_f = p_{1|f}\sigma^2_{1|f} + p_{2|f}\sigma^2_{2|f} + p_{3|f}\sigma^2_{3|f} + p_{1|f}p_{2|f}(\mu_{1|f} - \mu_{2|f})^2$$
$$+ p_{2|f}p_{3|f}(\mu_{2|f} - \mu_{3|f})^2 + p_{1|f}p_{3|f}(\mu_{1|f} - \mu_{3|f})^2 \tag{5.19}$$

If we examine closely equation (5.18), the reason for the relatively large variance of the loss from multiple failure modes becomes clear. In the equation, the variance of the distribution mixture has been decomposed into two major components. The first component $\sum_{k=1}^{M} p_{k|f}\sigma^2_{k|f}$ including the terms $p_{k|f}\sigma^2_{k|f}$ characterises only variation of the loss *within* the separate failure modes. The second component is the sum $\sum_{i<j} p_{i|f}p_{j|f}(\mu_{i|f} - \mu_{j|f})^2$

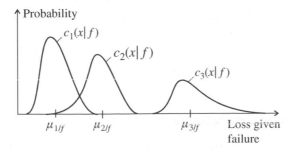

Figure 5.5 Distributions of the loss given failure from three different failure modes.

and characterises variation of the loss *between* the separate failure modes. These terms add up to form the total variance.

If the times to failure characterising the separate failure modes are unknown, the conditional probabilities $p_{k|f}$ are unknown and the variance σ_f^2 in equations (5.18) and (5.19) cannot be determined. Depending on the actual conditional probabilities $p_{k|f}$, the variance σ_f^2 may vary from the smallest variance characterising one of the failure modes to the largest possible variance obtained from a particular combination of failure modes activated with appropriate probabilities. Hence, an important question is to establish an exact upper bound for the variance of the conditional loss from mutually exclusive failure modes. This upper bound can be obtained by using the upper bound theorem proposed and proved in (Todinov, 2003).

Variance upper bound theorem: The exact upper bound of the variance of the conditional loss from multiple mutually exclusive failure modes is obtained from not more than two failure modes.

Mathematically, the upper bound variance theorem can be expressed as

$$\sigma_{\max|f}^2 = p_{\max|f}\sigma_{k|f}^2 + (1 - p_{\max|f})\sigma_{s|f}^2 + p_{\max|f}(1 - p_{\max|f})(\mu_{k|f} - \mu_{s|f})^2 \tag{5.20}$$

where k and s are the indices of the failure modes from which the upper bound of the variance is obtained; $p_{\max|f}$ and $1 - p_{\max|f}$ are the conditional probabilities for which the upper bound (the maximum variance) $\sigma_{\max|f}^2$ is attained. If $p_{\max|f} = 1$, the maximum variance is attained from sampling only the distribution of the loss characterising the kth failure mode.

For a large M, determining the upper bound variance by finding the global maximum of the right-hand side of equation (5.18) or (5.19) regarding the conditional probabilities $p_{k|f}$ is a difficult task which can be simplified significantly by using the upper bound variance theorem. The algorithm for finding the conditional probabilities which maximise the variance of the loss consists of checking the variances of all individual failure modes and the variances from sampling all possible pairs of failure modes (Todinov, 2003). As a result, determining the upper bound variance of the loss from M failure modes involves only $M(M+1)/2$ checks.

5.3.2 Variance of the Potential Loss

The variance of the potential loss will be determined from equation (5.5), which in fact describes a distribution mixture.

Indeed, by using the relationship $p_k = p_f \times p_{k|f}$, equation (5.5) can be presented as

$$c(x) = (1 - p_f)\,\delta(x) + p_f \sum_{k=1}^{M} p_{k|f} c_k(x|f) \tag{5.21}$$

where p_f is the probability of failure.

In fact, the probability density distribution $c(x)$ of the potential loss is obtained by sampling with probability $1 - p_f$ the delta function (corresponding to no failure) and the distribution mixture $\sum_{k=1}^{M} p_{k|f} c_k(x|f)$ which is the conditional distribution of the loss given failure. The variance of a distribution mixture of two distributions characterised by variances σ_1^2, σ_2^2 and means μ_1, μ_2 is (Todinov, 2003):

$$\sigma^2 = p\sigma_1^2 + (1 - p)\sigma_2^2 + p(1 - p)(\mu_1 - \mu_2)^2 \tag{5.22}$$

In equation (5.22), p is the probability of sampling the first distribution and $1 - p$ is the probability of sampling the second distribution. Since the mean and the variance of the delta-function $\delta(x)$ are zero, in equation (5.22) $\sigma_1^2 = 0$ and $\mu_1 = 0$. Next, the probability p in equation (5.22) of sampling the delta function $\delta(x)$ is equal to the probability of 'no failure' $1 - p_f$ while the probability of sampling the conditional distribution of the losses given failure is equal to the probability of failure p_f. After substituting these values in equation (5.22), the expression

$$\sigma^2 = p_f \sigma_f^2 + p_f(1 - p_f)\mu_f^2 \tag{5.23}$$

is obtained for the variance of the potential loss, where σ_f^2 is given by equation (5.18). The mean of the potential loss is

$$\mu = p_f \mu_f \tag{5.24}$$

where μ_f is the mean of the loss given failure determined from equation (5.17). Equation (5.24) is in fact the risk equation (4.1). If the variance σ^2 (or the uncertainty related to the potential loss) is large, the variability of the potential loss is high.

In some cases however, even determining the variance of the potential loss is not sufficient to discriminate correctly between two design solutions. This will be illustrated by the following numerical example. Suppose that two design solutions are compared. The loss given failure for the first solution is normally distributed with mean $\mu_{1|f} = \$320,000$ and standard deviation $\sigma_{1|f} = \$6000$ while the loss given failure for the second solution is normally distributed with mean $\mu_{2|f} = \$290,000$ and standard deviation $\sigma_{2|f} = \$65,000$. The probabilities of failure characterising the two design solutions are $p_1 = 0.55$ and $p_2 = 0.48$, respectively. The expected loss calculated from the risk equation is $K_1 = p_1 \times \mu_{1|f} = \$176,000$ for the first and $K_2 = p_2 \times \mu_{2|f} = \$139,200$ for the second design solution. Since the second design solution is characterised by a smaller expected loss, it will be preferred to the first solution if no further calculations are made.

This conclusion will also be confirmed if the variances of the potential loss characterising the two design solutions are calculated and compared by using equation (5.23). Indeed, after substituting the parameters values, equation (5.23) yields $\sigma_1 = 159,260$, $\sigma_2 = 151,721$ for the standard deviations of the potential loss characterising the two design solutions. In other words, the second solution is not only characterised by a smaller expected loss, but also by a smaller variance of the potential loss – yet another reason to be preferred to the first design solution.

Now let us calculate the probability that the potential loss will exceed the maximum tolerable limit of $x_{max} = \$330,000$. From equation (5.8), this probability is

$$P(X_1 \geq x_{max}) = p_1 \times \left[1 - \Phi \left(\frac{x_{max} - \mu_{1|f}}{\sigma_{1|f}} \right) \right] \approx 0.026$$

for the first design solution and

$$P(X_2 \geq x_{max}) = p_2 \times \left[1 - \Phi \left(\frac{x_{max} - \mu_{2|f}}{\sigma_{2|f}} \right) \right] \approx 0.13$$

for the second design solution, where $\Phi(\bullet)$ is the cumulative standard normal distribution. Clearly, the first design solution is associated with much smaller probability (2.6%) that the potential loss will exceed the maximum tolerable level compared to the second design solution which is characterised by a probability of 13%. The conclusion is that *if design solutions are compared in terms of the probability of the potential loss exceeding a particular threshold, the reliable selection criterion should be based on the actual distribution of the potential loss, not on the expected value or the variance of the potential loss.*

5.3.3 Upper Bound Estimate for the Variance of the Potential Loss in Case of Unknown Probability of Failure

Equation (5.23) is useful for determining the uncertainty associated with the potential loss but it works only if the probability of failure p_f can be determined. Here we show that even if the probability of failure p_f is unknown, the uncertainty associated with the potential loss can still be estimated by using an exact upper bound.

Suppose that only a single failure mode is present, characterised by expected loss given failure μ_f and standard deviation σ_f. The probability of failure p_f yielding the largest uncertainty in the potential loss can be determined by finding the maximum of the right-hand side of equation (5.23) in the interval $(0 \leq p_f \leq 1)$. The maximum variance of the potential loss is attained either at

$$p_f = \frac{\sigma_f^2 + \mu_f^2}{2\mu_f^2} \tag{5.25}$$

if $\sigma_f < \mu_f$ or at $p_f = 1$ if $\sigma_f \geq \mu_f$. This provides an opportunity to estimate an exact upper bound of the variance of the potential loss. If $\sigma_f \geq \mu_f$, the upper bound of the variance of the potential loss σ_{max}^2 coincides with the variance of the loss given failure $(\sigma_{max}^2 = \sigma_f^2)$. If $\sigma_f < \mu_f$, substituting p_f from equation (5.25) into equation (5.23) yields

$$\sigma_{max}^2 = \frac{(\sigma_f^2 + \mu_f^2)^2}{4\mu_f^2} \tag{5.26}$$

Now suppose that M failure modes are present, characterised by probabilities (absolute) $p_1, \ldots, p_M$ $(p_f = p_1 + \cdots + p_M)$. The delta function describing the probability density distribution corresponding to zero loss is characterised by probability $1 - p_f$. According to the upper bound variance theorem, the upper bound of the variance is obtained from sampling not more than two distributions. Since the delta function is the lower bound of the losses from all distributions it is certainly sampled (there are two kind of events only: 'failure' or 'no failure'). The upper bound of the variance of the potential loss is then obtained from sampling the delta function with some of the conditional distributions of the loss given failure characterising the separate failure modes. Consequently, in order to determine the upper bound of the variance of the potential loss expression

(5.26) should be calculated sequentially with each conditional distribution characterising the separate failure modes and the failure mode which yields the largest variance σ_{max}^2 should be used to determine the upper bound.

The upper bound variance σ_{max}^2 or the upper bound standard deviation σ_{max} of the potential loss permits a conservative, yet precise estimate to be made related to the uncertainty associated with the potential loss. Another advantage is that the uncertainty estimate is made without the need to know the probability of failure p_f or any of the conditional probabilities $p_{k|f}$ characterising the separate failure modes!

5.4   COUNTEREXAMPLES RELATED TO THE RISK OF FAILURE OF NON-REPAIRABLE SYSTEMS

The fact that a system with larger reliability does not necessarily mean a system with smaller expected loss from failure can be demonstrated on a very simple system containing only two components, logically arranged in series.

Let the two components building the system be characterised by constant hazard rates λ_1 and λ_2, respectively. Suppose that the expected loss given failure $\overline{C}_1$ associated with the first component is much greater than the expected loss given failure $\overline{C}_2$ associated with the second component $\overline{C}_1 >> \overline{C}_2$. Let $\lambda_1 = \lambda_2 = 0.012$ month^{-1}, $a = 60$ months, $\Delta = 0.004$, $\overline{C}_1 = 1000$ and $\overline{C}_2 = 100$. The probability of system failure before time a is then

$$p_f = (1 - \exp[-(\lambda_1 + \lambda_2)a]) \approx 0.763$$

and the expected value of the potential loss from failure before time a according to equation (5.13) is

$$K = (1 - \exp[-(\lambda_1 + \lambda_2)a]) \times \left(\frac{\lambda_1}{\lambda_1 + \lambda_2}C_1 + \frac{\lambda_2}{\lambda_1 + \lambda_2}C_2 \right) \approx 419.6$$

For a system with components characterised by hazard rates $\lambda_1 + \Delta$ and $\lambda_2 - 2\Delta$, the probability of failure becomes

$$p_f' = (1 - \exp[-(\lambda_1 + \lambda_2 - \Delta)a]) \approx 0.699$$

and the risk according to equation (5.13) is

$$K' = (1 - \exp\left[-(\lambda_1 + \lambda_2 - \Delta)a\right])$$

$$\times \left(\frac{\lambda_1 + \Delta}{\lambda_1 + \lambda_2 - \Delta}C_1 + \frac{\lambda_2 - 2\Delta}{\lambda_1 + \lambda_2 - \Delta}C_2\right) \approx 573$$

As can be verified from the numerical example, the reliability of the second system is larger ($p'_f < p_f$) compared to the first system yet the expected loss (the risk) is also larger ($K' > K$)! Consider now a system including components characterised by hazard rates $\lambda_1 - \Delta$ and $\lambda_2 + 2\Delta$. In this case, the probability of failure increases: $p''_f = 1 - \exp\left[-(\lambda_1 + \lambda_2 + \Delta)a\right] \approx 0.81$, while the risk decreases!

$$K'' = (1 - \exp\left[-(\lambda_1 + \lambda_2 + \Delta)a\right])$$

$$\times \left(\frac{\lambda_1 - \Delta}{\lambda_1 + \lambda_2 + \Delta}C_1 + \frac{\lambda_2 + 2\Delta}{\lambda_1 + \lambda_2 + \Delta}C_2\right) \approx 290.6$$

Clearly, a system with larger reliability may be associated with larger risk of failure. Components arranged in series whose failures are associated with different expected losses given failure can for example be found in subsea oil and gas production. Consider a simple production system consisting of a single well. Among other components arranged in series, the system also includes a topside control equipment (at the sea surface) and a production tree which is installed on the sea bed. Both components are logically arranged in series because oil production stops whenever a critical failure in the topside control equipment or in the production tree occurs. Since a critical failure of the topside control equipment is associated with relatively small downtime, the amount of lost production from this failure is relatively small. The cost of intervention is also relatively small because the topside equipment is easily accessible. Conversely, a critical failure in the difficult to access production tree located on the sea bed is associated with a mobilisation of an expensive oil rig with special equipment and trained crew and the repair takes a long time during which no oil is recovered. The cost of mobilisation of resources, together with the cost of intervention to recover and repair/replace the failed production tree could easily amount to millions of US dollars. The total lost production time could amount to several months which is associated with significant financial losses. Similar is the case of two components, for example a wing valve and a choke valve located on the same production tree. Failure of any of these components requires an immediate intervention for repair. While a failed choke valve could be repaired by an intervention involving a remotely operated vehicle (ROV) which does not require recovering the whole production tree to the surface, a failed wing valve would require such an intervention. This leads to a very high cost of intervention, associated with hiring an oil rig.

5.5 DETERMINING THE LIFE DISTRIBUTION AND THE RISK OF FAILURE OF A COMPONENT CHARACTERISED BY MULTIPLE FAILURE MODES

Suppose that a section from hydraulic fluid supply can fail in two failure modes: by clogging a filter with debris or by clogging a control valve. Suppose that event A_1 denotes no failure due to clogging the filter and event A_2 denotes no failure due to clogging the valve. The probability of no failure can be determined from the probability of the intersection

$$P(A_1 \cap A_2) = P(A_1)P(A_2 | A_1) \qquad (5.27)$$

Clearly, the probability of no failure due to clogging the valve $P(A_2)$ is different from the probability of no failure due to clogging the valve given that no clogging of the filter has occurred $(P(A_2) \neq P(A_2 | A_1))$. These two events are statistically dependent.

Suppose that a given component can fail in n statistically dependent failure modes. Let A_1, $A_2, \ldots, A_n$ be the events: no failure has been initiated from the first, the second, $\ldots$, the n-th failure mode. Let $P(A_k | A_1 A_2 \ldots A_{k-1})$ be the probability of no failure from the kth failure mode given that no failure will be initiated by the first (A_1), the second $(A_2), \ldots$, and the $(k-1)$-st failure mode (A_{k-1}).

The probability of no failure from the n failure modes is determined from the probability of the intersection: $A_1 \cap A_2 \cap \cdots \cap A_n$ that no failure will be initiated by the first, the second, $\ldots$, the nth failure mode. According to the formula related to the probability of an intersection of statistically dependent events (see any book on probability and statistics), this probability is

$$P(no\ failure) = P(A_1) \times P(A_2 | A_1) \times P(A_3 | A_1 A_2) \times \cdots$$
$$\times P(A_n | A_1 A_2 \ldots A_{n-1}) \qquad (5.28)$$

Accordingly, the probability of failure is

$$P(failure) = 1 - P(A_1) \times P(A_2|A_1) \times P(A_3|A_1A_2) \times \cdots$$
$$\times P(A_n|A_1A_2 \ldots A_{n-1}) \tag{5.29}$$

An application of this formula for determining the probability of failure from two statistically dependent failure modes depending on a common parameter will be discussed in Chapter 16.

In the case of statistically independent events, $P(A_k|A_1A_2 \ldots A_{k-1}) = P(A_k)$ and the probability of failure becomes

$$P(failure) = 1 - P(A_1) \times P(A_2) \times P(A_3) \times \cdots \times P(A_n) \tag{5.30}$$

Suppose now that the times to failure characterising the separate failure modes are given by the cumulative distribution functions $F_k(t)$, $k = 1, 2, \ldots, M$ and the expected losses from failure characterising each failure mode are $\overline{C}_1, \ldots, \overline{C}_M$, correspondingly. $F(t)$ which is the probability of failure before time t is simply $F(t) = 1 - R(t)$ where $R(t)$ is the probability that the system will survive time t. In order to survive time t, the system must survive all failure modes. Since $P(A_i) \equiv R_i(t)$ where $R_i(t) = 1 - F_i(t)$ is the probability of surviving the ith failure mode, substituting in equation (5.30) yields

$$F(t) = 1 - \prod_{i=1}^{n} [1 - F_i(t)] \tag{5.31}$$

for the cumulative distribution of the time to failure. Equation (5.31) is also valid for determining the distribution of the time to failure for a system with n components logically arranged in series.

Some of the cumulative distribution functions $F_k(t)$ may be empirical or so complicated that evaluating equation (5.31) may be a difficult task. Furthermore, some of the failure modes may be statistically dependent. A simple Monte Carlo simulation can easily avoid these predicaments and can be used to determine both: the life distribution of the component/system and the risk of failure before a specified time a. Here is the algorithm.

Algorithm 5.1

x[n]:   /* Global array containing the distribution of the lives of the component/system */
C[n]:   /* Global array containing the expected loss from failure associated
            with the separate failure modes */

procedure **Generate_life_from_failure_mode (*j*)**
{
  /* *Generates a time to failure from the j-th failure mode* */
}
/* *Main algorithm* */

S=0;
For *i* = 1 **to** Number_of_trials **do**
{
  /* *Samples the times to failure distributions characterising all failure modes
      and finds the failure mode associated with the smallest time to failure and its index* */

Min_life= **Generate_life_from_failure_mode (1)**; Failure_mode_indx=1;
For *j*=2 **to** *n* **do**
  {
  Life = **Generate_life_from_failure_mode (*j*)**;
  If (Life < Min_life) then {
                                Min_life=Life; Failure_mode_indx=j;
                                }
  }
x[i] = Min_life; /* *Stores the time to failure of the system in array x[ ]* */

if (Min_life < a) **then**
        S = S + C[Failure_mode_indx]; /* *Accumulates the loss if failure occurs
                                        before time a* */
}
Risk = S/Number_of_trials; /* *Calculates the risk of failure before time a* */

Sort array x[] with the times to failure, in ascending order;

At each simulation trial, the time to failure distributions characterising the separate failure modes are sampled sequentially. Since the failure mode associated with the minimum time to failure in fact fails the system, the minimum time to failure from each simulation trial is stored in array $x[ ]$.

Subsequently, a check is performed and if the system fails before time a, the expected loss associated with the failure mode which has failed the system is accumulated in the temporary sum S. At the end of the simulation, the risk of failure before the specified time a is determined by dividing S by the number of Monte Carlo simulation trials. This is effectively the expected value of the loss from all simulations.

After sorting array $x[ ]$ in ascending order, the cumulative distribution of the time to failure can be produced by plotting $i/(Number\_of\_trials + 1)$ against $x[i]$ (Todinov, 2005a).

5.6 UNCERTAINTY AND ERRORS ASSOCIATED WITH RELIABILITY PREDICTIONS

Reliability predictions are always associated with errors. These are is usually due to uncertainty associated with the model parameters and uncertainty associated with the models themselves.

5.6.1 Uncertainty Associated with the Reliability Parameters

The reliability parameters in the models are usually associated with uncertainty. A typical example is the uncertainty associated with the hazard rate of components.

Suppose that components exhibiting constant hazard rates have been tested for failures. After failure, the components are not replaced and the test is truncated on the occurrence of the kth failure, at which point T component-hours have been accumulated. In other words, the total accumulated operational time T includes the sum of the times to failure of all k components. The mean time to failure (MTTF) is estimated by dividing the total accumulated operational time T (the sum of all operational times) to the number of failures k:

$$\hat{\theta} = \frac{T}{k} \tag{5.32}$$

where $\hat{\theta}$ is the estimator of the unknown MTTF. It is a well-established fact that if θ denotes the true value of the MTTF, the expression $2k\hat{\theta}/\theta$ follows a χ^2-distribution with $2k$ degrees of freedom, where the end of the observation time is at the kth failure (Meeker and Escobar, 1998). Consequently, if the number of failures is small, the MTTF can be associated with large uncertainty.

Suppose that the distributions representing the uncertainties of the reliability parameters are known. Then the following two-step Monte Carlo simulation algorithm can be used to handle uncertainty in the reliability parameters.

Algorithm 5.2

function Is_premature_failure()
  { /* Checks for premature failure before a specified time 'a' by using the current values
    of the reliability parameters. Returns '1' if failure before time a has occurred
    and '0' otherwise */
  }

function Generate_random_parameter(*j*)
{
  /* *Generates an instance of the j-th reliability parameter by a random sampling from*
    its distribution */
}
/* *Main algorithm* */
No_failure_counter = 0;
For *i* = 1 **to** Number_of_trials **do**
{
  /* *Generate values for all n reliability parameters* */
  For *j*=1 **to** *n* **do**
    Generate_random_parameter(*j*);
    Failure = **Is_premature_failure()**; /* *Checks for a premature failure by using the*
                                  current values of the reliability parameters */
  If (**not** Failure) **then** No_failure_counter = No_failure_counter + 1;
}
Reliability = No_failure_counter / Number_of_trials.

This algorithm will be verified by a simple example related to calculating the reliability of a system involving two components logically arranged in series, characterised by constant hazard rates. Suppose that the uncertainty associated with the hazard rates of the two components is specified by

$$\lambda_{1\,\min} \leq \lambda_1 \leq \lambda_{1\,\max} \tag{5.33}$$

$$\lambda_{2\,\min} \leq \lambda_2 \leq \lambda_{2\,\max} \tag{5.34}$$

where the component hazard rates are uniformly distributed in the specified uncertainty intervals. The hazard rate of the first component is then characterised by a uniform distribution with probability density function

$$f_1(\lambda) = 1/(\lambda_{1\,\max} - \lambda_{1\,\min}) \tag{5.35}$$

while the hazard rate of the second component is characterised by a uniform distribution with probability density function

$$f_2(\lambda) = 1/(\lambda_{2\,\max} - \lambda_{2\,\min}) \tag{5.36}$$

Since a system failure is present if at least one of the components fails, the probability that the system will survive a time interval of length a is equal to the product of the probabilities $P(T_1 > a)$ and $P(T_2 > a)$ that the first and the second component will survive a time interval of length a. The probability that the hazard rate of the first component will be in the infinitesimally small interval $(\lambda, \lambda + d\lambda)$ and the component will survive time a is given by the product

$$\frac{d\lambda}{\lambda_{1\,max} - \lambda_{1\,min}} \exp(-\lambda a).$$

According to the total probability theorem, the probability that the first component will survive time a, irrespective of where the hazard rate is in the uncertainty interval $(\lambda_{1\,min}, \lambda_{1\,max})$ is given by the integral

$$P(T_1 > a) = \int_{\lambda_{1\,min}}^{\lambda_{1\,max}} \frac{1}{\lambda_{1\,max} - \lambda_{1\,min}} \exp(-\lambda a)\, d\lambda$$

$$= \frac{\exp(-\lambda_{1\,min}a) - \exp(-\lambda_{1\,max}a)}{a(\lambda_{1\,max} - \lambda_{1\,min})} \quad (5.37)$$

Similarly, the probability that the second component will survive time a, irrespective of where the hazard rate is in the interval $(\lambda_{2\,min}, \lambda_{2\,max})$ is given by the integral

$$P(T_2 > a) = \int_{\lambda_{2\,min}}^{\lambda_{2\,max}} \frac{1}{\lambda_{2\,max} - \lambda_{2\,min}} \exp(-\lambda a)\, d\lambda$$

$$= \frac{\exp(-\lambda_{2\,min}a) - \exp(-\lambda_{2\,max}a)}{a(\lambda_{2\,max} - \lambda_{2\,min})} \quad (5.38)$$

The probability that the system will survive time a is then given by

$$P(T_1 > a \cap T_2 > a) = \frac{[\exp(-\lambda_{1\,min}a) - \exp(-\lambda_{1\,max}a)]}{a^2(\lambda_{1\,max} - \lambda_{1\,min})(\lambda_{2\,max} - \lambda_{2\,min})} \quad (5.39)$$

An algorithm for calculating the reliability by incorporating the uncertainty associated with the reliability parameters was used to determine the reliability of a system consisting of two components in series, with hazard rates of the components varying uniformly in the intervals

$$\lambda_{1\,min} = 0.012 \text{ year}^{-1} \leq \lambda_1 \leq \lambda_{1\,max} = 0.045 \text{ year}^{-1} \quad (5.40)$$

$$\lambda_{2\,min} = 0.036 \text{ year}^{-1} \leq \lambda_2 \leq \lambda_{2\,max} = 0.089 \text{ year}^{-1} \quad (5.41)$$

The implementation of Algorithm 5.2 for two components is given below.

Algorithm 5.3

Generate_exp_time_to_failure(lambda)
{/* Generates a time to failure following the negative exponential distribution */
    u = **real_random**();
    time_to_failure = -(1/lambda) x ln(u);
    return time_to_failure;
}
function **Generate_random_hazard_rate**($\lambda_{min}, \lambda_{max}$);
{
    // Generates uniformly distributed hazard rate in the interval λ_{min}, λ_{max}
    u = **real_random**();
    random_lambda=$\lambda_{min} + (\lambda_{max} - \lambda_{min})u$;
    return random_lambda
}
No_failure_counter = 0;
a = 3; *//the specified time interval is three years*
$\lambda_{1\,min} = 0.012\ year^{-1}$; $\lambda_{1\,max} = 0.045\ year^{-1}$; */*specified uncertainty interval*
                                for the first hazard rate */
$\lambda_{2\,min} = 0.036\ year^{-1}$; $\lambda_{2\,max} = 0.089\ year^{-1}$; */*specified uncertainty interval*
                                for the second hazard rate */
For $i = 1$ **to** Number_of_trials **do**
{
    /* Generate values for the hazard rates */
    lambda1 = **Generate_random_hazard_rate**($\lambda_{1\,min}, \lambda_{1\,max}$);
    lambda2 = **Generate_random_hazard_rate**($\lambda_{2\,min}, \lambda_{2\,max}$);
    Time1 = **Generate_exp_time_to_failure**(lambda1);
    Time2 = **Generate_exp_time_to_failure**(lambda2);
If (Time1>a **and** Time2>a) **then** No_failure_counter = No_failure_counter+1;
}
Reliability = No_failure_counter / Number_of_trials.

By using a simulation based on 10 million Monte Carlo simulation trials, a reliability value of 0.762 that the system will survive 3 years of operation has been calculated.

The same probability of 0.762 has been calculated from the theoretical equation (5.39) which illustrates the presented algorithm.

A useful technique for reducing the uncertainty in reliability parameters by combining past information and experimental data is the Bayesian updating technique (Lee, 1997).

It can be illustrated by the following simple example. Suppose that the reliability r of a particular component is determined through testing. Before the start of the test, a uniform prior distribution

$$f(r) = \begin{cases} 1 & for\ 0 \leq r \leq 1 \\ 0 & otherwise \end{cases} \qquad (5.42)$$

is assumed for the unknown reliability. In other words, the unknown reliability value could reside anywhere in the interval $(0, 1)$. Suppose that x components in a particular set of n components have survived the test. In a specified sequence of outcomes, the probability of exactly x components surviving the test is $P(x|r) = r^x(1-r)^{n-x}$. Thus, if surviving the test is denoted by '1' and failing it by '0', the probability of the sequence '01101' containing five tests is $r^3(1-r)^2$. The prior distribution $f(r)$ of the reliability r can then be updated by using the Bayes' theorem (Ang and Tang, 1975):

$$f(r|x) = \frac{r^x(1-r)^{n-x}}{\int_0^1 r^x(1-r)^{n-x}dr}, \qquad for\ 0 \leq r \leq 1 \qquad (5.43)$$

and $f(r|x) = 0$, otherwise. The posterior distribution $f(r|x)$ related to the unknown reliability is the Beta probability distribution (Abramowitz and Stegun, 1972). If $n = 5$ components were tested, three of which survived, the equation for the posterior distribution $f(r|x)$ of the unknown reliability becomes

$$f(r|x) = \frac{r^3(1-r)^{5-3}}{\int_0^1 r^3(1-r)^{5-3}dr} \qquad (5.44)$$

Since $\int_0^1 r^3(1-r)^2dr = 1/60$, the posterior distribution of the reliability representing the associated uncertainty becomes

$$f(r|x) = 60\ r^3(1-r)^2 \qquad (5.45)$$

5.6.2 Uncertainty in Reliability Predictions Associated with Using a Constant Hazard Rate Estimated from Aggregated Data

Poor quality data can be associated with large errors which could give rise to large errors of all subsequent analyses and decisions.

Assuming a constant hazard rate when the hazard rate is not constant, for example, can be a significant source of errors in reliability predictions. Particularly dangerous is the case where early-life failure data or wearout failure data are aggregated with constant failure rate and a common 'constant' failure rate is calculated and used for reliability predictions. This can be demonstrated by the following numerical example. Assume that within a period of 12 years, a particular equipment is characterised by 8 early-life failures within the first 2 years with times to failure 0.2, 0.35, 0.37, 0.48, 0.66, 0.95, 1.4 and 1.9 years, 6 failures in the useful-life period between the start of the 2nd year and the start of the 10th year, with times to failure 2.45, 3.9, 5.35, 7.77, 8.37, 9.11 years, and 12 wearout failures between the start of the 10th year and the start of the 12th year, with times to failure 10.05, 10.46, 10.52, 10.65, 10.82, 11.23, 11.41, 11.53, 11.65, 11.72, 11.84 and 11.98 years. If a constant hazard rate is calculated for the useful-life region, by excluding the early-life failures and the wearout failures, the estimate $\hat{\lambda}_u = 6/\sum_{i=1}^{6} t_i = 0.16$ year^{-1} where $(t_1 = 2.45, \ldots, t_6 = 9.11)$ will be obtained. If the early-life failures are aggregated with the failures in the useful-life region, the estimate $\hat{\lambda}_{e,u} = 14/\sum_{i=1}^{14} t_i = 0.32$ year^{-1} $(t_1 = 0.2, \ldots, t_{14} = 9.11)$ will be obtained. If the early-life failures are excluded but the wearout failures are aggregated with the failures in the useful-life region, the estimate $\hat{\lambda}_{u,w} = 18/\sum_{i=1}^{18} t_i = 0.1$ year^{-1} $(t_1 = 2.45, \ldots, t_{18} = 11.98)$ will be obtained.

If the reliability associated with a time interval of $t = 7$ years is now calculated, given that the equipment has survived 2 years of continuous operation, these three different hazard rates will result in three very different estimates regarding the reliability:

$$R_u = \exp(-\hat{\lambda}_u t) \approx 0.33$$

$$R_{e,u} = \exp(-\hat{\lambda}_{e,u} t) \approx 0.11$$

$$R_{u,w} = \exp(-\lambda_{e,u,w} t) \approx 0.5$$

Let us have a look at the expression used to produce the hazard rate estimate in the useful-life region:

$$\hat{\lambda}_u = \frac{n_u}{\sum_{i=1}^{n_u} t_{i,u}}$$

where n_u is the number of failures in this region and $t_{i,u}$ are the times to failure. Suppose that we aggregate the same number $n_e = n_u$ of failures from the early-life region. Since for the accumulated service time characterising the early-life region $\sum_{i=1}^{n_u} t_{i,e} < \sum_{i=1}^{n_u} t_{i,u}$ is fulfilled, it is clear that

$$\hat{\lambda}_{e,u} = \frac{n_u + n_u}{\sum_{i=1}^{n_u} t_{i,u} + \sum_{i=1}^{n_u} t_{i,e}} > \frac{2n_u}{2\sum_{i=1}^{n_u} t_{i,u}} = \frac{n_u}{\sum_{i=1}^{n_u} t_{i,u}} = \hat{\lambda}_u$$

Alternatively, let us aggregate the same number $n_w = n_u$ of failures from the wearout region. Since for the accumulated service time characterising the wearout region $\sum_{i=1}^{n_u} t_{i,w} > \sum_{i=1}^{n_u} t_{i,u}$ is fulfilled, it is clear that

$$\hat{\lambda}_{u,w} = \frac{n_u + n_u}{\sum_{i=1}^{n_u} t_{i,u} + \sum_{i=1}^{n_u} t_{i,w}} < \frac{2n_u}{2\sum_{i=1}^{n_u} t_{i,u}} = \frac{n_u}{\sum_{i=1}^{n_u} t_{i,u}} = \hat{\lambda}_u$$

which is confirmed by the numerical example. In short, aggregating only early-life failures results in a constant hazard rate larger than the one estimated solely from failures in the useful-life region while aggregating only wearout failures results in a smaller constant hazard rate. As a result, aggregating early-life failures and useful-life failures results in pessimistic estimates of the reliability during the useful-life while aggregating wearout failures and useful-life failures results in optimistic reliability estimates. The effects from aggregating failures from the early-life region and the wearout region have opposite signs and compensate to some extent. Indeed if we aggregate failures from the three regions, the constant hazard rate estimate

$$\hat{\lambda}_{e,u,w} = \frac{26}{\sum_{i=1}^{26} t_i} = 0.146 \text{ year}^{-1}$$

is obtained which is relatively close to the estimate

$$\hat{\lambda}_u = \frac{6}{\sum_{i=1}^{6} t_i} = 0.16 \text{ year}^{-1}$$

characterising the useful-life region. The constant hazard rate assumption has been widely used because of its simplicity. Results from examining real data sets from some well-known data bases, for example, indicated that the useful-life failure data are commonly mixed with early-life or wearout failure data. As demonstrated earlier, calculations based on constant hazard rate when it is non-constant could result in significant errors in the reliability estimates. The example also shows that reliability predictions based on constant hazard rate models estimated from databases which aggregate failure data should be used with caution.

5.6.3   Uncertainty Associated with the Selected Model

If the functions of the system or the underlying physical mechanism of the process have not been well understood, inappropriate models could be selected. Selecting an inappropriate model is associated with large uncertainty in the model predictions. Model selection should not be dictated only by the desire to get the best fit to the observed data. Model selection is about constructing a model which is *consistent with the underlying physical mechanism* of failure.

The capability of a model to give a very good fit to a single data set may indicate little because a good fit can be achieved simply by making the model more complicated (Chatfield, 1998), by *over-parameterising* for example. As a rule, over-parameterised models have a poor predictive capability and are associated with large uncertainty in the model predictions.

Often, the selection of an appropriate model is suggested by the structure of the described quantity. Suppose that a quantity X is a sum $X = \sum_{i=1}^{n} Y_i$ of a large number of statistically independent quantities Y_i, none of which dominates the distribution of the sum. For example, the distribution of a geometrical design parameter (e.g., length) incorporating the additive effects of a large number of factors. If the number of separate contributions (additive terms) is relatively large, and if none of the contributions dominates the distribution of their sum, according to the central limit theorem (DeGroot, 1989), the distribution of the quantity X can be approximated by a normal distribution with mean equal to the sum of the means, and variance equal to the sum of the variances of the separate contributions. The Gaussian distribution will be the appropriate model for X.

Predictions regarding the quantity X will be associated with significant errors if X is modelled by a model different from the Gaussian. Due to the inevitable statistical fluctuation in the data (especially if their number is small) it may occur that an alternative model (different from Gaussian) gives a better fit. This is not a reason however, to select the alternative model. For another, perhaps a larger data set, the alternative model will be associated with poor predictions.

If, on the other hand, the data clearly suggest that the distribution describing its behaviour is multimodal, the Gaussian model (being unimodal) is obviously inappropriate. Thus, if the system is characterised by multiple, mutually exclusive failure modes, the distribution of the loss given failure characterising the system cannot be Gaussian in general, even if all of the individual failure modes are characterised by Gaussian distributions of the loss given failure. The distribution of the loss given failure characterising the system is a distribution mixture.

Now suppose that the quantity X is a product $X = \prod_{i=1}^{n} Y_i$ of a large number of statistically independent quantities Y_i, none of which dominates the distribution of the product. Such is for example the common model $X = Y^n$, where Y is a random variable. According to the central limit theorem, for a large number n, the logarithm of the product $\ln X = \sum_{i=1}^{n} \ln Y_i$ is approximately normally distributed regardless of the probability distributions of Y_i. Consequently, if $\ln X$ is normally distributed, according to the definition of the log-normal distribution, the random variable X is log-normally distributed and this is the appropriate model for X. A model different from the log-normal model here would obviously be inappropriate and associated with errors.

The selected model should also be robust which means that the model predictions should not be sensitive to small variations in the input data. For example, the estimated system reliability should not be overly sensitive to small variations of the reliabilities of the components. As discussed earlier, the small errors associated with the reliabilities of the separate components are often due to insufficient number of observations, errors associated with the recorded times to failure, aggregating wearout data and constant hazard rate data, etc. As a result, non-robust models are often associated with large uncertainty in the model predictions.

5.7   POTENTIAL LOSS AND POTENTIAL OPPORTUNITY

Suppose now that apart from failure events leading to potential losses there exist also opportunity events associated with potential gains. Suppose that the loss events are associated with a probability density function $f(t)$ and cumulative distribution function $F(t)$ of the time to occurrence. Similarly, the opportunity events are characterised by a probability density function $g(t)$ and a cumulative distribution function $G(t)$ of the time to occurrence.

Similar to the way the equations related to the potential losses from failure have been derived, equations can also be derived for the potential gain from M different types of opportunity events.

Following the analogy with the potential loss, the equation regarding the cumulative distribution of the potential gain is

$$G(x) = (1 - p_g) H(x) + p_g \sum_{k=1}^{M} p_{k|g} G_k(x \mid o) \tag{5.46}$$

where p_g is the probability of an opportunity event, $p_{k|g}$ is the conditional probability that given an opportunity event, it is the kth type of opportunity event that will generate the gain. $G_k(x|o)$ is the conditional cumulative distribution characterising the kth type of opportunity events. This is the probability $P(X \le x) \equiv G_k(x|o)$ that the gain X from the kth type of opportunity will not exceed a specified value $x(x > 0)$ given that the opportunity has occurred ('o' in equation 5.46 stands for 'opportunity').

Differentiating equation (5.46) with respect to x yields

$$g(x) = (1 - p_g)\,\delta(x) + p_g \sum_{k=1}^{M} p_{k|g} g_k(x|o) \qquad (5.47)$$

for the probability density of the gain ($\delta(x)$ is the Dirac's delta function which is the derivative of the Heaviside function $H(x)$). For a single type of opportunity events, multiplying by 'x' and integrating both sides of equation (5.47) yields

$$\overline{G} = p_g G_g \qquad (5.48)$$

for the expected value $\overline{G}$ of the potential gain which is a product of the probability of an opportunity and the expected value of the gain given the opportunity.

Equation (5.46) can be used for determining the probability that the potential gain will exceed a specified quantity x. This probability is

$$P(X > x) = 1 - G(x) = 1 - (1 - p_g)\,H(x) - p_g \sum_{k=1}^{M} p_{k|g} G_k(x|o)$$

which, for $x > 0$, becomes

$$P(X > x) = p_g \sum_{k=1}^{M} p_{k|g}[1 - G_k(x|o)] \qquad (5.49)$$

The sum

$$P(X > x|o) = \sum_{k=1}^{M} p_{k|g}[1 - G_k(x|o)]$$

can be interpreted as the conditional probability that given an opportunity, the gain will be greater than x.

The probability that the potential gain will exceed a specified quantity is always smaller than the probability that the conditional gain will exceed the specified quantity. Similar to the measure 'potential loss', the measure 'potential gain' also incorporates the uncertainties in the gains associated with the separate opportunity events.

Suppose now that both failure events and opportunity events can occur during the finite time interval $(0, a)$, where the expected loss given that a loss event has occurred is $\overline{C}$ while the expected gain given that an opportunity event has occurred is $\overline{G}$. It is important to determine the probability of a gain or a loss in the time interval $(0, a)$, and to assess the value of the expected net profit. The probability of a gain can be determined from the following argument.

Consider the probability that an opportunity event will arrive first in the infinitesimal time interval $(t, t + \mathrm{d}t)$. This probability can be expressed as a product $g(t)[1 - F(t)]\mathrm{d}t$ of the probabilities of the following statistically independent events: (i) an opportunity event will arrive in the time interval $(t, t + \mathrm{d}t)$, the probability of which is $g(t)\mathrm{d}t$ and (ii) a failure event will not initiate failure before time t, the probability of which is $[1 - F(t)]$. According to the total probability theorem, the probability that an opportunity event will arrive first in the time interval $(0, a)$ is

$$p_g = \int_0^a g(t)\,[1 - F(t)]\,\mathrm{d}t \tag{5.50}$$

Similarly, for the probability that a failure event will occur first, the expression

$$p_f = \int_0^a f(t)\,[1 - G(t)]\,\mathrm{d}t \tag{5.51}$$

is obtained. The expected net profit $\overline{N}_p$ can then be determined from

$$\overline{N}_p = p_g \overline{G} - p_f \overline{C} \tag{5.52}$$

where the probabilities p_g and p_f are determined from equations (5.50) and (5.51). During selecting an alternative from a set of competing solutions however, the variation of the expected net profit $\overline{N}_p$ is an important factor which must also be considered. Thus, if two alternative solutions are characterised by similar expected net profits but with different variances characterising the loss from failure, the solution associated with the smaller variance of the loss should be preferred. This will be the alternative characterised by a smaller risk of excessive losses.

For opportunity and loss events following a homogeneous Poisson process and characterised by constant rates of occurrence λ_g and λ_f, respectively, $g(t) = \lambda_g \exp(-\lambda_g t)$ and $F(t) = 1 - \exp(-\lambda_f t)$. Substituting these in equation (5.50) and integrating results in

$$p_g = \int_0^a \lambda_g \exp\left[-(\lambda_g + \lambda_f)t\right] dt = \frac{\lambda_g(1 - \exp\left[-(\lambda_g + \lambda_f)a\right])}{\lambda_g + \lambda_f} \quad (5.53)$$

for the probability that an opportunity event will occur first. Similarly, substituting in equation (5.51) and integrating results in

$$p_f = \int_0^a \lambda_f \exp\left[-(\lambda_f + \lambda_g)t\right] dt = \frac{\lambda_f(1 - \exp\left[-(\lambda_f + \lambda_g)a\right])}{\lambda_f + \lambda_g} \quad (5.54)$$

for the probability that a failure event will occur first. Consequently, the expression related to the expected net profit from equation (5.52) becomes

$$\overline{N}_p = [1 - \exp\left(-\lambda_f + \lambda_g\right)a]\left(\frac{\lambda_g}{\lambda_g + \lambda_f}\overline{G} - \frac{\lambda_f}{\lambda_g + \lambda_f}\overline{C}\right) \quad (5.55)$$

Equation (5.55) has been verified. For example, for numerical values: $a = 5$ years; $\lambda_g = 0.55$ year^{-1}, $\lambda_f = 0.35$ year^{-1}, $\overline{C} = \$900$ and $\overline{G} = \$1200$, equation (5.55) yields $\overline{N}_p = \$379$ for the expected net profit, which has been confirmed by a Monte Carlo simulation based on 100,000 trials.

6

LOSSES FROM FAILURES FOR REPAIRABLE SYSTEMS WITH COMPONENTS LOGICALLY ARRANGED IN SERIES

Consider the common case of a repairable system with M components logically arranged in series $(M = 1, 2, \ldots)$ operating during the finite time interval $(0, a)$. The system fails if at least one of its components fails, and in this sense the system is equivalent to a component with M mutually exclusive failure modes. After each system failure, only the failed component is returned to as good as new condition by replacing it with an identical new component and the system is put back in operation. The downtimes after each system failure are neglected for the purposes of the presentation in this chapter.

The expression for the cumulative distribution $L(x)$ of the potential losses from failures in the finite time interval $(0, a)$ is

$$L(x) = \sum_{k=0}^{\infty} p(k)\, C(x|k) \qquad (6.1)$$

where $p(k)$ is the probability of k failures in the time interval $(0, a)$ and $C(x|k)$ is the probability that the losses X will be smaller than x given that exactly k failures have occurred in the time interval $(0, a)$ $(C(x|0) = 1)$. The cumulative distribution function $C(x|k)$ can be determined by solving convolution integrals (intractable approach for a large number of failures) or by a Monte Carlo simulation.

The Monte Carlo algorithm in pseudo-code, yielding the potential losses from a random number of failures in a specified time interval is as follows:

Algorithm 6.1

function Loss_given_failure()
{
Returns a random loss given failure by using the inverse transformation method
(Algorithm A2 from Appendix A)
}

function Number_random_failures()
{
*Returns the number of random failures in the time interval by sampling the point
process describing the failure occurrences*
}

For i=1 to Number_of_trials **do**
{
Generates the number of random failures k in the time interval 0,a.
k = Number_random_failures(); Current_loss = 0;

For j=1 to k **do**
{
     X = Loss_given_failure();
     Current_loss=Current_loss+X; *//Accumulates the loss from the current failure*
}
   Potential_losses[i]=Current_loss; *//Saves the loss from the current simulation trial*
}

Sort the array containing the potential losses in ascending order and build the distribution of the potential losses from failures.

The algorithm stores the number of random failures in the variable k after sampling the point process (e.g. homogeneous Poisson process) modelling the failure occurrences. Subsequently, by using the inverse transformation method (DeGroot, 1989) the distribution of the loss given failure $C(x|f)$ is sampled k times and the sum of the sampled values is stored in the variable Current_loss. At the end of each simulation trial, the potential loss is stored in the array Potential_losses[] which, at the end of the simulation, is sorted in ascending order. The empirical distribution of the potential losses is built by plotting the values Potential_losses[i] against $i/(n+1)$ where $n =$ Number_of_trials is the number of simulation trials.

In all of the derivations below, for the sake of simplicity, we shall be working with the expected value of the losses given failure. The expected loss given k failures $\overline{C}(k)$ in the time interval $(0, a)$ is equal to k times the expected loss $\overline{C}$ given a single failure ($\overline{C}(k)=k\,\overline{C}$). This follows immediately from the additive property of the expected values $\overline{C}(k) = E[X_1 + \cdots + X_k] = E[X_1] + \cdots + E[X_k] = k\overline{C}$, where X_i is a random variable denoting the loss from the ith failure.

The probability that the potential losses from failures will be smaller than or equal to a critical quantity L_{max} ($X \leq L_{max}$) can be determined from the probability of having not more than $r = [L_{\max}/\overline{C}]$ failures in the finite time interval $(0, a)$, where $[L_{\max}/\overline{C}]$ denotes the greatest integer part of the ratio $L_{\max}/\overline{C}$. Consequently, the probability becomes

$$P(X \le L_{\max}) = \sum_{y=0}^{r} p(y) \tag{6.2}$$

where $p(y)$ is the probability of exactly y failures in the interval $(0, a)$. Hence, the probability that the losses from failures will exceed $L_{\max}$ is

$$P(X > L_{\max}) = 1 - \sum_{y=0}^{r} p(y) \tag{6.3}$$

Similar to the potential loss for non-repairable systems, a maximum potential loss L_a at a pre-set level can be defined. The pre-set level α is the probability that the actual loss will be as extreme or more extreme than the maximum potential loss L_a ($\alpha = P(L > L_a)$). Once a probability level α has been specified, the maximum potential losses $L_{a,i}$ characterising alternative design solutions can be compared.

From equation (6.1), an expression for the expected losses from failures in the finite time interval $(0, a)$ is obtained:

$$\overline{L} = \sum_{k=0}^{\infty} p(k) \, k \, \overline{C} \tag{6.4}$$

For systems with components logically arranged in series, all sequential component failures define a point process on the time axis. The system failures comprise all point processes associated with the separate components and can be regarded as a *superimposed point process* (Thompson, 1988). The number of system failures $N(t)$ until time t is

$$N(t) = N_1(t) + \cdots + N_M(t) \tag{6.5}$$

where $N_1(t), \ldots, N_M(t)$ are random variables giving the number of system failures caused by the separate components. Since the intensity of system failures is defined as $\lambda(t) = \lim_{\Delta t \to 0} P[N(t + \Delta t) - N(t) \ge 1]/\Delta t$, and the intensities of component failures are $\lambda_i(t) = \lim_{\Delta t \to 0} P[N_i(t + \Delta t) - N_i(t) \ge 1]/\Delta t$ $(i = 1, 2, \ldots, M)$, $\lambda(t)\Delta t$ gives the probability of at least one system failure in the infinitesimal time interval $(t, t + \Delta t)$, and $\lambda_i(t)\Delta t$ gives the probabilities of at least one system failure caused by the ith component. Since the component failures are mutually exclusive, and there are no simultaneous failure occurrences, at least one system failure is present in the infinitesimally small time interval $(t, t + \Delta t)$ if either at least one system failure caused by the first mode is present or at least one system failure caused by the second failure mode is present and so on. As a result, the probability $\lambda(t)\Delta t$ of at least one system failure in the infinitesimal time interval $(t, t + \Delta t)$, is a sum of the probabilities $\lambda_i(t)\Delta t$ of at least one system failure characterising the separate components (as sum of probabilities of mutually exclusive events): $\lambda(t)\Delta t = \lambda_1(t)\Delta t + \cdots + \lambda_M(t)\Delta t$. Dividing both sides by Δt yields:

$$\lambda(t) = \lambda_1(t) + \cdots + \lambda_M(t) \tag{6.6}$$

for the total system failure density $\lambda(t)$ and the failure densities $\lambda_i(t)$, $i = 1, \ldots, M$ associated with the separate components. We will also refer to $\lambda(t)$ as rate of occurrence of failures (ROCOF) for the system (Ascher and Feingold, 1984).

6.1 LOSSES FROM FAILURES FOR REPAIRABLE SYSTEMS WHOSE COMPONENT FAILURES FOLLOW A NON-HOMOGENEOUS POISSON PROCESS

An important property of the non-homogeneous Poisson process reported for example in Thompson (1988) is that its intensity is equal to the hazard rate of the first arrival time.

Consequently, if the subsequent failures of a component with non-constant hazard rate $\lambda(t)$ define a non-homogeneous Poisson process, the ROCOF for the system will be also $\lambda(t)$, equal to the hazard rate of the component.

Consider now a system consisting of M components characterised by hazard rates $h_1(t), h_2(t), \ldots, h_M(t)$. The subsequent failures and replacements of each component follows a non-homogeneous Poisson process. If only the time to the first failure for the system is considered (the system is considered non-repairable), its hazard rate will be equal to the sum of the hazard rates of the components. For statistically independent components, the system failures will be a superposition of non-homogeneous Poisson processes characterising the failures of the separate components and will also be a non-homogeneous Poisson process. According to the property of the non-homogeneous Poisson process mentioned at the beginning of this section, the system failure intensity $\lambda(t)$ (the ROCOF when the system is considered repairable) will be equal to the system's hazard rate when the system is considered non-repairable. Considering that the hazard rate of the non-repairable system is equal to the sum of the hazard rates of the separate components:

$$\lambda(t) = h_1(t) + h_2(t) + \cdots + h_M(t) \tag{6.7}$$

is fulfilled for the hazard rate of the system. Since the ROCOF for the repairable system is equal to the hazard rate $\lambda(t)$ of the non-repairable system, it is given by equation (6.7).

For a non-homogeneous Poisson process, the probability $p(k)$ of k failures in the finite time interval $(0, a)$ is given by the Poisson distribution

$$p(k) = \frac{1}{k!} \left(\int_0^a \lambda(t)\, dt \right)^k \exp\left(-\int_0^a \lambda(t)\, dt \right) \tag{6.8}$$

where $\lambda(t)$ is the non-constant ROCOF. For a constant cost of failure $\overline{C}$ characterising all components, substituting this expression in equation (6.4) yields

$$\overline{L} = \overline{C} \exp\left(-\int_0^a \lambda(t)\, dt \right) \times \left(\int_0^a \lambda(t)\, dt \right) \left(1 + \frac{1}{1!} \left(\int_0^a \lambda(t)\, dt \right)^1 \right.$$
$$\left. + \frac{1}{2!} \left(\int_0^a \lambda(t)\, dt \right)^2 + \cdots \right) = \overline{C} \int_0^a \lambda(t)\, dt \tag{6.9}$$

for the expected losses from failures in the interval $(0, a)$. The integral $\overline{N} = \int_0^a \lambda(t)\, dt$ is the expected number of failures in the finite time interval $(0, a)$. For components characterised by different, but time-independent costs of failure $\overline{C}_1, \overline{C}_2, \ldots, \overline{C}_M$, the expected losses from failures can be presented as

$$\overline{L} = \overline{C}_1 \left(\int_0^a h_1(t)\, dt \right) + \cdots + \overline{C}_M \left(\int_0^a h_M(t)\, dt \right) \tag{6.10}$$

Equation (6.10) can also be presented as

$$\overline{L} = \overline{N}_1 \overline{C}_1 + \cdots + \overline{N}_M \overline{C}_M \tag{6.11}$$

where $\overline{N}_i = \int_0^a h_i(t)\, dt$ is the expected number of failures associated with the ith component.

Now suppose that the expected losses from failure characterising the components are time-dependent. The expected losses from failures caused by the ith component in the infinitesimal interval $(t, t+dt)$ are given by $dN = h_i(t)\overline{C}_i(t)\,dt$ where $h_i(t)\,dt$ is the expected number of failures in the infinitesimal time interval and $\overline{C}_i(t)$ is the expected loss given failure at time t. The total expected losses from failures from the ith component are then given by the integral $\overline{L}_i = \int_0^t h_i(t)\overline{C}_i(t)\,dt$. The total expected losses from failures for the whole system are then

$$\overline{L} = \int_0^t [h_1(v)\overline{C}_1(v) + \cdots + h_M(v)\overline{C}_M(v)]\,dv \qquad (6.12)$$

Equation (6.9) has important applications. Consider for example early-life failures of a particular system following a non-homogeneous Poisson process during a finite time interval $(0, a)$ (Fig. 6.1(a)). The non-constant ROCOF for the system $\lambda(t)$ has been reduced to a new smaller ROCOF

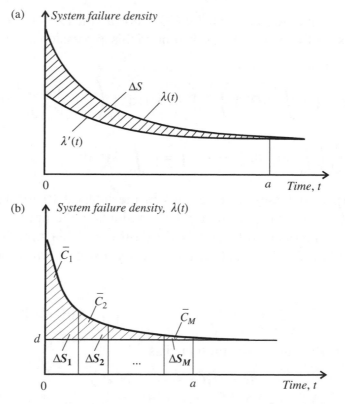

Figure 6.1 (a) Reducing the initial system failure density $\lambda(t)$ to a failure density $\lambda'(t)$; (b) Determining the expected losses from prevented early-life failures.

$\lambda'(t)$, Fig. 6.1(a). This is commonly achieved by measures aimed at reducing early-life failures: for example, by better design, manufacturing, assembly, material, quality control and inspection. If the expected cost of a single failure is $\overline{C}$, according to equation (6.9), the expected losses from failures before the reliability improvement are $\overline{L} = \overline{C} \int_0^a \lambda(t)\,dt = \overline{C}S$, where $S = \int_0^a \lambda(t)\,dt$ is the area S beneath the failure density curve $\lambda(t)$. Similarly, the expected losses from failures after the reliability improvement are $\overline{L}' = \overline{C} \int_0^a \lambda'(t)\,dt = \overline{C}S'$, where $S' = \int_0^a \lambda'(t)\,dt$ is the area beneath the final failure density curve $\lambda'(t)$. Because the integral $\Delta S = \int_0^a [\lambda(t) - \lambda'(t)]\,dt$ is the area between the initial failure density $\lambda(t)$ and the final failure density $\lambda'(t)$ (the hatched area in Fig. 6.1(a)), the prevented expected losses from failures are $\Delta \overline{L} = \overline{C} \Delta S$.

The hatched area ΔS in Fig. 6.1(a) can be interpreted as the expected number of prevented failures from reducing the initial system failure density. The reliability investment creates value if the expected losses $\Delta \overline{L} = \overline{C} \Delta S$ from prevented failures exceeds the costs towards this prevention. An important application of this model is in investigating the effect of eliminating early-life failures on the financial revenue.

The model can also be generalised for the common case where the losses given failure vary with time. In this case, the expected losses given failure are $\overline{C}_1, \overline{C}_2, \ldots, \overline{C}_M$ and depend on the actual time subinterval where the early-life failure occurs, (Fig. 6.1(b)). Let $\Delta S_1, \Delta S_2, \ldots, \Delta S_M$ be the areas of the hatched regions corresponding to each time subinterval. Since ΔS_i equals the expected number of prevented early-life failures in the ith time subinterval, the expected losses $\Delta \overline{L}$ from prevented early-life failures in the time interval $(0, a)$ are determined from

$$\Delta \overline{L} = \Delta S_1 \overline{C}_1 + \Delta S_2 \overline{C}_2 + \cdots + \Delta S_M \overline{C}_M \tag{6.13}$$

For a continuous time dependence $\overline{C}(t)$ of the expected loss given failure, $\Delta \overline{L}$ becomes

$$\Delta \overline{L} = \int_0^a [\lambda(t) - d]\overline{C}(t)\,dt \tag{6.14}$$

where d is the constant hazard rate characterising the useful life region (Fig. 6.1(b)).

6.2   LOSSES FROM FAILURES FOR REPAIRABLE SYSTEMS WHOSE COMPONENT FAILURES FOLLOW A HOMOGENEOUS POISSON PROCESS

Consider now the case where all individual components are characterised by constant hazard rates λ_i. Since after each failure of a component, it is replaced by an identical component, the system failures are a superposition of the components' failures. The subsequent failures and replacements of component i, however, is a homogeneous Poisson process with density numerically equal to its hazard rate λ_i. Since a superposition of several homogeneous Poisson processes with densities λ_i is a homogeneous Poisson process with density

$$\lambda = \sum_{i=1}^{M} \lambda_i \tag{6.15}$$

equation (6.15) also holds for the ROCOF λ of a system with M components logically arranged in series. Combining equation (5.14) related to the expected loss given failure and equation (6.9), results in

$$\overline{L} = \lambda a \overline{C} = \lambda a \sum_{k=1}^{M} \frac{\lambda_k}{\lambda_1 + \cdots + \lambda_M} \overline{C}_k \tag{6.16}$$

for the expected losses from failures in the finite time interval $(0, a)$, $(\lambda = \sum_{k=1}^{M} \lambda_k)$, which can be transformed into

$$\overline{L} = \sum_{k=1}^{M} \lambda_k a \, \overline{C}_k \tag{6.17}$$

If $L_{\max}$ are the maximum acceptable expected losses and the losses $\overline{C}$ associated with failure of each component are the same, solving equation (6.16) with respect to the system ROCOF λ gives

$$\lambda^* = \frac{L_{\max}}{\overline{C} a} \tag{6.18}$$

for the upper bound of the ROCOF which guarantees that if for the system's ROCOF, $\lambda \leq \lambda^*$ is fulfilled, the expected losses from failures $\overline{L}$ will be within the maximum acceptable level $L_{\max}$. This is, in effect, setting reliability requirements to limit the expected losses from failures below a maximum acceptable limit. In cases where the expected losses $\overline{C}_k$ given failure associated with the different failure modes/components are different, the hazard rates which satisfy the inequality

$$\overline{L} = \sum_{i=1}^{M} \lambda_i \overline{C}_i \leq \frac{L_{\max}}{a} \tag{6.19}$$

limits the expected losses from failures below the maximum acceptable level $L_{\max}$.

Setting reliability requirements of the type specified by equation (6.18) can be illustrated by the following example. Suppose that for a particular unit, the losses from failures associated with the separate components are approximately the same. This is common in cases where, in order to replace a failed component, the whole unit has to be replaced. In this case, despite variations in the costs of the separate components belonging to the unit, the losses from failures are determined by the cost of mobilisation of resources and the cost of intervention and replacement of the whole unit, and are the same for all components in the unit. Suppose that the cost of intervention and replacement of the unit is approximately $100,000. If the maximum acceptable losses from failures during 15 years is $600,000, the hazard rate λ of the unit should be smaller than or equal to 0.4 year^{-1}.

$$\lambda \leq \lambda_L = \frac{L_{\max}}{\overline{C}a} = \frac{600000}{15 \times 100000} = 0.4$$

This is an example of setting reliability requirements to limit the expected losses from failures. In cases where the losses from failure of the components are different, the hazard rates of the components which satisfy inequality (6.19) will limit the expected losses from failures below the maximum acceptable level $L_{\max}$.

If the actual losses from failures are of importance, and not the expected losses, the approach to setting reliability requirements is different. For a system whose failures follow a homogeneous Poisson process with density λ, and the cost given failure is constant $\overline{C}$ (irrespective of which component has failed), according to equation (6.2), the probability that the losses from failures X will be smaller than a maximum acceptable value $L_{\max}$ is

$$P(X \leq L_{\max}) = \sum_{y=0}^{r} \frac{(\lambda a)^y}{y!} \exp(-\lambda a) \tag{6.20}$$

where $r = [L_{max}/\overline{C}]$, and the right-hand side of equation (6.20) is the cumulative Poisson distribution. L_{max} could be for example the maximum tolerable losses from failures (e.g. the budget allocated for unscheduled maintenance).

Equation (6.20) can be used to verify that the potential losses from failures do not exceed a critical limit and will be illustrated by a simple example. Suppose that the available amount of repair resources for the first 6 months ($a = 6$ months) of operation of a system is 2000 units. Each system failure requires 1000 units of resources for intervention and repair. Suppose also that the system failures follow a homogeneous Poisson process with intensity $\lambda = 0.08$ month^{-1}. The probability that within the first 6 months of operation, the potential losses X will exceed the critical limit of 2000 units can be calculated by subtracting from unity the probability of the complementary event: 'the potential losses within the first 6 months will not exceed 2000 units':

$$P(X > L_{max} = 2000) = 1 - P(X \leq L_{max} = 2000)$$

Since $r = [L_{max}/\overline{C}] = 2$, applying equation (6.20) results in

$$P(X \leq L_{max} = 2000) = \sum_{y=0}^{2} \frac{(\lambda a)^y}{y!} \exp(-\lambda a) = 0.987$$

for the probability that the potential losses will not exceed the critical value. Hence, the probability that the potential losses will exceed the critical value of 2000 units is

$$P(X > L_{max} = 2000) = 1 - P(X \leq L_{max} = 2000) = 0.013$$

If we require the upper bound of the system failure density which guarantees with confidence $q\%$ that the potential losses in the interval $(0, a)$ will not exceed the maximum acceptable limit L_{max}, the equation

$$\frac{q}{100} = \sum_{y=0}^{r} \frac{(\lambda a)^y}{y!} \exp(-\lambda a) \tag{6.21}$$

where $r = [L_{max}/\overline{C}]$ must be solved numerically with respect to λ.

The solution λ_L gives the upper bound of the ROCOF which guarantees that if for the system ROCOF $\lambda \leq \lambda_L$ is fulfilled, the potential losses L will be within the available resources L_{max} ($L \leq L_{max}$):

$$\overline{L} = \lambda a \overline{C} = \lambda a \sum_{k=1}^{M} \frac{\lambda_k}{\lambda_1 + \cdots + \lambda_M} \overline{C}_k \qquad (6.22)$$

Suppose that a branch in a reliability network, with components logically arranged in series, is characterised by the property that no two components from the branch can be in a failed state at the same time. Since $\overline{C} = \sum_{k=1}^{M} [\lambda_k/(\lambda_1 + \cdots + \lambda_M)]\overline{C}_k$ is the loss given failure of the branch (lost production time, cost, etc.) equation (5.14) can also be used to simplify complex reliability block diagrams by replacing branches with multiple components in series with a single equivalent block (Fig. 6.2(a)) characterised by an equivalent hazard rate

$$\lambda_e = \sum_{i=1}^{M} \lambda_i \qquad (6.23)$$

and equivalent expected loss given failure (e.g. equivalent downtime given failure)

$$\overline{C}_e = \sum_{k=1}^{M} \frac{\lambda_k}{\lambda_1 + \lambda_2 + \cdots + \lambda_M} \overline{C}_k \qquad (6.24)$$

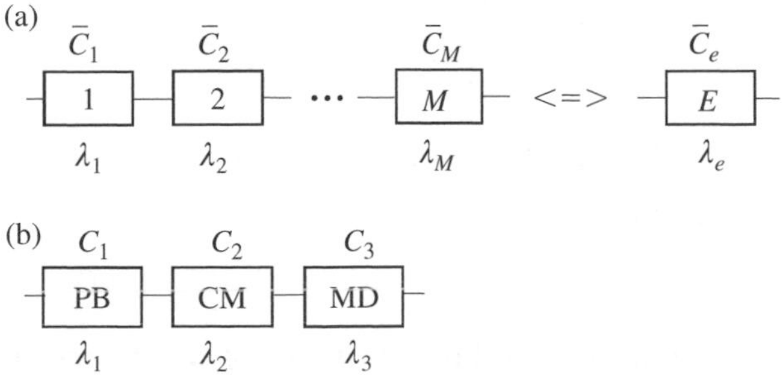

Figure 6.2 (a) Simplifying complex reliability networks by replacing branches with components arranged in series with equivalent single components. (b) A common system in series containing a power block (PB); control module (CM) and mechanical device (MD).

As a result, the speed of the algorithms tracking the losses from failures for complex reliability networks can be increased significantly.

6.3 COUNTEREXAMPLE RELATED TO REPAIRABLE SYSTEMS

The fact that a larger reliability of a repairable system does not necessarily mean lower expected losses from failures can be demonstrated on the simple system in Fig. 6.2(b) composed of a power block (PB), control module (CM) and a mechanical device (MD). Two systems of this type are compared, whose components' hazard rates and losses from failure are listed in Table 6.1.

Table 6.1 Reliability parameters of two repairable systems with components arranged in series (Fig. 6.2(b)).

Component	Hazard rate [year^{-1}]		Losses from failure
	1st system	2nd system	
PB	$\lambda_1 = 0.2$	$\lambda_1' = 0.1$	$C_1 = C_1' = \$2500$
CM	$\lambda_2 = 0.3$	$\lambda_2' = 0.1$	$C_2 = C_2' = \$1500$
MD	$\lambda_3 = 0.21$	$\lambda_3' = 0.26$	$C_3 = C_3' = \$230,000$

The reliability of the first system for $t = 2$ years is

$$R(t) = \exp[-(\lambda_1 + \lambda_2 + \lambda_3)\,t] = 0.24$$

while the expected losses from failures during $t = 2$ years of operation are given by equation (6.17):

$$\overline{L} = \sum_{i=1}^{3} t\lambda_i C_i = \$98500$$

Correspondingly, the reliability of the second system associated with a time interval of $t = 2$ years is

$$R(t) = \exp[-(\lambda_1' + \lambda_2' + \lambda_3')t] = 0.40$$

while the expected losses from failures during $t = 2$ years of operation are

$$\overline{L} = \sum_{i=1}^{3} t\lambda_i' C_i \approx \$120400$$

As can be verified, although the second system has superior reliability, it is also associated with larger expected losses!

This counterexample shows that for a system containing components associated with different losses from failure, a larger system reliability does not necessarily mean smaller losses from failures.

Now, let us assume that the losses given failure characterising all components in the system are the same: $C_1 = C_2 = C_3 = C$. In this case, the expected losses become

$$\overline{L} = \sum_{i=1}^{3} \lambda_i t C = Ct \sum_{i=1}^{3} \lambda_i$$

for the first system and

$$\overline{L}' = \sum_{i=1}^{3} \lambda_i' t C = Ct \sum_{i=1}^{3} \lambda_i'$$

for the second system. Clearly, in this case, the smaller the system hazard rate $\lambda = \sum_{i=1}^{3} \lambda_i$, the larger the reliability of the system, the smaller the expected losses. This example shows that for a system which consists of components associated with the same loss given failure, a larger system reliability always means smaller losses from failures.

6.4 FAILURE AND OPPORTUNITY

Suppose now that on a finite time interval $(0, t)$, failure events follow a homogeneous Poisson process with density λ_f and each realisation of a failure event is associated with expected loss $\overline{C}$. Suppose also that during the same time interval opportunity events also arrive, following a homogeneous Poisson process with density λ_s, each of which yields expected gain $\overline{G}$. The probability that there will be no net loss during the finite time interval $(0, t)$ can be determined from the following argument. The probability that there will be no net loss is equal to the sum of the probabilities of the following compound events: (i) no opportunity events and no loss events in the interval $(0, t)$; (ii) exactly one opportunity event and number of loss events equal or smaller than $k_1 = [\overline{G}/\overline{C}]$ (which is the largest integer not exceeding $\overline{G}/\overline{C}$);

(iii) the probability of exactly two opportunity events and number of loss events equal or smaller than $k_2 = [2\overline{G}/\overline{C}]$ and so on. Expressing this as series results in

$$P(no\ net\ loss) = \exp[-(\lambda_s + \lambda_f)t]\left[1 + \sum_{i=1}^{\infty}\frac{(\lambda_s t)^i}{i!}\left(\sum_{j=0}^{k_i}\frac{(\lambda_f t)^j}{j!}\right)\right] \quad (6.25)$$

for the probability that there will be no net loss in the finite time interval $(0, t)$.

7

RELIABILITY ANALYSIS OF COMPLEX REPAIRABLE SYSTEMS BASED ON CONSTRUCTING THE DISTRIBUTION OF THE POTENTIAL LOSSES

7.1 RELIABILITY NETWORKS OF TWO COMPETING PRODUCTION SYSTEMS

The case study related to determining the potential losses featured in this chapter involves a comparison of the economic performance of a single-control-channel production system based on eight production units (Fig. 7.1) and a dual-control-channel production system also based on eight production units (Fig. 7.2). Each production unit has a production capacity of 200 volume units per day. All production units contribute equally to the total production, with a constant production profile during a 15-year life cycle. The purpose of the reliability value analysis is to

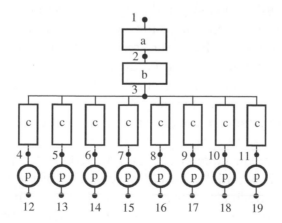

Figure 7.1 A reliability network of an eight-unit single-control-channel production system.

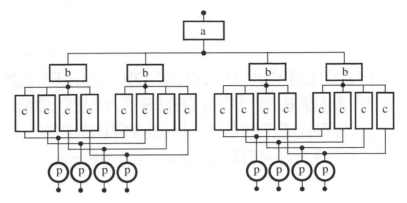

Figure 7.2 A reliability network of an eight-unit dual-control-channel production system.

determine the potential losses for the two systems and by comparing them to determine which production system is associated with smaller potential losses.

The reliability networks related to the production systems are presented in Figs 7.1 and 7.2, respectively, where the production units have been denoted by open circles and identical components have been denoted by identical letters.

There is a distinct difference between the architectures of the two production systems. Indeed, while each production unit from the single-control system in Fig. 7.1 is controlled by a single channel, each production unit from the dual-control system in Fig. 7.2 is controlled from two channels, only one of which is sufficient to maintain production from the production unit. We distinguish two kind of failures in these systems. *Critical failures* cause one or more production units to stop production and require immediate intervention for repair. Failures of redundant components which do not cause any production unit to stop production are *non-critical*. While any component failure in the single-control system is critical and associated with losses, many component failures in the redundant control channels of the dual-control system are not critical and are not associated with losses (Fig. 7.3).

A breakdown repair policy has been adopted, which means that intervention for repair is initiated only if at least one of the production units stops production. Such maintenance policy is common in cases where the cost of intervention is particularly high, such as in deep-water oil and gas production.

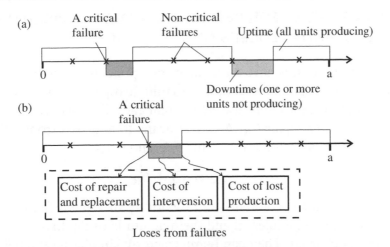

Figure 7.3 (a) Critical and non-critical failures. Only critical failures are associated with losses. (b) Basic components of the losses from critical failures.

A critical failure in the dual-control system is present only if component failures block both control channels to a production unit or the production unit fails itself. This means a lack of link between one of the end nodes (12–19 in Fig. 7.1) (connection) with the start node (node '1' in Fig. 7.1) through working components. For the single-control system, a critical failure is present if any of the eight paths in Fig. 7.1: (1,2,3,4,12), (1,2,3,5,13), (1,2,3, 6,14), (1,2,3,7,15), (1,2,3,8,16), (1,2,3,9,17), (1,2,3,10,18) or (1,2,3,11,19) is blocked (broken). The nodes of the reliability network in Fig. 7.1 (the filled circles) have been numbered by 1, 2, ..., 19.

The last failed component is repaired first. Because this is the component which actually fails the system, it is assumed that this failed component will be identified and repaired first. By repairing the last failed component, the production system is restored to an operational state without having to repair all failed components. (If the dual-control system was put in operation after repairing all failed components, to a large extent, the benefit from the redundant control channels would have been lost.)

No intervention for repair is initiated for non-critical failures which do not cause a production unit to stop production. Non-critical failures are left until a critical failure occurs when, with a single intervention, the last failed component and all previously failed components are replaced. Only then, the costs of intervention and the cost of replacement of all failed components are accumulated.

7.2 AN ALGORITHM FOR RELIABILITY ANALYSIS BASED ON THE POTENTIAL LOSSES FROM FAILURES

The losses from failures have three major components: (i) cost of lost production C_{LP}, (ii) cost of intervention C_I to initiate repair which also includes the cost of mobilisation of resources for repair and (iii) cost of replaced components and cost of repair C_R. As a result, the total losses from failures L can be presented as a sum of these three components

$$L = C_{LP} + C_I + C_R \qquad (7.1)$$

The potential losses combine the probability of critical failures occurring in a specified time interval and the magnitude of the losses given that failures have occurred. They can be revealed by simulating the behaviour of production systems during their life cycles. For this purpose, a discrete-event-driven simulator can be designed, capable of tracking the potential losses for systems with complex topology, composed of a large number of components. Variation of the number of failures and their time occurrences during a specified time interval (e.g. the design life) causes a variation in the potential losses.

The calculated values for the potential losses are subsequently used to build a cumulative distribution which provides an opportunity to determine the probability that the potential losses will exceed a specified critical threshold. Taking the average of the simulated potential losses, from all simulation histories, yields the expected losses from failures.

By using the discrete-event simulator, the potential losses related to alternative solutions can be compared and the solution associated with the smallest potential losses selected.

The average production availability and the cost of lost production are calculated from equation (2.15) as a ratio of the average actual production time and the maximum possible production time. For a life cycle of 15 years and eight production units with equal production capacities, the maximum possible production time is $M_d = 8 \times 15 \times 365 = 43{,}800$ days (1 year $\approx$ 365 days has been assumed), and equation (2.15) yields

$$A_P = 1 - L_d/43{,}800 \qquad (7.2)$$

where $L_d = \sum_{i=1}^{8} l_{d,i}$ is the expected total number of production unit-days lost for all eight production units during the life cycle of the production system.

The expected cost of lost production C_{LP} is calculated from

$$C_{LP} = L_d \times V_d \times P_V \qquad (7.3)$$

where L_d is the expected number of lost production unit-days, V_d is the volume of production per day, per production unit and P_V is the selling price per unit volume production.

A discrete-event simulation was used to reveal the distribution of the potential losses from failures. The simulation is event-driven, with time increments determined by the failure times of the separate components. Four principal blocks form the core of the simulator: (i) event handler, (ii) block for system reliability analysis, (iii) block for generating the lives of the repaired/replaced components, (iv) block for accumulating the losses from failures.

The event-handler tracks all failures (both critical and non-critical) in the system and increments the system time by intervals equal to the time intervals between the component failures. Since no failure occurs within these time intervals, the system time can be increased in large steps, equal to the lengths of these intervals. As a result, the computational speed can be increased significantly compared to methods based on time slicing and checking for failures in each 'time slice'.

After the minimum time to a component failure has been determined, the failed component is saved in a stack and a check is performed whether the component failure has caused a critical failure. A critical failure is indicated by a non-existence of a path to at least one of the production components.

If the analysed component failure is a critical failure, an intervention for repair is simulated and the losses from the critical failure are accumulated. Simulating a repair consists of taking all failed components from the stack and after delays determined by the downtimes associated with repair/replacement of the separate components, new lives for the repaired/replaced components are generated and the components are 'put back' in operation.

A large number of simulations of critical failure histories during the system's life cycle reveals the variation of the potential losses. After finishing all Monte Carlo simulation trials, the cumulative distribution of the potential losses is built. The expected losses from failures are also calculated, by dividing the sum of all potential losses obtained from the separate trials to the number of simulation trials. An outline of the simulation algorithm in pseudo-code is given in the next page.

Algorithm 7.1

For i = 1 to Num_trials **do**
{
Generate new lives for all components and place them in ascending order in a list of component lives;
  Repeat
    {
      Current_time = *The time of the next failure. This time is taken from the head of the list of component lives (the minimum component life)*;
      If (Current_time > *the length of the life cycle*) **then break**;
      Save the component with minimum life into a stack;
      Delete the time to failure of the component from the list of component lives;
    All_paths_exist = **paths**(); /* *checks whether paths exist to all production components* */
  If (All_paths_exist = 0) **then** /* *critical failure* */
    {
      Accumulate the losses from the critical failure in a record with index i, related to the current simulation trial;
    /* *Initiate repair* */
    Take out one by one all failed components from the stack. After delays determined by the downtimes for repair of the components, new lives are generated for the new components which are 'put back in operation';

    Place the new lives of the components into the list of component lives;
    Restore the system's connectivity;
    }
Until (*a break-statement is encountered in the loop*);
}
Sort the records containing the losses from failures obtained in the simulation trials;
Plot the cumulative distribution of the potential losses;

Divide the sum of the potential losses from all simulation trials to the number of trials and determine the expected losses from failures.

The loop with control variable '*i*' executes the block of statements in the braces Num_trials number of times. If a statement '**break**' is encountered in the body of the Repeat-Until loop, the execution continues with the next simulation trial, by skipping all statements between the statement '**break**' and the end of the loop.

A key part of this algorithm is determining after each failure of a component whether there are paths through working components in the reliability network to each production node. If a path to at least one of the production nodes does not exist, a critical failure is registered and repair is simulated.

The existence of paths through working components to all production nodes is determined by using Algorithm 3.5.

7.3 INPUT DATA AND RESULTS RELATED TO THE POTENTIAL LOSSES FOR TWO COMPETING PRODUCTION SYSTEMS

The input data related to the values of the reliability parameters characterising the separate components and the costs associated with intervention and repair are listed in Table 7.1.

Table 7.1 Input data for both production systems.

Component	MTTF (years)	Downtime for intervention and replacement (days)	Cost of replacement ($)
a	10.2	10	2500
b	4.2	16	6500
c	2.0	28	7500
p	12.0	40	9000

All components in the system are characterised by their cumulative distributions of the time to failure. The algorithms described here are capable to handle any specified distribution for the time to failure and the repair time characterising a particular component, including empirically defined distributions. Defining the components with their time to failure distribution rather than with their hazard rate is convenient for one major reason. While the concept hazard rate has no meaning for repairable systems, by using their time to failure distributions, repairable sub-systems can be incorporated in any repairable system and conveniently treated as single components.

Here, for the purposes of the illustration, the hazard rates characterising the separate components have been assumed to be constant. As a consequence, the time to failure of any specified component 'i' is described by the negative exponential distribution:

$$F(t) = 1 - \exp(-t/\mathrm{MTTF}_i) \qquad (7.4)$$

Since this time to failure distribution is fully defined if the mean time to failure MTTF_i of component 'i' is known, the second column of the input data table (Table 7.1) lists the MTTF_i values of the separate components.

Usually, the downtime required for intervention and repair of any of the failed components follows a distribution (e.g. a log-normal distribution).

Table 7.2 Results obtained from running the discrete-event simulator tracking the losses from failures for the dual-control and the single-control system.

Calculated result	Dual-control system	Single-control system
Maximum potential losses at 5% pre-set level	$LD_{0.05} \approx \$27.5 \times 10^6$	$LS_{0.05} = \$59 \times 10^6$
Production availability	97.5%	94.1%
Total expected number of lost production unit-days	1096.9	2580.17
Expected cost of intervention	$\$16.36 \times 10^6$	$\$36.3 \times 10^6$
Expected cost of replacement	$\$0.92 \times 10^6$	$\$0.55 \times 10^6$
Expected cost of lost production	$\$5.27 \times 10^6$	$\$12.38 \times 10^6$
Expected total losses from failures	$\$22.55 \times 10^6$	$\$49.23 \times 10^6$
Standard deviation of the total losses from failures	$\$2.88 \times 10^6$	$\$5.71 \times 10^6$

Here, for the sake of simplicity, these times have also been assumed to be constant.

The values (in days) of the downtimes have been listed in the third column of Table 7.1. We must point out that the described algorithm can be used with any specified type of distribution of the times to failure of the components and the downtimes for repair. For the sake of simplicity, the cost of intervention if a critical failure is present was assumed to be constant: $500,000. The costs of replacement of failed components are listed

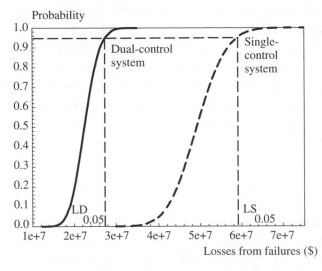

Figure 7.4 Cumulative distributions of the potential losses for the dual-control and the single-control eight-unit production systems from Figs 7.1 and 7.2.

in the fourth column of Table 7.1. For both systems, the selling price of the 200 volume units produced by each production component per day was taken to be $24.

The cumulative distribution of the potential losses has been tracked by simulating the behaviour of the two production systems during their design life of 15 years. All results were obtained from running a discrete-event simulator implementing in C++ Algorithm 7.1 described earlier.

The expected values and the standard deviations of the calculated parameters are listed in Table 7.2. The distributions of the potential losses for both production systems, obtained from 10,000 simulations histories, are presented in Fig. 7.4.

7.4 ANALYSIS OF THE RESULTS

The results in Table 7.2 and Fig. 7.4 show that the eight-unit dual-control system possesses a superior production availability compared to the eight-unit single-control production system. During 15 years of continuous operation, the single-control production system is characterised by approximately 1483 more lost production unit-days compared to the dual-control system. This corresponds to a production availability increase of 3.4% for the dual-control system.

This increase can be partly attributed to the architecture of the dual-control system based on two control channels where a critical failure is present only if component failures block both control channels to a production unit or the production unit fails itself.

At first glance, it appears that the dual-control production system in Fig. 7.2 is a duplicate of a four-unit dual-control production system or can be treated as series of linked one-unit dual-control production systems. Because of the symmetry, it also seems that conclusions related to the production availability characterising a one-unit or a four-unit dual-control production systems could also be stated for the eight-unit dual-control production system. Such statements, however, would be erroneous. A dual-control production system is characterised by enhanced production availability which increases with increasing the number of production units in the system.

Indeed, let us have a look at the reliability network of the eight-unit dual-control production system in Fig. 7.2. The essential difference from the eight-unit single control system in Fig. 7.1 is that when, for example, a production unit from the right-hand group of four production units stops production, *all of the failed components in the system are repaired.* Consequently, critical failures affecting one or more production units from one of the groups of four production units cause failed redundant components associated with the other group of four production units to be repaired/replaced too. The sub-sections including four production units behave like stand-alone production systems (Fig. 7.5(a)) subjected to 'periodic inspection' and repair of failed redundant components. As a result, sub-sections combining four production units in a larger system will exhibit larger production availability compared to stand-alone four-unit systems such as in Fig. 7.5(a).

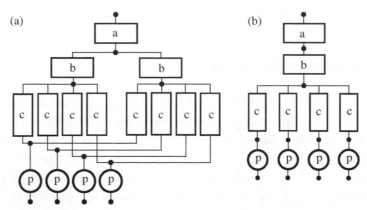

Figure 7.5 Reliability networks of (a) a four-unit dual-control production system and (b) a four-unit single-control production system.

The more production units exist in the system, the more frequently a critical failure will be present, the more frequent will be the 'inspections' and repair of failed redundant components. As a result, a production unit from a multi-unit dual-control production system will have enhanced production availability compared to the case where it works alone! The more production units exist in the system, the greater the increase in the availability of the production unit.

This argument shows that no predictions regarding the production availability of dual-control systems containing a large number of production units should be made by stating the availability characterising dual-control systems containing a smaller number of production units or a single production unit. Separate computer simulations are necessary to determine the production availability in each particular case.

For single-control systems, however, the production availability of a system based on a single production unit is equal to the production availability of a system composed of multiple production units. Indeed, provided that a particular production unit (e.g. the first of the eight units) from the system in Fig. 7.1 is producing, component failures affecting other production units have no effect on it because of the following reasons:

- If any of the other units stops production because of failure of components 'a' or 'b', the first unit (to which corresponds node '12') will also be in a failed state. As soon as the failed component is repaired, the first production unit, with the rest of the units will all be returned in operational state.
- If any other production unit stops production because of failure of a component denoted by 'c' or because of failure of the production unit itself, intervention and repair of the failed components are initiated without interrupting the production from the first unit. Due to the single-control channel, no redundant failed components are repaired upon critical failure; therefore, the reliability of the channels controlling the first production unit is not affected. The availability of the first production unit is equal to its availability as if it works alone, as a stand-alone system such as in Fig. 7.6(b). Because of the symmetry, the production availability of all production units is the same – equal to the production availability of a stand-alone system composed of a single production unit (Fig. 7.6(b)). Hence, the production availabilities of all single-control systems including different number of production units will be the same.

In order to confirm these conclusions, computer simulations have been performed, determining the production availabilities of a one-, four- and

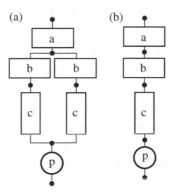

Figure 7.6 Reliability networks of (a) a single-unit dual-control production system and (b) a single-unit single-control production system.

Table 7.3 Production availabilities calculated for one-, four- and eight-unit production systems.

Number of production units	Type of control	
	Dual control (%)	Single control (%)
One	96	94.1
Four	97	94.1
Eight	97.5	94.1

eight-unit dual-control and single-control systems. The results listed in Table 7.3 confirm that if a breakdown repair policy is adopted, with increasing the number of production units, the availability of dual-control production systems increase while the availability of single-control systems remains the same.

The average production availability however is an expected value which does not reveal the variation of the actual availability and from it, the variation of the losses from failures. The combined variation of the potential losses including the variations of the availability, the number of critical failures and the number of failed components have been captured by the maximum potential losses at a pre-set level. The maximum potential losses at the pre-set level determine the necessary capital reserve for covering losses.

The single-control system is characterised by maximum potential losses $LS_{0.05} = \$59 \times 10^6$ at a 5% pre-set level ($\alpha = 0.05$) while the for the dual-control system the maximum potential losses at the same pre-set level was $LD_{0.05} = \$27.5 \times 10^6$. In other words, at a pre-set level of 5% for both systems, the necessary reserve for covering the potential losses associated with the single-control system is more than twice the necessary reserve for the dual-control system.

Compared to the single-control production system, the expected total losses from failures of the dual-control system are by \$26.68 million smaller. The cost of intervention and the production losses associated with the dual-control system are smaller than the corresponding losses characterising the single-control system. The large reduction in the cost of lost production is due to the significant increase of the production availability (by 3.4%). The losses due to intervention also decreased because the redundant control channels built in the dual-control system made critical failures less frequent, which resulted in a less frequent intervention for repair and correspondingly smaller total intervention costs. The replacement costs are higher for the dual-control system because, compared to the single-control system, it includes more components entailing a greater number of component failures and replacements. Despite the greater total number of component failures however, the dual-control system is characterised by fewer critical failures and this is the reason why the intervention losses and the cost of lost production are smaller compared to the single-control production system.

As can be verified from the distributions of the losses from failures characterising the two competing production systems (Fig. 7.4), the variation of the losses from failures is significant for both systems. However, Table 7.2 and the graphs in Fig. 7.4 show that the dual-control production system is characterised by a smaller variation of the losses from failures compared to the single-control system.

Because the single-control production system is associated with larger losses, its corresponding distribution of the potential losses is located to the right from the distribution characterising the dual-control production system (Fig. 7.4). For any specified level of the losses from failures, the probability that the single-control system will yield larger losses is larger than the corresponding probability characterising the dual-control system. In other words, the single-control system is associated with larger risk that the potential losses will exceed a specified value.

The advantage of comparing two alternative solutions built with the same type of components (*a*, *b*, *c* and *p*, see Figs 7.1 and 7.2) is that the effect related to inaccuracies and uncertainties in the input data is eliminated and the effect solely attributable to the system architecture revealed. *In this sense, alternative design architectures can be compared and selected at the design stage even if no reliable data are available for any of the components building the systems.*

7.5 INFLUENCE OF THE SYSTEM TOPOLOGY ON THE LOSSES FROM FAILURES

The influence of the system topology on the expected losses from failures will be illustrated by the three system topologies in Fig. 7.7 based on a single production component (component p in Fig. 7.7).

Suppose again, that all components are characterised by constant hazard rates with MTTF listed in the second columns of Tables 7.4 and 7.5. Again, in case of a critical failure, a constant expected cost of $100,000 per single intervention has been assumed, for all system topologies.

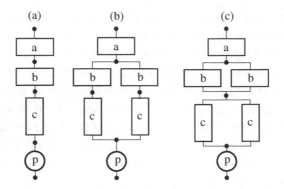

Figure 7.7 Three system topologies based on a single production component.

Table 7.4 Input data set 1.

Component	MTTF (years)	Downtime (days)	Cost of replacement ($)
a	5.2	5	25,000
b	12.4	11	65,000
c	18.3	15	75,000
p	7.5	45	90,000

Table 7.5 Input data set 2.

Component	MTTF (years)	Downtime intervention + repair (days)	Cost of replacement ($)
a	5.2	5	25,000
b	0.3	11	65,000
c	0.5	15	75,000
p	7.5	45	90,000

The downtimes in days are listed in the third columns of Tables 7.4 and 7.5. The expected losses from failures have been simulated during a period of 15 years by using the two data sets given in Tables 7.4 and 7.5. According to equation (2.15), the production availability is determined from

$$A_P = 1 - \frac{L_d}{t \times 365} \qquad (7.5)$$

where t is the number of years ($t = 15$) (1 year $\approx$ 365 days has been assumed) and L_d is the total number of production days lost.

For all system topologies (1, 2 and 3), the expected production availability, expected cost of intervention and expected cost of repair/replacement during 15 years, calculated on the basis of 10,000 simulation trials, have been presented in Figs 7.8–7.10, respectively.

The comparison shows clearly that for production systems with hierarchy, availability is improved significantly by introducing redundancies in sections with inferior reliability compared to the rest of the system. Introducing redundancies in sections with superior reliability compared to

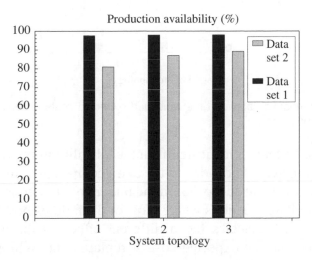

Figure 7.8 Comparison between the simulated production availabilities for the three system topologies in Fig. 7.7.

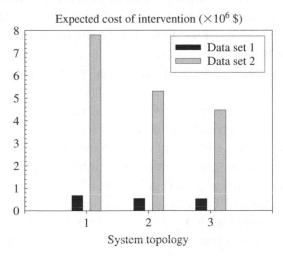

Figure 7.9 Comparison between the simulated expected intervention costs for the three system topologies in Fig. 7.7.

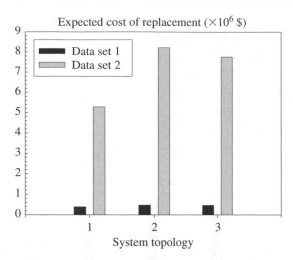

Figure 7.10 Comparison between the simulated expected costs of replacement for the three system topologies in Fig. 7.7.

the rest of the system yields insignificant availability improvement, which is achieved at increased cost due to the extra components. In this case, the reliability improvement due to the redundant sections is compromised by the remaining unreliable sections where most of the failures now concentrate.

Introducing redundancies has a different effect on the expected cost of intervention and the expected cost of replacement. While introducing redundancies increases the cost of replacement, it reduces significantly the intervention costs by reducing the number of critical failures. If the losses from failures are dominated by the cost of intervention (such is the case in deep-water oil and gas production), introducing redundancies reduces significantly the losses from failures. Introducing redundant control modules, for example, decreases significantly the system downtime and the lost production.

An interesting feature in Fig. 7.10 is the decrease of the expected cost of replacement for system topology '3' with respect to topology '2'. Despite that the two systems have identical components, the cross-link (additional connection) in topology 3 makes it more reliable compared to topology 2. While the system topology '2' in Fig. 7.7(b) will fail each time a component 'b' in one of the control branches and component 'c' in the other control branch fail, the system topology '3' (Fig. 7.7c) will still be operating. Consequently, compared to system topology '2', the number of critical failures and the number of replacements for system topology '3' are smaller, which causes the slight decrease of the replacement costs.

8

RELIABILITY VALUE ANALYSIS FOR COMPLEX SYSTEMS

8.1 DERIVING THE VALUE FROM DISCOUNTED CASH-FLOW CALCULATIONS

In order to reveal the net present values (NPV) of competing design solutions, discounted cash-flow models (Wright, 1973; Mepham, 1980; Vose, 2000; Arnold 2005) have been proposed. These cash-flow models have the form

$$\text{NPV} = -C_p + \sum_{i=1}^{n} \frac{\bar{I}_{F,i} - \bar{O}_{F,i}}{(1+r)^i} \qquad (8.1)$$

where n is the number of years, r is the discount rate, C_p is the capital investment, $\bar{I}_{F,i}$ is the expected value of the inflow in the i-th year (the positive cash flow), and $\bar{O}_{F,i}$ is the expected value of the outflow (the total expenditure) in the i-th year (the negative cash flaw). The correct estimation of the losses from failures is at the heart of the correct determination of the outflow. The problem with the classical NPV models related to their application for reliability value analysis is that they are based on the expected values of the outflow. This feature of the models does not permit tracking the variation of the NPV due to variation of the number of failures per year and the variation of their times of occurrence.

Indeed, let us consider a production system with components logically arranged in series, characterised by a constant hazard rate λ. If a period comprising the first year only ($a = 1$) is considered, the expected losses from failures are $\bar{L}_1 = \lambda a \bar{C}_1$ where $\bar{C}_1$ are the expected losses given failure, for the first year only. If the value $\bar{L}_1$ was used to determine the expected outflow $\bar{O}_{F,i}$ in equation (8.1) which is due to losses from failures, the variation of the *NPV* at the end of the first year due to a different number of failures during the year would have been lost completely. Indeed, during the year, there may not be any failures, or there may be a single, two, ..., many failures.

Depending on the actual failure times, the losses from failures vary significantly. Thus, for a deep-water production system, a failure which occurs at the peak of the production profile where the amount of produced oil per day is large, the impact is much greater compared to a failure which occurs towards the end of the system's life cycle, where the amount of recovered oil per day is relatively small.

Furthermore, as shown in Chapter 5, the financial impact associated with early-life failures is significantly larger compared to the financial impact associated with failures occurring later in life because of the time value of money. Furthermore, early-life failures also entail losses due to warranty payments.

In view of the drawbacks of the classical models, a dynamic discounted cash-flow model based on a direct Monte Carlo simulation for determining the variation of the NPV as a function of the failure pattern, appears to be an attractive alternative. The model incorporates the losses from failures, the capital costs and the income generated from selling the product. The dynamic NPV cash-flow model has the form

$$\text{NPV} = -C_p + \sum_{i=1}^{n} \frac{I_{F,i} - O_{F,i}}{(1 + r)^i} \tag{8.2}$$

where NPV is the net present value, n is the number of years, r is a risk-free discount rate, C_p is the capital investment, $I_{F,i}$ is the actual inflow in the i-th year (the positive cash flow) and $O_{F,i} = O_{m,i} + L_i$ is the actual outflow in the i-th year (the negative cash flaw). The actual outflow $O_{F,i}$ is composed of the actual running and maintenance costs $O_{m,i}$ (the maintenance costs due to losses from failures are not incorporated in the term $O_{m,i}$) and the actual losses from failures L_i in year i. At the heart of the NPV model is the model for tracking the potential losses from failures described in Chapter 7.

In accordance with the losses from failures model, the losses from failures

$$L_i = C_{\text{LP},i} + C_{\text{I},i} + C_{\text{R},i} \tag{8.3}$$

in the i-th year are a sum of the cost of lost production $C_{\text{LP},i}$, the cost of intervention $C_{\text{I},i}$ and the cost of repair/replacement $C_{\text{R},i}$ in the i-th year.

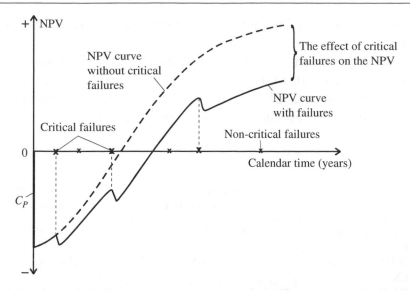

Figure 8.1 Schematic variation of the NPV with time, caused by losses from critical failures.

NPV can be tracked by simulating the behaviour of competing design solutions during their life cycle. Variation of the number of critical failures and their time occurrences during a specified time interval (e.g. the design life) causes a variation in the losses from failures and subsequently, a variation of the NPV (Fig. 8.1).

The calculated values for the losses from failures and the NPV values from each simulation trial (critical failures history) are subsequently used to build a cumulative distribution of the NPV. The average of the calculated NPV from all simulation histories yields the expected value of the NPV. By using the discrete-event simulator, the NPV of two alternative solutions can be compared and the solution associated with the larger NPV selected.

8.2 INPUT DATA FOR THE RELIABILITY VALUE ANALYSIS

The value analysis example here involves a comparison of the economic performance of an eight-unit dual-control-channel production system with an eight-unit single-control-channel production system (Figs 7.2 and 7.1). Each production unit has a production capacity of 200 volume units per day. All production units contribute equally to the total production, with constant production profile during the 15-year life cycle. The purpose of the reliability value analysis is to determine the NPV for the two competing systems and by comparing them, to establish which production system is more beneficial.

In order to determine the distribution of the NPV, in addition to the reliability and maintainability data from Table 7.1, necessary to determine the losses from failures, NPV-specific data related to the competing systems are also necessary. These have been listed in Table 8.1.

Table 8.1 NPV-related input data for calculating the NPV of the two production systems.

	Dual-control production system	Single-control production system
Capital expenditure (CAPEX)	25×10^6	22×10^6
Operating expenditure (OPEX)	0.75×10^6	0.65×10^6

For both systems the selling price of the 200 volume units produced by each unit, each day, was taken to be \$24. All NPV calculations have been performed at a risk-free discount rate of 6%.

8.3 DETERMINING THE DISTRIBUTION OF THE NPV

Losses from failures and NPV have been tracked by simulating the behaviour of the two production systems during their design life of 15 years. The discrete-event-simulator described in Chapter 7 was used for this purpose, with an additional block for calculating the NPV. An outline of the simulation algorithm in pseudo-code is given below.

Algorithm 8.1

For i = 1 **to** Num_trials **do**
{
Generate new lives for all components and place them in a list, in ascending order;

Repeat
        Current_time = *The time of the next failure. This time is taken from the head of the list of component lives (the minimum component life);*
    If (Current_time > *the length of the life cycle*) **then break**;
    Save the component with minimum life into a stack;

Delete the time to failure of the component from the list of component lives;

All_paths_exist = **paths**(); /* *checks whether paths exist to all production components* */

If (All_paths_exist = 0) **then** /* *critical failure* */
{

    Accumulate the losses from the critical failure;
    Record the losses for the particular year;

/* *Initiate repair* */
    Takes out one by one all failed components from the stack. After a delay determined by the corresponding downtimes for replacement/repair, new lives are generated for the new components which are put back in operation;

    Place the new lives of the components into the list of component lives;
    Restore the system's connectivity;
}

Until (*a break-statement is encountered in the loop*);
Calculate the net present value NPV[i] related to the current failure history;
}

Sort the records containing the NPV values obtained in the simulation trails;
Plot the cumulative distribution of the NPV values;

Divide the accumulated NPV-values to the number of trials and determine the expected net present value.

A large number of simulations of critical failure histories during the entire life cycle (Fig. 8.2) reveals the variation of the NPV.

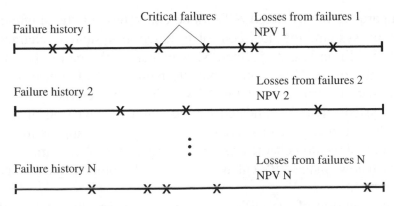

Figure 8.2 A large number of simulated critical failure histories reveals the variation of the losses from failures and the NPV.

After terminating all Monte Carlo simulation trials, the expected NPV is calculated by dividing the accumulated NPV-values to the total number of simulation trials.

The NPV cash-flow model based on discrete-event simulation has significant advantages to cash-flow models based on the expected value of the losses from failures. Unlike these models, the model based on simulation reveals the variation of the NPV caused by the variation of the number of critical failures and their actual times of occurrence during the system's life cycle.

8.4   RESULTS AND ANALYSIS RELATED TO THE NPV

The results were obtained from running the discrete-event simulator, used for tracking the NPV for the dual-control and single-control production system. The expected NPV values and their standard deviations are listed in Table 8.2. Figure 8.3 features the distribution of the NPV values for both production systems, obtained from 10,000 simulations histories.

Table 8.2 Results obtained from running the discrete-event simulator for tracking the NPV and the probability of surviving an MFFOP of 6 months.

Calculated result	Dual-control system	Single-control system
Expected NPV value	$\$89.28 \times 10^6$	$\$75.9 \times 10^6$
Standard deviation of the NPV value	$\$1.93 \times 10^6$	$\$3.83 \times 10^6$
Empirical probability of surviving an MFFOP of 6 months	35.3%	8.23%

MFFOP: minimum failure-free operation period.

Compared to the expected NPV value characterising the single-control production system, the dual-control production system is characterised by $13.38 million larger expected NPV value. Furthermore, the variation (uncertainty) of the NPV values characterising the dual-control system is smaller than the variation of the NPV values characterising the single-control system. This is indicated by the standard deviation of the NPV values characterising the dual-control system which is approximately half the standard deviation characterising the single-control system.

All cash flows in the NPV calculations were discounted by a 6% risk-free discount rate.

The distribution of the NPV values characterising the two competing production systems is presented in Fig. 8.3. Clearly, the variation of the NPV

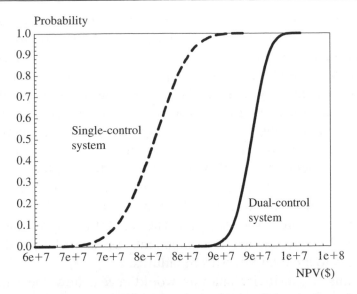

Figure 8.3 Distribution of the NPV for the dual-control and the single-control eight-unit production systems.

solely due to the failure pattern (history) is significant for both production systems. The dual-control production system, however, is characterised by larger NPV values and its corresponding NPV-distribution curve is located to the right of the NPV-distribution curve characterising the single-control system. Compared to the single-control system, for any specified level of the NPV, the dual-control system is characterised by a smaller probability that the actual NPV will fall below the specified level. These comparisons indicate clearly that the dual-control production system is more beneficial compared to the single-control system and should be preferred.

The simulation results demonstrate that the dual-control production system is characterised by a superior availability and smaller losses from failures. An inspection for determining only the status of the components (working or failed) conducted at regular intervals will have little effect on the single-control production system but will have a profound impact on the dual-control production system. Such an inspection will increase the availability of the dual-control system and decrease its losses from failures. Indeed, if failure of a redundant component has been discovered during any of the inspections, repair will be initiated and the failed redundant components will be restored to as-good as-new condition. As a result, critical failures will be delayed. If status inspections however, also monitor the degree of deterioration of the components (e.g. due to corrosion) and excessively deteriorated components are timely replaced, the inspections will have a profound impact on the availability of both systems.

An important step would be to conduct a cost-benefit analysis on the effect of inspections at regular intervals on the losses from failures and the NPV of a dual-control system. For the single-control system, failure of any component is a critical failure. Because there are no redundant control components, determining the status of the components (only verifying whether the components are working or failed) conducted at regular intervals will not have an impact on the availability and losses from failures.

The analysis capability could be improved by developing a module which ranks the separate blocks of the production system according to the losses from failures they are associated with. The reliability improvement efforts should then be directed to the components associated with the largest contributions to the total losses from failures.

Conducting a sensitivity analysis would reveal how the variation of different factors like selling price, operational costs, capital costs and discount rate affects the NPV. The proposed models have been successfully applied and tested for reliability value analyses of productions systems in deep-water oil and gas production.

8.5   ANALYSIS OF THE RESULTS RELATED TO THE PROBABILITY OF EXISTENCE OF THE MFFOP

The probability of surviving a specified minimum failure-free operation period (MFFOP), for example 6 months, for the dual-control and single-control production systems was determined to be 35.3% and 8.23%, correspondingly. The required MFFOP relates to the first critical failure – the first failure which is associated with loss of production from any of the eight production units. For the single-control system, the probability of surviving a period of specified length can easily be determined analytically, which creates an opportunity to validate the empirically obtained probability from the simulator. Since any component failure in the reliability network of the single-control system causes one or more production units to stop production, all component failures are critical failures. The probability of surviving the specified operating period of 0.5 years is equal to the probability of not having a critical failure within this period, which is determined from

$$p_{\text{MFFOP, SC}} = \exp\left[-0.5 \times \sum_{i=1}^{18} \frac{1}{\text{MTTF}_i} \right] \approx 0.082$$

where $MTTF_i$ is the mean time to failure in years, characterising the i-th component. The theoretical probability of surviving 6 months confirms the empirical probability of 8.23% obtained from the discrete-event simulator.

The probability that an MFFOP of length 6 months will exist, is relatively small for both the dual-control and the single-control production systems. Such a low probability of surviving the required period is due mainly to the relatively low MTTF characterising the components building the systems. Despite the relatively large availability of 97.5% characterising the dual-control production system, the probability of surviving a modest period of 6 months is only 35.3%; in other words, a high-production availability can be associated with relatively small reliability. This is yet another confirmation that reliability requirements solely based on availability targets do not necessarily guarantee high reliability and small losses from failures.

8.5.1 A Method for Determining the MFFOP Corresponding to a Pre-set Level

Instead of determining the probability of surviving a specified MFFOP, a desired probability level α can be specified $(0 < \alpha < 1)$ and the $MFFOP_\alpha$ corresponding to this level can be determined.

MFFOP$_\alpha$ is an alternative reliability measure. It is the maximum operating interval, the probability of a critical failure within which does not exceed the pre-set level α.

Unlike the MTTF, which for non-constant rate of occurrence of failures is misleading (as explained in Chapter 2), the MFFOP corresponding to a pre-set confidence level is a powerful reliability measure which *does not depend on the variation of the rate of occurrence of failures with time*.

An MFFOP of 1.5 years at a pre-set level of $\alpha = 0.05$ essentially states that with probability 95%, the system/component will survive 1.5 years of continuous operation without a critical failure (associated with losses). The larger the reliability of the system, the larger the MFFOP at a pre-set level. If a common pre-set level α is specified, the $MFFOP_{\alpha,i}$ characterising different systems can be compared. This creates the possibility to select at the design stage the system architecture/solution characterised by the largest MFFOP.

Determining the MFFOP corresponding to a pre-set confidence level can be done by using an algorithm based on a Monte Carlo simulation. The cumulative distribution of the time to a critical failure of the system (Fig. 8.4) is built first by using the algorithms for system reliability analysis

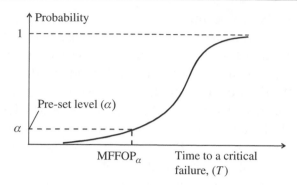

Figure 8.4 Determining the MFFOP corresponding to a pre-set level α from the cumulative distribution of the times to a critical failure.

described in Chapter 3. Suppose that num_times times to a critical failure of the system, obtained from the simulations, are stored in ascending order in the array cumul_array[]. The MFFOP corresponding to a pre-set value of α is obtained by determining a cut off point (in the array) which specifies a fraction of α times to failures. As a result, the MFFOP_α at the pre-set level α is obtained from the dependence

$$\text{Index} = [\alpha \times \text{num\_times}] \tag{8.4}$$

where $[\alpha \times \text{num\_times}]$ denotes the largest integer which does not exceed $\alpha \times \text{num\_times}$ (Fig. 8.4):

$$\text{MFFOP}_\alpha = \text{cumul\_array}\,[\text{index}] \tag{8.5}$$

A numerical example based on new concept can be given with the reliability network in Fig. 3.3 where for each of the six identical components, it is assumed that the time to failure follows the negative exponential distribution

$$F(t) = 1 - \exp(-t/20) \tag{8.6}$$

where time t is measured in years.

For a pre-set probability level $\alpha = 0.05$, the method described earlier yields an MFFOP of 6.65 years ($\text{MFFOP}_{0.05} = 6.65$). The distribution of the time to a critical failure (no path between nodes '1' and '4' (Fig. 3.3)) is given in Fig. 8.5.

8.5.2   Determining the Probability of a Rolling MFFOP

An important characteristic of repairable production systems is the so-called rolling warranty period (a rolling MFFOP). This is a period, free from

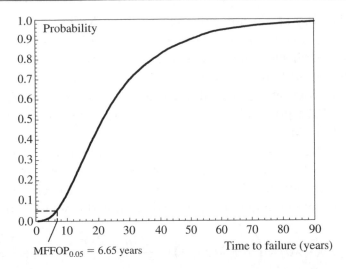

MFFOP$_{0.05}$ = 6.65 years

Figure 8.5 Determining an MFFOP of 6.65 years, guaranteed with 95% confidence level, from the cumulative distribution of the times to a critical failure of the reliability network in Fig. 3.3.

critical failures, which exists *before each* critical system failure (Fig. 8.6). The existence of such a rolling MFFOP guarantees with certain probability that there will be no a situation where a critical failure is followed by another critical failure or more critical failures within a specified time interval.

Figure 8.6 A rolling MFFOP of length *s* (a rolling warrantee) before each critical failure in a specified time interval (0, *a*).

The existence of such a failure-free period is of critical importance to supply systems for example, which accumulate the supplied resource before it is dispatched for consumption (e.g. compressed gaseous substances). Suppose that after a critical failure followed by a repair, the system needs the MFFOP of specified length to restore the amount of supplied resource to the level existing before the critical failure, or to deliver the spare components which have been consumed by the critical failure. In this case, the probability of disrupting the supply equals the probability of clustering of two or more critical failures within the critical recovery period.

The probability of existence of such a rolling MFFOP can be obtained from a Monte Carlo simulation of the performance of the system throughout its life cycle and checking the number of simulation histories N_{MFFOP} during which such an MFFOP exists before each critical failure. Dividing this number to the total number of Monte Carlo simulation trials N_{tr} provides an estimate of the probability ($P_{\text{MFFOP}} \approx N_{\text{MFFOP}}/N_{tr}$) of existence of the specified rolling MFFOP.

In cases where the downtimes are negligible and the systems are characterised by a constant rate of occurrence of critical failures λ, the probability p_{MFFOP} of existence of the MFFOP of length s before each critical failure during a life cycle with length a can be determined from (Todinov, 2004b)

$$p_{\text{MFFOP}} = \exp(-\lambda a) \times \left(1 + \lambda(a - s) + \frac{\lambda^2(a - 2s)^2}{2!} + \cdots + \frac{\lambda^r(a - rs)^r}{r!} \right)$$

$$(8.7)$$

where $r = [a/s]$; $[a/s]$ denotes the greatest integer part of the ratio a/s which does not exceed it.

For systems characterised by a constant rate of occurrence of critical failures λ, determining the MFFOP corresponding to a pre-set level α is reduced to solving numerically the equation

$$1 - \alpha = \exp(-\lambda a) \times \left(1 + \lambda(a - \text{MFFOP}_\alpha) + \frac{\lambda^2(a - 2\text{MFFOP}_\alpha)^2}{2!} \right.$$

$$\left. + \cdots + \frac{\lambda^r(a - r\,\text{MFFOP}_\alpha)^r}{r!} \right)$$

$$(8.8)$$

with respect to MFFOP_α.

9

RELIABILITY ALLOCATION BASED ON MINIMISING THE TOTAL COST

9.1 MINIMISING THE TOTAL COST: VALUE FROM THE RELIABILITY INVESTMENT

Decreasing the probability of failure p_f of a component or system can only be achieved by increasing its reliability. Increasing reliability however requires resources and an optimisation procedure is necessary for minimising the sum of the losses from failure and the cost of resources invested in reliability improvement. The total costs (expenditure) can be presented as:

$$G = Q + K \tag{9.1}$$

where G is the total cost, Q is the cost towards reliability improvement and K is the risk of failure. For multiple failures in a specified time interval, the equation related to the total cost becomes

$$G = Q + \overline{L} \tag{9.2}$$

where again G is the total cost and Q is the cost towards reliability improvement. The difference from equation (9.1) is the term $\overline{L}$ which is the expected losses from multiple failures.

While an item can be 100% free of defects, it can never be characterised by 100% reliability. There are a number of reasons why such a reliability level cannot be achieved: the current state of technology which has certain limitations regarding the strength variability of the produced items, the human errors, which account for a significant number of failures, the existing uncertainty regarding the likely load the product will endure during service, the lack of full control on random failures caused by external overloads, the lack of full control on deterioration processes, the variation of strength which is due to the natural variation of material properties, the presence of software faults, the presence of latent faults in components, etc. Although the variation of the bulk material properties (such as yield strength) can be reduced, it can never be eliminated. For local material properties which control the resistance to fracture and fatigue however (e.g. fracture toughness), variation is always present and cannot be reduced because it is a function of local microstructural features, texture, inclusions, lattice orientation, most of which are random quantities.

In the sections to follow, we will show that improving reliability is associated with reducing the variability of material properties, monitoring the condition of the operating equipment, including redundancy, removing faults, conducting inspections related to the status of the operating components, using better quality materials, using corrosion inhibitors and corrosion protection, increasing the connectivity of the system, more efforts during the design towards eliminating potential failure modes.

Similar considerations apply to reducing losses given failure. Reducing the consequences from failure requires significant investments in fail-safe devices, protection systems, damage arrestors, evacuation equipment, systems for early warning, equipment containing the spread of fire, etc.

Clearly, reducing the risk of failure K in equation (9.1) and the expected losses from failures in equation (9.2) require substantial investment Q. Suppose that an investment Q^* reduces the risk of failure to a particular tolerable level K^*. Clearly, an investment Q significantly beyond this particular level Q^* cannot be justified – the marginal decrease of the risk level cannot outweigh the resources towards the risk reduction.

For the special case of a non-repairable system characterised by a constant hazard rate λ, the total cost $G(\lambda - x)$, associated with decreasing the hazard rate λ by x, can be written in the form

$$G(\lambda - x) = Q(x) + K(\lambda - x) \qquad (9.3)$$

where $K(\lambda - x)$ is the risk associated with decreasing the current system hazard rate λ by x, $Q(x)$ is the cost towards decreasing the current system hazard rate λ by x $(Q(0) = 0)$ and $K(\lambda - x)$ is the risk of failure after decreasing the current hazard rate λ by x. $K(\lambda - x) = p_f(\lambda - x) \times \overline{C}$, where $p_f(\lambda - x)$ is the probability of failure associated with system hazard rate $\lambda - x$. The difference

$$\Delta G = G(\lambda - x) - G(\lambda) = Q(x) + K(\lambda - x) - K(\lambda) \qquad (9.4)$$

gives the relative total cost from decreasing the hazard rate λ by a value x. The hazard rate λ_{opt} which yields the smallest value of the total cost

$G(\lambda - x)$ can be obtained by minimising $G(\lambda - x)$ in equation (9.3) with respect to x in the interval $(0, x_{max})$ $(0 \leq x \leq x_{max} < \lambda)$. If x^* is the value minimising $G(\lambda - x)$, $\lambda_{opt} = \lambda - x^*$ is the optimal value for the hazard rate, which minimises the total cost.

The difference ΔG taken with a negative sign measures the *value from the reliability investment*:

$$V = -\Delta G = G(\lambda) - G(\lambda - x) = K(\lambda) - K(\lambda - x) - Q(x) \qquad (9.5)$$

According to the definitions presented earlier, the optimal hazard rate $\lambda_{opt} = \lambda - x^*$, which minimises the total cost maximises the value from the reliability investment.

A positive sign of V in equation (9.5) indicates that the reliability investment $Q(x)$ creates value: the risk reduction exceeds the cost towards achieving this reduction. The larger V is, the bigger is the value from the reliability investment. A negative sign of V indicates that the amount $Q(x)$ spent on reliability improvement is not justified by the risk reduction.

The amount $Q(x)$ includes the cost of all activities which reduce the losses from failures. $Q(x)$ includes the cost of:

- More reliable materials and components.
- Redundant components.
- Corrosion and erosion protection (coatings, anodes, corrosion inhibitors and corrosion resistant alloys).
- Quality control checks and inspections.
- Cleaner, more homogeneous materials with reduced defect content.
- Protection against intensive wearout.
- Condition monitoring devices, sensors and early warning detectors.
- Preventive and protective risk reduction measures.
- Safety devices.
- Components reducing the spread of damage given that failure occurs.
- Design modifications improving the reliability of the initial design.
- Reliability analyses determining the level of reliability incorporated in the design.
- Reliability design reviews.
- Reliability testing and control after manufacturing.
- Statistical process control during manufacturing.

Often, a decision whether to replace an existing equipment with a more reliable but also a more expensive one is required. Assume that the existing equipment costs Q_0 and the risk of failure in a specified time interval is K_0. The alternative equipment costs Q_1 and the expected loss (the risk) from failure in the specified time interval is K_1. For the specified time interval, the total costs associated with the existing and the alternative equipment are $G_0 = Q_0 + K_0$ and $G_1 = Q_1 + K_1$, respectively. The difference $\Delta V = G_0 - G_1 = Q_0 + K_0 - (Q_1 + K_1)$, which can also be presented as

$$\Delta V = (Q_0 - Q_1) + (K_0 - K_1) \tag{9.6}$$

measures the *value of the alternative solution*. In this case, $K_0 - K_1$ in equation (9.6) is the risk reduction associated with implementing the alternative solution and $|Q_0 - Q_1|$ is the required extra cost at which this risk reduction is achieved.

The alternative solution adds value if $\Delta V > 0$, in other words if the extra cost towards buying alternative equipment is outweighed by the risk reduction.

9.2   RELIABILITY ALLOCATION TO MINIMISE THE TOTAL COST

The strategy to achieve efficient reliability allocation at a component level for a production configuration with a specified number of production units, during a specified time interval, can be outlined as follows.

Suppose that some of the components in the production configuration can be replaced by alternative components, each characterised with particular reliability and cost. Furthermore, redundancies can be selected for the components, or the system topology can be altered in different ways by keeping only the number and the production capacity of the production components.

For a given production capacity, selling price of the product and a discount rate, the net present value obtained from a particular production configuration remains a function of the cost of the equipment, the maintenance and running costs and the losses from failures. The sum of the cost of the equipment, the maintenance costs and the running costs we will refer to as capital costs.

Maximising the net present value involves minimising the sum $G = Q + \overline{L}$ which combines the capital costs Q related to the equipment and the expected losses from failures $\overline{L}$ in the specified time interval. The set of selected alternatives for the components and the modifications in the system topology which minimise the objective function $G = Q + \overline{L}$ is the optimal solution which yields the maximum net profit/value. This optimisation task can be solved numerically, for example, by using a hybrid optimisation method combining a local optimisation and random search. The local optimisation permits descending to a local minimum whilst the random search permits the search to continue in other parts of the space of alternatives so that different prospective local minima are explored.

Let us consider the important special case where for each component in the production system there exist alternatives. Usually, (but not always) the larger the reliability of the alternative, the larger its price, the larger the capital costs. Suppose that for each component i in a system composed of M components, there are n_i available alternatives, each characterised by a different time to failure distribution F_{ij}, and capital costs q_{ij}. The index i stands for the i-th component ($i = 1, 2, \ldots, M$) while index j stands for the j-th alternative ($j = 1, 2, \ldots, n_i$). Usually, the real engineering systems contain a relatively large number of components M with relatively small number of alternatives n_i for each component $i (i = 1, \ldots, M)$ or no alternatives at all ($n_i = 1$). The total number of possible alternatives NA is then equal to

$$NA = n_1 \times n_2 \times \cdots \times n_M \tag{9.7}$$

Let $\mathbf{a} = \{a_1, a_2, \ldots, a_M\}$ be a vector containing the selected alternatives for the components ($1 \leq a_k \leq n_k$), $Q(\mathbf{a}) = q_{a_1} + q_{a_2} + \cdots + q_{a_k}$ be the sum of the total capital costs associated with the selected alternatives and $\bar{L}(\mathbf{a})$ be the corresponding expected losses from failures.

The problem reduces to determining the optimal $\mathbf{a}^* = \{a_1^*, a_2^*, \ldots, a_M^*\}$ alternatives which yield the smallest sum $G(\mathbf{a}) = Q(\mathbf{a}) + \bar{L}(\mathbf{a})$ for a repairable system or $G(\mathbf{a}) = Q(\mathbf{a}) + K(\mathbf{a})$ for a non-repairable system. The optimal selection of alternatives can be obtained by minimising $G(\mathbf{a})$ with respect to $\mathbf{a}$.

Since, the expected losses from failures are a product of the expected number of failures $\bar{N}(\mathbf{a})$ and the expected loss given failure, the objective function to be minimised is

$$G(\mathbf{a}) = Q(\mathbf{a}) + \bar{N}(\mathbf{a}) \times \bar{C} \tag{9.8}$$

in case of a repairable system

$$G(\mathbf{a}) = Q(\mathbf{a}) + p_f \times \bar{C} \tag{9.9}$$

in case of a non-repairable system.

By expressing $Q(\mathbf{a})$ in equations (9.8) and (9.9), the problem is reduced to selecting M alternatives for the separate components $(a_1, a_2, \ldots, a_M)$ such that the sum

$$G = \sum_{i=1}^{M} q_{i,a_i} + \overline{L}(a_1, a_2, \ldots, a_M) \tag{9.10}$$

in case of multiple failures (repairable system) or

$$G = \sum_{i=1}^{M} q_{i,a_i} + K(a_1, a_2, \ldots, a_M) \tag{9.11}$$

in case of a single failure (non-repairable system) are minimised, where $\sum_{i=1}^{M} q_{i,a_i}$ are the capital costs associated with the selected alternatives, $\overline{L}(a_1, a_2, \ldots, a_M)$ are the expected losses from failures associated with them and $K(a_1, a_2, \ldots, a_M)$ is the risk of failure.

Equation (9.10) can also be presented as

$$G = \sum_{i=1}^{M} q_{i,a_i} + \overline{N}(a_1, a_2, \ldots, a_M) \times \overline{C} \tag{9.12}$$

where $\overline{N}(a_1, a_2, \ldots, a_M)$ is the expected number of failures in the specified time interval and $\overline{C}$ is the expected loss given failure.

9.3   RELIABILITY ALLOCATION TO MINIMISE THE TOTAL COST FOR A SYSTEM WITH COMPONENTS LOGICALLY ARRANGED IN SERIES

9.3.1   Repairable Systems

Consider now the important special case of a repairable system composed of M sub-systems logically arranged in series. This means that the sub-system's failures are mutually exclusive and no two sub-systems can be in a failed state at the same time. The total expected losses from failures are therefore a sum of the expected losses from failures generated by the separate sub-systems. Equation (9.12) therefore becomes

$$G = \sum_{i=1}^{M} [q_i + \overline{N}_i \times \overline{C}_i] \tag{9.13}$$

where q_i, N_i and $\overline{C}_i$ are all related to the i-th sub-system.

G in equation (9.13) can be minimised if the sums $q_i + \overline{N}_i \times \overline{C}_i$ are minimised individually for each sub-system. Consequently, a reliability allocation which maximises the net profit for a system composed of sub-systems logically arranged in series is achieved by determining the component alternatives which minimise the sum of the capital costs and the expected losses from failures for each sub-system.

For the important special case of sub-systems which are single components characterised by constant hazard rates λ_i, $i = 1, 2, \ldots, M$, the expected number of failures of the i-th component is $\overline{N}_i = \lambda_i a$, where a is the length of the time interval. Equation (9.13) then becomes

$$G = \sum_{i=1}^{M} [q_i + \lambda_i a \overline{C}_i] \tag{9.14}$$

where $\overline{C}_i$ are the expected losses given failure of the i-th component. The term $\lambda_i a \overline{C}_i$ in equation (9.14) gives the expected losses from failures associated with the i-th component. The total cost G in equation (9.14) can be minimised if $q_i + \lambda_i a \overline{C}_i$ are minimised individually, for each component. Consequently, a reliability allocation which maximises the value for a system with components logically arranged in series is achieved by determining the alternatives which minimise the total cost – the sum of the capital costs and the losses from failures for each component in the system. Equation (9.14) and the reliability allocation algorithm described are also valid for single-control production systems with hierarchy, similar to the system in Fig. 7.1. A characteristic feature of these systems is that failure of any component immediately incurs losses. Similar to the systems with components arranged in series, the total losses from failures are a sum of the losses generated by failures of the separate components.

9.3.2 Non-Repairable Systems

If the cost of failure $C(t)$ is a discrete function accepting constant values $C_1, C_2, \ldots, C_N$ in N years, equation (9.3) regarding the total cost becomes:

$$G(\lambda - x) = Q(x) + \sum_{i=1}^{N} \frac{C_i}{(1+r)^i} \{\exp[-(i-1)(\lambda - x)] - \exp[-i(\lambda - x)]\}$$

$$\tag{9.15}$$

where $K = \sum_{i=1}^{N} \frac{C_i}{(1+r)^i} [\exp(-(i-1)(\lambda - x)) - \exp(-i(\lambda - x))]$ gives the risk of failure within N years; the index 'i' denotes the i-th year and C_i is the loss given failure in the i-th year, combining the cost of the component and the cost of the consequences from failure. In equation (9.15) r is the discount rate.

The right-hand side of equation (9.15) can be minimised numerically with respect to x. If x^* is the value minimising $G(\lambda - x)$, $\lambda_{opt} = \lambda - x^*$ is the optimal value for the hazard rate. The problem is from one-dimensional non-linear optimisation and can for example be solved by using standard numerical methods. It is possible that a decrease in $G(\lambda - x)$ may follow after some initial increase. In other words, the investment $Q(x)$ must go beyond a certain value before a decrease in the total cost $G(\lambda - x)$ can be expected. This is for example the case where a certain amount of resources is being invested into developing a more reliable design. If the investment does not continue with implementing the design however, no reduction of the total cost will be present despite the initial investment in developing the design. In fact, the total cost will be larger.

Often, only information regarding the costs $Q(x_i)$ of N alternatives $(i = 1, \ldots, N)$ is available. In this case, the total costs $G(\lambda - x_i)$ characterising all alternatives can be compared and the alternative $k(1 \leq k \leq N)$ characterised by the smallest total cost $G(\lambda - x_k)$ selected.

An equation for the total cost can also be constructed for a system with M components logically arranged in series. Assume that for each component, different alternatives exist, characterised by different reliabilities and costs. For any selected vector of alternatives characterised by times to failure distributions $\{F_{a_1}, F_{a_2}, \ldots, F_{a_M}\}$ corresponds a cost $Q(\mathbf{a}) = q(a_1) + \cdots + q(a_M)$, where $q(a_i)$ is the cost of the selected alternative for the i-th component. The total cost is

$$G(a) = Q(a) + \sum_{i=1}^{N} \left\{ P(i-1 \leq T \leq i) \times \frac{1}{(1+r)^i} \sum_{k=1}^{M} p_{k|f} \overline{C}_{k|f} \right\} \quad (9.16)$$

where $P(i-1 \leq T \leq i)$ is the probability that the time to failure of the system will be between the i-1st and the i-th year; $p_{k|f}$ are the conditional probabilities that given failure, it is the k-th component which initiated it first (see equation 5.2).

By minimising the right-hand side of equation (9.16) numerically with respect to the separate alternatives $\{a_1, \ldots, a_M\}$, an optimal vector

$\{a_1^*, \ldots, a_M^*\}$ of alternatives can be determined which minimises the total cost and maximises the value from the reliability investment.

9.4 RELIABILITY ALLOCATION BY EXHAUSTIVE SEARCH THROUGH ALL AVAILABLE ALTERNATIVES

Within each sub-system, provided that the number of different alternatives NA is not too large, the sums $q_i + \overline{N}_i \times \overline{C}_i$ can be minimised by using a full exhaustive search through all possible combinations of alternatives. An algorithm based on a full exhaustive search serves as a benchmark for all heuristic algorithms, designed for optimisation of systems including a large number of components and alternatives. Usually, such heuristic algorithms combine random selections in the space of alternatives and local minimisation.

Suppose that a particular sub-system is composed of M components $(i = 1, M)$, and for each component i there are n_i available alternatives, each characterised by a distribution of the time to failure F_{ij}, and capital cost q_{ij}. The index i stands for the i-th component $(i = 1, 2, \ldots, M)$ and index j stands for the j-th alternative $(j = 1, 2, \ldots, n_i)$. A recursive algorithm which generates all possible alternatives is given below.

Algorithm 9.1

na[M] = {n1, n2, ..., nM}; // *Contains the numbers of the alternatives n1, ..., nM*
available for all M components

cur[M];   // *Contains the current alternative*;
M;   // *Contains the number of components in the subsystem*;
Gmin = **total cost from all first alternatives**;   // *Contains the minimal current total cost;*
Initially set to be equal to the total cost
associated with all first alternatives
of the components;
cur_min[M]; // *Contains the indices of the alternatives yielding the current minimal*
total cost

procedure **combin** (beg)
{
  if (beg = M+1) **then** {
            // *The current combination of alternatives is in the array cur[M];*
            Use the current combination of alternatives to calculate the
            expected losses from failures L associated with the subsystem;
            Calculate the cost Q of the alternatives stored in the array cur[M];

Calculate the current total cost G = Q + L for the subsystem;

If (G < Gmin) **then** {Gmin=G; **Save the current alternatives in**
cur_min[M];}
}
else
  {
    for j=1 to na[beg] **do**
    {
      cur[beg]=j; combin(beg+1); // *Recursive call*
    }
  }
}
}
// *A call from the main routine:*
combin(1);
}

Procedure combin() is called recursively from the loop 'j' which scans all possible alternatives for component with index 'beg'. Initially, procedure combin() is called with parameter beg=1. The size of the initial task related to generating all possible alternatives for the sub-system containing M components has been reduced by decomposing it to two simpler tasks: going through all possible alternatives (whose number is stored in na[1]) of the first component, and combining these with all possible alternatives for the remaining $M - 1$ components gives all possible alternatives. In turn, finding all possible alternatives for the remaining $M - 1$ components is obtained by going through all possible alternatives (whose number is stored in na[2]) for the second component and combining them with all possible alternatives for the remaining $M - 2$ components and so on. Following this algorithm, finding all possible alternatives is organised with recursive calls. A return from a recursive call is executed if no more components exists (if the value of the variable beg equals $M + 1$). At that point, a full set of alternatives exists in the array cur[M]. The capital cost Q associated with these alternatives is then calculated as well as the losses from failures associated with them. The total cost G is obtained as a sum of these two quantities and is subsequently compared with the current minimum value G_{min} of the total cost obtained so far. If the current total cost G is smaller than the current minimum value G_{min}, G replaces G_{min} and the indices of the current alternatives yielding the minimum total cost are saved. This process continues, until the recursive procedure exhausts all possible sets of alternatives. At the end, the set of alternatives yielding the minimum total cost will be obtained.

The same algorithm can also be used for minimising the total cost $G = Q + K$ for a non-repairable system. The only modification is that instead of the expected losses from failures L, the risk of failure K as a function of the selected alternatives is calculated.

If the total number of alternatives is very large, minimising the total cost can be done by using heuristic algorithms. Such is for example the implementation of genetic algorithms for determining the set of optimal alternatives which minimises the total cost (Hussain and Todinov, 2007).

9.5 NUMERICAL EXAMPLES

The described algorithms will be illustrated by a simple numerical example. Suppose that for the system in Fig. 6.2(b) composed of three components logically arranged in series, three alternatives exist for the power block (PB), control module (CM) and mechanical device (MD), with hazard rates (year^{-1}) specified by the matrix:

$$\lambda = \begin{pmatrix} 0.5 & 0.15 & 0.34 \\ 0.25 & 0.51 & 1.1 \\ 0.44 & 0.001 & 0.11 \end{pmatrix} \tag{9.17}$$

and prices specified by the matrix

$$q = \begin{pmatrix} \$370 & \$596 & \$421 \\ \$328 & \$211 & \$48 \\ \$680 & \$950 & \$800 \end{pmatrix} \tag{9.18}$$

where λ_{ij} and q_{ij} give the hazard rate and the cost of the j-th alternative of the i-th component (PB, $i = 1$; CM, $i = 2$; MD, $i = 3$). Failure of any component causes a system failure whose cost, for simplicity, has been assumed to be constant: $C = \$1000$.

During 2 years of continuous operation ($a = 2$ years), the minimum total cost is attained for alternatives ($a_1 = 2$, $a_2 = 1$, $a_3 = 2$): the second alternative of the first component, the first alternative of the second component and the second alternative of the third component. These alternatives have been obtained by minimising $q + \lambda a C$ for each component separately, according to the method discussed earlier. The minimum total cost during 2 years of operation is

$$G = q_{1,a_1} + q_{2,a_2} + q_{3,a_3} + \lambda_{1,a_1} a C + \lambda_{2,a_2} a C + \lambda_{3,a_3} a C = \$2676$$

It is interesting to point out that only alternatives characterised by the smallest hazard rates for the three components have been selected by the algorithm. This is because, for repairable systems, the accumulated losses from failures of components dominate the costs of the components. Consequently, selecting alternatives associated with small hazard rates reduces the total cost, especially for long life cycles.

Suppose now, that the system is non-repairable, and the focus is on the first and only failure before $a = 2$ years. The cost of system failure is again $C = \$1000$, the hazard rates and the costs of the alternatives are specified again by matrices (9.17) and (9.18). Since the system is now non-repairable, the loss from failure is the risk of failure before $a = 2$ years. Alternatives λ_{1,a_1}, λ_{2,a_2} and λ_{3,a_3} are now sought for the components which minimise the total cost:

$$G = q_{1,a_1} + q_{2,a_2} + q_{3,a_3} + C \times (1 - \exp[-(\lambda_{1,a_1} + \lambda_{2,a_2} + \lambda_{3,a_3})a]) \quad (9.19)$$

The discount rate r has been assumed to be zero.

The exhaustive search algorithm described earlier, yields alternatives $a_1 = 1$, $a_2 = 3$ and $a_3 = 1$ for the first, the second and the third component, respectively, which yield the minimum total cost $G_{\min} = \$2081$. Clearly, the alternatives which minimise the total cost for a repairable system are not necessarily the ones which minimise the total cost for a non-repairable system.

By analysing equation (9.19) we can conclude that provided that the hazard rates of some of the components in series are large (e.g. hazard rates λ_{1,a_1} and λ_{2,a_2}), the hazard rate of the remaining components (in our case the hazard rate λ_{3,a_3}) has a little impact on the total cost, while the impact of the cost of the remaining components on the total cost is significant. This is the reason behind the selection of the first alternative for the third component. Instead of the very reliable second alternative which is more expensive, the significantly less reliable but at the same time less expensive first alternative has been selected. The reason is that the relatively large hazard rates characterising the alternatives of the first two components already yield a relatively large probability of failure before $a = 2$ years. The reliability of a system in series is smaller than the reliability of the least reliable component. Little risk reduction is gained by selecting a very reliable component if there is at least a single component in the system whose reliability is low. In fact selecting a high-reliability alternative in this case will increase the total cost because it costs more. This is why the second alternative for the third component was not selected.

9.6 APPLICATIONS

9.6.1 Reliability Allocation Which Minimises the Risk of Failure

Reliability allocation to minimise the risk of failure will be illustrated by an underwater assembly composed of three components (a valve block (1), a control umbilical (2) and a control unit (3)). They are logically arranged in series, which means that failure of any component causes the system to stop production which requires an intermediate intervention for repair. Failure of the valve block requires the whole installation to be retrieved to the surface which is a very expensive and lengthy intervention involving a large specialised intervention vessel. During the intervention, the system is not producing, which incurs extra cost of lost production.

Failure of the control umbilical or the control unit requires only a remotely operated vehicle (ROV), the cost of whose deployment is significantly smaller compared to the cost of the large intervention vessel required for the valve. The times for repair of the control umbilical and the control unit are also significantly smaller compared to the time for repair of the valve. Suppose that the expected cost of failure associated with the valve is $C_1 = £930,000$, the cost of failure of the umbilical is $C_2 = £112,000$ while the cost of failure of the control unit is $C_3 = £58,000$. Suppose also that three alternatives exist for each component, with hazard rates (year^{-1}) specified by the matrix:

$$\lambda = \begin{pmatrix} 0.31 & 0.22 & 0.15 \\ 0.15 & 0.09 & 0.03 \\ 0.27 & 0.18 & 0.11 \end{pmatrix} \tag{9.20}$$

and prices specified by the matrix

$$q = \begin{pmatrix} \$90,000 & \$135,000 & \$165,000 \\ \$88,000 & \$96,000 & \$160,000 \\ \$39,000 & \$49,000 & \$78,000 \end{pmatrix} \tag{9.21}$$

The time interval of operation was set to be $a = 10$ years.

The alternatives $a_1 = 3$, $a_2 = 1$ and $a_3 = 1$ for the components, selected by the optimisation algorithm yielded a total cost $556,582. For the valve (1), the third (the most reliable) alternative has been selected, which was also the most expensive. For the umbilical (2) and the control unit (3), the least expensive alternatives rather than the most reliable ones have been selected. This selection guaranteed a minimum sum of the cost of the alternatives and the risk of failure before $a = 10$ years.

9.6.2 Redundancy Optimisation

An important application of the proposed reliability allocation methods is in topology optimisation where for a system with a specified topology, a decision has to be made on where to introduce redundancy so that the total cost is minimised. Consider for example a non-repairable system containing M components logically arranged in series, which is required to survive a particular number of years (a) of operation without failure. Two alternatives exist for each component in the system: (i) 'no redundancy' alternatives characterised by reliabilities $r_1, r_2, \ldots, r_M$ and costs $q_1, q_2, \ldots, q_M$ and (ii) 'full active redundancy' alternatives characterised by reliabilities $1 - (1 - r_1)^2$, $1 - (1 - r_2)^2, \ldots, 1 - (1 - r_M)^2$ and costs $2q_1, 2q_2, \ldots, 2q_M$. The problem is to select appropriate alternative for each component so that the total cost, which is a sum of the cost of the components and the risk of failure before the specified operating time of a years, is minimised.

Consider again the generic system from Fig. 6.2(b) containing three components: power block (1), control module (2) and mechanical device (3).

Suppose that the reliabilities of the components associated with 2 years of operation are $r_1 = 0.6$, $r_2 = 0.7$ and $r_3 = 0.54$, with costs $q_1 = \$180$, $q_2 = \$42$ and $q_3 = \$190$.

The optimal alternatives selected by the exhaustive search algorithm were $a_1 = 1$, $a_2 = 2$ and $a_3 = 1$ corresponding to no redundancy for the first component, a redundancy for the second component and no redundancy for the third component. This combination of alternatives yields a total cost $G = \$1159$. At first glance, this selection seems counter-intuitive, because a redundancy alternative was not selected for the third component which is the least reliable component ($r_3 = 0.54$). The reason for such a selection is the cost of the redundant component which cannot be outweighed by the risk reduction provided by the redundancy. If a redundancy was selected for the third component (alternatives combination $a_1 = 1$, $a_2 = 1$ and $a_3 = 2$) the total cost will be $G = \$1271$. If a combination of alternatives ($a_1 = 1$, $a_2 = 2$ and $a_3 = 2$) was selected, the total cost will be $G = \$1213$. Both of these combinations yield total cost greater than the total cost from the optimal combination of alternatives ($a_1 = 1$, $a_2 = 2$ and $a_3 = 1$).

9.6.3 Optimal Selection of Suppliers and a System Topology

Another important application of the proposed allocation method is for optimising the selection of suppliers. Suppose that for exactly k components in a system, there are more than one supplier. Let $n_1, n_2, \ldots, n_k$ be the number of suppliers available for the first, second,..., k-th component. Each alternative supplier is characterised by two parameters: the time to failure distribution of the delivered component and its cost. The task is to select a set of k alternatives $(a_1, a_2, \ldots, a_k; \ 1 \le a_1 \le n_1; \ 1 \le a_2 \le n_2; \ 1 \le a_k \le n_k)$ for the suppliers which minimise the total cost $G = Q + K$ or $G = Q + \bar{L}$.

An application of significant practical importance is the following.

Similar to the case considered earlier, for exactly k components in the system, there is more than one supplier. At the same time, for exactly m components in the system there is a possibility for introducing redundancy (full active or a standby redundancy). The problem is to select a set of alternative suppliers and to decide on where to incorporate redundancy in order to minimise the total cost.

9.7 RELIABILITY ALLOCATION TO LIMIT THE EXPECTED LOSSES FROM FAILURES BELOW A MAXIMUM ACCEPTABLE LEVEL

The process of reducing the expected losses from failures can be illustrated by a single component with M mutually exclusive failure modes, each characterised by a constant hazard rate. Triggering any failure mode causes system failure. According to equation (6.17), the expected losses from failures are given by $\bar{L} = \sum_{k=1}^{M} a \lambda_k \bar{C}_k$.

Now, the failure modes can be ranked in descending order according to the relative contribution $\lambda_k a \bar{C}_k / \sum_{i=1}^{M} \lambda_i a \bar{C}_i$ of each failure mode $(k = 1, 2, \ldots, M)$ to the total expected losses from failures $\sum_{i=1}^{M} \lambda_i a \bar{C}_i$, associated with all failure modes.

Next, a Pareto chart can be built on the basis of this ranking and from the chart (Fig. 4.12), the failure modes accountable for most of the expected losses from failures during the specified time interval can be identified.

Consequently, the limited resources for reliability improvement (time, money, people) should be concentrated on the failure modes accountable for most of the losses. This approach allocates in the most efficient way the limited reliability improvement budgets in order to get maximum gains from

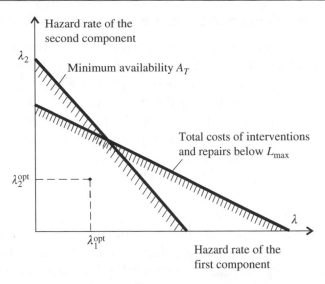

Figure 9.1 Setting reliability requirements as an intersection of hazard rate envelopes for a system composed of two components, logically arranged in series.

reducing the losses from failures. Furthermore, an importance measure M based on the expected losses from failures can be introduced:

$$M = \partial \overline{L} / \partial \lambda_i \tag{9.22}$$

where $\partial \overline{L}$ is the change of the expected losses from failures and $\partial \lambda_i$ is an increment of the hazard rate for the i-th component.

Considering equation (6.11), in order for the expected losses $\overline{L}$ from failures to be smaller than a maximum specified limit $\overline{L}_{max}$, the inequality

$$\overline{L} = E(N_1) \times \overline{C}_{1|f} + E(N_2) \times \overline{C}_{2|f} + \cdots + E(N_M) \times \overline{C}_{M|f} \leq \overline{L}_{max} \tag{9.23}$$

must be satisfied, where $E(N_i)$ and $\overline{C}_{i|f}$ are the expected number of failures and the cost given failure, associated with the individual components. This inequality can also be presented as

$$\frac{E(N_1)}{\overline{L}_{max}/\overline{C}_{1|f}} + \cdots + \frac{E(N_M)}{\overline{L}_{max}/\overline{C}_{M|f}} \leq 1 \tag{9.24}$$

Expected losses from failures smaller than the maximum acceptable limit $\overline{L}_{max}$ are guaranteed if the expected numbers of failures $E(N_i)$ associated with the separate components satisfy inequality (9.24) and the conditions

$$0 \leq E(N_1) \leq \overline{L}_{max}/\overline{C}_{1|f}, \ldots, 0 \leq E(N_M) \leq \overline{L}_{max}/\overline{C}_{M|f} \tag{9.25}$$

For inequality (9.24) to be fulfilled, the expected number of failures $E(N_i)$ for the i-th component cannot go beyond the upper bound $\overline{L}_{max}/\overline{C}_{i|f}$, $i = 1, 2, \ldots, M$. Equations (9.24) and (9.25) confirm the basic principle for a risk-based design: the larger the losses given failure for a component, the smaller the upper bound of the expected number of failures for the component, the larger the required minimum reliability level from the component. The maximum expected numbers of failures which still guarantee losses from failures not larger than the maximum acceptable limit $\overline{L}_{max}$ are obtained for a set of $E(N_i)$ which satisfy equality (9.24) (i.e. which lie on the hyper-plane defined by equation (9.24)).

For the important special case of components characterised by constant hazard rates λ_i, the expected number of failures are $E(N_i) = \lambda_i a$ and the constant hazard rates must satisfy the inequality

$$\frac{\lambda_1}{\overline{L}_{max}/(a\overline{C}_{1|f})} + \cdots + \frac{\lambda_M}{\overline{L}_{max}/(a\overline{C}_{M|f})} \leq 1 \qquad (9.26)$$

and the constraints

$$0 < \lambda_1 \leq \overline{L}_{max}/(a\overline{C}_{1|f}), \ldots, 0 < \lambda_M \leq \overline{L}_{max}/(a\overline{C}_{M|f}) \qquad (9.27)$$

The maximum hazard rates which still guarantee losses from failures not larger than the maximum acceptable limit L_{max} are obtained for a set of hazard rates $\lambda_1, \lambda_2, \ldots, \lambda_M$ which satisfy equality (9.26), that is which lie on the hyper-plane defined by equation (9.26). Clearly, equality (9.26) is satisfied for an infinite number of combinations for the hazard rate values. An optimisation procedure can then be employed to select among these, the hazard rates associated with the smallest reliability investment.

An important application of the described method is in guaranteeing that the availability of a production system with components arranged in series will be greater than a specified minimum level. The average availability for the system is $A_P = 1 - \overline{L}/a$, where $\overline{L}$ is the expected lost production time and a is the maximum available production time. Guaranteeing availability greater than A_T ($A > A_T$) means $1 - \overline{L}/a > A_T$, which is equivalent to $\overline{L} < \overline{L}_{max} = a(1 - A_T)$, where $\overline{L}_{max}$ is the maximum acceptable expected lost production time (downtime). If $\overline{C}_{k|f}$ in equations (9.23)–(9.26) denote the lost production time due to failure of the k-th component, solving equation (9.26) will yield a set of hazard rates $\lambda_1, \lambda_2, \ldots, \lambda_M$ which guarantee that the expected lost production time $\overline{L}$ will be smaller than the maximum acceptable expected downtime $\overline{L}_{max}$. In other words, the set of hazard rates $\lambda_1, \lambda_2, \ldots, \lambda_M$ will guarantee the specified availability target.

10

GENERIC APPROACHES TO REDUCING THE LIKELIHOOD OF CRITICAL FAILURES

Central to the risk management of the technical systems considered are the following two fundamental steps:

(i) Identifying components or sets of components with a large contribution to the total losses from failures.
(ii) Directing the risk-reduction efforts towards these components.

Losses from failures can be reduced by different measures, broadly divided into three basic categories: (i) measures reducing the likelihood of critical failures; (ii) measures reducing the consequences given that failure has occurred and (iii) measures which simultaneously reduce the likelihood of a critical failure and the consequences given failure.

The old adage 'prevention is better than cure' applies fully to management of technical risk. In cases where the intervention for repair is very difficult or very expensive (e.g. deep-water oil and gas production), preventive measures should always be preferred to protective measures. While protective measures reduce or mitigate the consequences from failure, preventive measures exclude failures altogether or reduce the possibility of their occurrence.

10.1 REDUCING THE LOSSES FROM FAILURES BY IMPROVING THE RELIABILITY OF COMPONENTS

An important way of reducing the likelihood of critical failures is to build the systems with very reliable components or to improve the reliability of the existing components. Component reliability is increased by strengthening the components against their failure modes which can be achieved by a careful study of the failure modes, the underlying failure mechanisms and the failure promoting factors.

A thorough root cause analysis provides a solid basis for reliability improvement. Knowledge regarding the circumstances and processes which contribute to the failure events is the starting point for a real reliability improvement. In this respect, a good formal failure reporting system and subsequent failure analysis are important tools for reducing the number of failure modes. The main purpose of the *root cause analysis* is to identify the factors promoting the failure mode and determine whether the same or related factors are present in other parts of the system. Identifying the root causes initiates a process of preventing the failure mode from occurring by appropriate modifications of the design, the manufacturing process or the operating procedures.

A typical example of reliability improvement by a root cause analysis can be given with improving the reliability of hot-coiled Si–Mn suspension springs suffering from premature fatigue failure. Typically, automotive suspension springs are manufactured by hot winding. The cut-to-length cold-drawn spring rods are austenitised, wound into springs, quenched and tempered. This is followed by warm pre-setting, shot peening, cold pre-setting and painting (Heitmann et al., 1996).

The initial step of the analysis is conducting rig tests inducing fatigue failures of a large number of suspension springs under various conditions. Fracture surfaces are then preserved and scanning electron microscopy is employed to investigate the fatigue crack initiation sites. If large size inclusions are discovered at the fatigue crack origin, a possible fatigue life improvement measure would involve changing to a supplier of cleaner spring steel.

Optical metallography of sections from the failed springs must also be made, in order to make sure that there is no excessive decarburisation. If the depth of the decarburised layer is significant, its fatigue resistance is low and care must be taken to control the carbon potential of the furnace atmosphere, in order to avoid excessive decarburisation. Alternatively, the chemical composition of the steel can be altered by microalloying, in order to make it less susceptible to decarburisation. The grain size at the surface of the spring wire must also be examined because microstructures with excessively large grain size are characterised by reduced toughness and fatigue resistance. Correspondingly, the austenitisation temperature and the duration of the austenitisation process must guarantee that the grain size remains relatively small.

The spring surface after quenching must also be examined in order to make sure that there are no excessive tensile residual stresses or quenching microcracks. Tempering must guarantee optimal hardness and yield strength which maximise the failure life. Finally, after shot peening, the residual stresses at the surface of the spring wire should be measured (e.g. by an X-ray diffractometer) to make sure that they are of sufficient magnitude and uniformly distributed over the circumference of the spring wire. If for example, the residual stresses are found to be highly non-uniform or of small magnitude, they would offer little resistance against fatigue crack initiation and propagation. Changes in the shot-peening process must then be implemented to guarantee a sufficient magnitude and uniformity of the residual stresses.

The more complex the system, the higher the reliability required from the separate components. Indeed, for the sake of simplicity, suppose that a complex system is composed of N identical components, arranged logically in series. If the required system reliability is R_s, the reliability of a single component should be $R_0 = (R_s)^{1/N}$. Clearly, with increasing the number of components N, the reliability R_0 required from the separate components to guarantee the specified reliability R_s for the system approaches unity. In other words, in order to guarantee the required system reliability R_s for a large system, the components must be highly reliable. At the same time, the number of defective components must be very small. In this respect, *the six-sigma quality philosophy* (Harry and Lawson, 1992) is an important approach based on a production with very small number of defective items (zero defect levels). Modern electronic systems, in particular, include a large number of components. Adopting a six-sigma process guarantees no more than two defective components out of a billion manufactured and this is an efficient approach for reducing failures in complex systems.

According to the discussion in Chapter 6, for a component or a system whose critical failures follow a non-homogeneous Poisson process, the area beneath the hazard rate curve within a specified time interval is equal to the expected number of critical failures. In the time interval $(0, a)$, in Fig. 10.1, reducing the rate of occurrence of failures from curve '1' to curve '2' results in a reduction of the expected number of failures equal to the hatched area S.

10.1.1   Improving Reliability by Removing Potential Failure Modes

In order to improve the reliability of a product, design analysis methods such as FMEA (failure mode and effect analysis; MIL-STD-1629A, 1977)

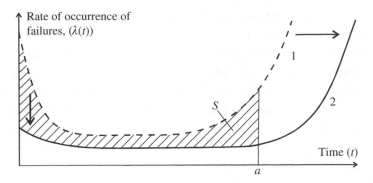

Figure 10.1 Reducing the losses from failures by decreasing the rate of occurrence of failures.

and its extension FMECA (failure modes, effects and criticality analysis) including criticality analysis can be used (Andrews and Moss, 2002). These ensure that as many as possible potential failure modes have been identified and their effect on the system performance assessed. The objective is to identify critical areas where design modifications can reduce the probability of failure or the consequences of failure. In this way, potential failure modes and weak spots which need attention are highlighted and the limited resources for reliability improvement are focused there.

10.1.2 Reliability Improvement by Reducing the Likelihood of Early-life Failures

Most component failures occurring early in life are quality-related failures caused by substandard items which find their way into the final products. Early-life failures are usually caused by poor design, manufacturing, quality control, assembly and workmanship, leaving latent faults in the components.

An important factor promoting early-life failures is also the variability associated with critical design parameters (e.g. material properties and dimensions) which leads to variability associated with the strength.

In the infant mortality region, the rate of occurrence of failures can be decreased (curve 1 in Fig. 10.1) by improving the reliability of components through better design, materials, manufacturing quality control and assembly. A significant reserve in decreasing the rate of occurrence of failures at the start of life is decreasing the uncertainty associated with the actual loads experienced during service. The number of defective components which cause early-life failures can be reduced by quality control, environmental stress screening (ESS) and accelerated testing whose purpose is to trap and eliminate latent faults.

Early-life failures affect particularly strongly the net present value and this has already been demonstrated in Chapter 4. Early-life failures usually occur during the pay-back period of the installed equipment and they are also associated with substantial losses due to warranty payments.

10.1.3 Reliability Improvement by Reducing the Likelihood of Wearout Failures

In the wearout region, the rate of occurrence of failures can be decreased significantly by preventive maintenance consisting of replacing worn-out components. This delays the wearout phase and, as a result, the rate of occurrence of failures decreases which means a smaller expected number of wearout failures (Fig. 10.1). Design changes which result in a significantly reduced rate of accumulation of damage (e.g. reduced rates of fatigue and corrosion damage) are important measures for reducing wearout failures. Reducing the rate of accumulation of fatigue damage, for example, can be achieved by appropriate design modifications avoiding stress concentrators; appropriate treatment of the surface layers; reduced loading amplitudes, selecting materials with increased fatigue resistance, free from surface defects, etc. Reducing the rate of accumulation of corrosion damage can, for example, be achieved by appropriate material selection, various corrosion protection measures and appropriate design.

10.2 MEASURES GUARANTEEING A SMALL LIKELIHOOD OF A CRITICAL FAILURE DURING A SPECIFIED MINIMUM FAILURE-FREE OPERATING PERIOD

Reducing the likelihood of critical failures during a specified time interval (minimum failure-free operating period, MFFOP) can be achieved in three principal ways: (i) *Reducing the likelihood of failure modes during the specified MFFOP*; (ii) *preventing failure modes from occurring during the specified MFFOP* (failure modes are designed out, blocked or prevented in some way so that their occurrence in the specified time interval is extremely unlikely) and (iii) *delaying failure modes* (failure modes are delayed to such an extent that they are more likely to appear beyond the end of the specified MFFOP rather than within it; Fig. 2.2). Guaranteeing with a high probability

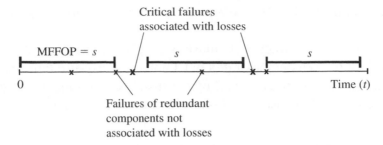

Figure 10.2 Guaranteeing with a high probability of an MFFOP of specified length s before each critical failure during the life cycle of the system.

of an MFFOP of specified length is central to reducing the losses by reducing the failure occurrences.

Often, a rolling MFFOP needs to be guaranteed not only before the first critical failure, but also before each subsequent critical failure (Todinov, 2005a). The rolling warranty is a typical example, where before each critical failure, the same warranty period of minimum length s is required (Fig. 10.2).

Reducing the losses from failures by guaranteeing a small likelihood of failure within the specified MFFOP implies no critical failures or a small likelihood of critical failures associated with losses. The larger the losses from failures C, the smaller the maximum acceptable probability p_{max} of failure within the specified MFFOP should be. This is summarised by the equation

$$p_{f max} = \frac{K_{max}}{C} \tag{10.1}$$

As a result, the maximum acceptable probability of failure within the specified MFFOP is a function of the losses from failure.

Failure occurrences can be suppressed within the specified MFFOP by delaying failure modes. Again, this is achieved by measures which result in a significantly reduced rate of damage accumulation. Protection from the harmful influence of the environment such as encapsulation into inert gas atmosphere and other forms of corrosion protection, erosion protection and protection against wearout are all examples of measures delaying failure modes. The rate of damage accumulation can also be decreased significantly by removing defects and imperfections from the material and by reducing the amplitude and frequency of the acting loads.

10.3 PREVENTIVE BARRIERS FOR REDUCING THE LIKELIHOOD OF FAILURE

Figure 10.3 features a generic example of preventive barriers designed to reduce the likelihood of a release of toxic substance during handling containers for storage. Reducing the likelihood of a release of toxic substance relies on physical barriers (e.g. metal container, second protective shield, redundant fixtures during handling), non-physical barriers (e.g. instructions for handling toxic substances) and human actions barriers (e.g. quality control, regular inspections, etc.).

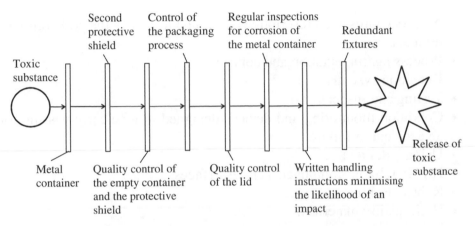

Figure 10.3 Different types of preventive barriers for reducing the likelihood of a release of toxic substance during handling containers for storage.

Studying the pathways for pathogenic contamination of consumer products, for example, is an essential step towards reducing the likelihood of infection. In this respect, analysing the critical points of the production process where contamination with pathogenic microorganisms can occur is vital. Setting preventive barriers at these critical points reduces significantly the possibility of food contamination and food-borne illnesses.

10.3.1 Passive Preventive Barriers and Their Functions

The catastrophic event is avoided by adding a physical or non-physical barrier or by removing the hazard altogether. The likelihood of a leakage from a valve can, for example, be reduced by including a physical barrier (a high-quality seal), a non-physical barrier (control and inspection) and a mixed barrier consisting of pressure testing.

A typical example of a passive preventive barrier is the physical separation or increasing the distance between sources of hazards and sources of triggering conditions. Such a separation prevents hazards and triggering conditions from interaction, thereby creating a barrier against failures.

Protection against the harmful influence of the environment such as encapsulating into inert gas atmosphere, corrosion protection, erosion protection and protection against temperature variations are all examples of passive preventive barriers.

Other examples of preventive barriers reducing the likelihood of accidents or failures are:

- All procedures and prescriptions designed to reduce the probability of human errors.
- Process instrumentation and control.
- Detection systems.
- Testing and inspection.
- Condition monitoring and devices designed to give early warning of an accident or failure.
- Using redundancy.
- Highly reliable components and interfaces.
- Robust designs.
- High-quality materials.
- Derating.
- Design to avoid unfavourable stress states.

Preventing failure modes from occurring in a specified time interval can be achieved by designing them out through appropriate modifications. Preventing failure modes caused by a wrong sequence or order of actions being taken can be achieved by designing *failure prevention interlocks*. These make the occurrence of failure modes practically impossible.

Physical interlocks are devices and circuits which block against a wrong action or a sequence of actions being taken. A physical interlock, for example, will prevent an aeroplane to take off without setting properly all flight controls for a successful take off or if all boarding doors have not been latched firmly into closed position. If, for example, starting a machine under load will cause failure, a built-in interlock device could make it impossible to start the machine if it is under load. Failures are often caused by exceeding the operational or environmental envelope. Efficient failure prevention interlocks for this type of failures are usually circuits which prevent operation during conditions of extreme heat, cold, humidity, vibrations, etc. Such an interlock can be designed for the common coupling fan-cooled device. If the fan fails, the power supply is automatically disconnected in order to prevent an overheating failure of the cooled device.

Logic interlocks eliminate the occurrence of erroneous actions. Preventing the hand of an operator from being in the cutting area of a guillotine can, for example, be made if the cutting action is activated only by a simultaneous pressure on two separate knobs/handles which engage both hands of the operator.

Time interlocks work by separating tasks and processes in time so that any possibility of collisions or mixing dangerous types of processes and actions is excluded. Suppose that a supply system fails if two or more demands follow within a critical interval needed for the system to recover. If the operation of the system is resumed only after a built-in delay has elapsed, equal to this minimum critical period, a time interlock will effectively be created excluding the possibility for overloading from sequential demands. In another example, a structural failure due to a premature removal of the scaffolding is prevented by a built-in minimum time delay allowing the concrete to set and acquire a particular minimum strength.

10.3.2 Active Preventive Barriers and Their Functions

The catastrophic event is avoided by detecting and avoiding failure or accident. Active preventive barriers follow the sequence Detect–Diagnose–Act and involve a combination of hardware, software and human action. Detection and monitoring is only part of the function. The collected/measured information needs to be processed and interpreted, after which an appropriate action must be taken. For example, a measured trend of increasing the temperature and the vibrations from a bearing indicates intensive wearout and incipient failure, which can be prevented by a timely replacement of the worn-out bearing.

10.4 INCREASING THE RELIABILITY OF COMPONENTS IN PROPORTION WITH THE LOSSES FROM FAILURES ASSOCIATED WITH THEM

If the cost given failure of a particular component is C, the minimum reliability R_{min} that needs to be designed in the component is

$R_{min} = 1 - K_{max}/C$, where K_{max} is the maximum tolerable level of risk (the expected potential loss). The larger the cost of failure C, the larger the minimum reliability required from the component. Even identical components should be designed to different reliability levels if their failures are associated with different losses.

This principle which is an underlying theme in this book is often overlooked in engineering designs. Usually, the same type of bolts or fixtures used for general purpose applications where the cost of failure is insignificant are also used in applications where the cost of failure is significant. Surely, the reliability of the components used must reflect the cost of failure and components associated with increased cost of failure must be designed to a higher-reliability level in order to bring down the risk within the tolerable level.

For a single failure mode, the risk-based design principle guaranteeing an expected loss smaller than the maximum tolerable level of risk K_{max} can be formulated as:

$$K = p_f C \leq K_{max} \qquad (10.2)$$

where p_f is the probability of a critical system failure. For M mutually exclusive failure modes, the risk-based design principle can be derived from the requirement of the expected potential loss (the risk) from all failure modes not to exceed the maximum tolerable level K_{max}:

$$K = p_{f1}C_1 + p_{f2}C_2 + \cdots + p_{fM}C_M \leq K_{max} \qquad (10.3)$$

where p_{fi} is the probability that the i-th failure mode will cause a critical failure within the specified MFFOP and C_i is the cost of failure given the i-th failure mode. Equation (10.3) can also be presented as

$$\frac{p_{f1}}{K_{max}/C_1} + \frac{p_{f2}}{K_{max}/C_2} + \cdots + \frac{p_{fM}}{K_{max}/C_M} \leq 1 \qquad (10.4)$$

where

$$0 \leq p_{fi} \leq \frac{K_{max}}{C_i}, \quad i = 1, \ldots, M \qquad (10.5)$$

Equations (10.3) and (10.4) are in fact a formulation of the risk-based design principle for multiple failure modes: to limit the risk below the maximum acceptable level K_{max}, the probabilities of the separate failure mode must satisfy inequalities (10.4) and (10.5).

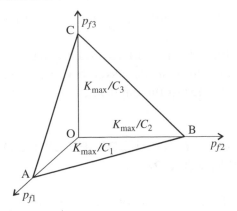

Figure 10.4 The risk of failure will not exceed the maximum tolerable level K_{max} if the point defined by the probabilities of activating the separate failure modes lies inside or on the tetrahedron OABC.

For the special case of three failure modes, conditions (10.4) and (10.5) become

$$\frac{p_{f1}}{K_{max}/C_1} + \frac{p_{f2}}{K_{max}/C_2} + \frac{p_{f3}}{K_{max}/C_3} \le 1 \qquad (10.6)$$

$$0 \le p_{fi} \le \frac{K_{max}}{C_i}, \quad i = 1, 2, 3. \qquad (10.7)$$

In a coordinate system defined by the three probabilities p_{f1}, p_{f2} and p_{f3}, equation (10.6) describes the plane ABC in Fig. 10.4. This, together with the planes defined by conditions (10.7), defines the tetrahedron OABC. In order for the risk K to be smaller than or equal to the maximum tolerable level K_{max}, the probabilities of activating the separate failure modes must define a point which lies on or inside the tetrahedron OABC (Fig. 10.4).

Given the expected losses associated with the separate failure modes, reducing the sum in equation (10.3) can be done by removing failure modes ($p_{fi} = 0$) or reducing their likelihoods. Suppose that Δp_{fi} are the reductions in the likelihoods of the separate failure modes. From the sum expressing the risk of failure we get

$$\Delta K = \Delta p_{f1} C_1 + \Delta p_{f2} C_2 + \cdots + \Delta p_{fM} C_M \qquad (10.8)$$

In equation (10.8), ΔK is the risk reduction corresponding to the reduction of the likelihoods of the separate failure modes. Since C_i vary substantially with the failure modes, the sensitivity of the risk $\Delta K/\Delta p_{fi}$ to the different failure modes varies. The most efficient reduction of the risk of failure is achieved by reducing the likelihood of the failure modes associated with the largest losses.

Reducing the likelihood of the failure modes can be done by better design, material processing, manufacturing, assembly, maintenance and protection from the harmful influence of the environment.

10.5   LIMITING THE POTENTIAL LOSSES BY REDUCING THE LENGTH OF EXPOSURE

Typical examples of limiting the potential losses by reducing the risk exposure are:

- Reducing the length of operation in order to reduce the probability of encountering an overstress load. Indeed, if the overstress load follows a homogeneous Poisson process with density ρ and the length of the time interval is a, the probability of encountering an overstress load during the time interval $(0, a)$ is $p_f = 1 - \exp(-\rho a)$. This probability can be reduced by reducing the length of the time interval a.
- Reducing the length of operation in dusty or humid environments in order to reduce the risk of degradation failure.
- Reducing the length of stay in dangerous zones in order to reduce the likelihood of poisoning, infection, hypothermia, or other health damage.
- Limiting the amount of flammable material or a toxic substance handled at a time.
- Limiting individual exposures to obligors to limit the credit risk to a bank, etc.

11

SPECIFIC PRINCIPLES FOR REDUCING THE LIKELIHOOD OF FAILURES

11.1 REDUCING THE RISK OF FAILURE BY BUILDING IN REDUNDANCY

As production systems become more complex, their analysis becomes increasingly difficult. Complexity increases the risks of both *random component* failures and *design-related* failures. Incorporating redundancy in the design is particularly effective where random failures predominate. *Redundancy* is a technique whereby one or more components of a system are replicated in order to increase reliability (Blischke and Murthy, 2000). Since a design fault would usually be common to all redundant components, design-related failures may not be reduced by including redundancy. In other words, a fault-free design is an important prerequisite for the redundancy to have a significant impact.

Including redundancy reduces significantly the number of interventions and the associated losses. This can be demonstrated on the basis of two active redundant components characterised by constant hazard rates λ_A and λ_B correspondingly (Fig. 11.1(a)).

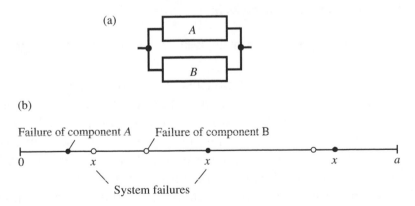

Figure 11.1 Reducing the number of system failures by including redundancy.

193

Assume for simplicity the same cost of intervention per system failure and the same hazard rate for both components $(\lambda_A = \lambda_B = \lambda)$. A system failure is present if both components A and B fail. Since each intervention is characterised by the same cost, the expected cost of intervention is proportional to the expected number of system failures in the specified time interval. If the system consists of a single component, the expected number of system failures in a time interval with length a will be λa. For the dual-redundant system in Fig. 11.1(a), the expected number of system failures in a time interval with length a can be estimated from a simple Monte Carlo simulation, the algorithm of which is given below.

Algorithm 11.1

```
system_failures_counter =0;
For i=1 to Number_of_trials do
  {
  current_system_time=0;
  Repeat
    {
      generate time_to_failure_A;
      generate time_to_failure_B;

      if (time_to_failure_A > time_to_failure_B) then
          current_system_time = current_system_time + time_to_failure_A;
      else current_system_time= current_system_time + time_to_failure_B;

      if (current_system_time > a) then break;
      else system_failures_counter = system_failures_counter +1;

    } until 'break' is executed in the loop;
  }
Expected_number_of_failures = system_failures_counter / Number_of_trials;
```

In Fig. 11.1(b), failures of components A and B are denoted by filled and open circles, correspondingly, while the system failures are denoted by 'x'. After a system failure, both components are replaced with new components. Clearly, each system failure coincides with the failure of the component characterised by the greater time to failure and this is the basis of the algorithm described earlier. The current system time is kept in the variable current_system_time. After each system failure, both components are replaced and new times to failure are generated for each component.

Since the next system failure will occur after a time interval equal to the largest time to failure of the components, the current system time is always incremented by this time interval. Subsequently, a check is performed whether the specified design life *a* has been exceeded. If it has been exceeded by the current system time, a *brake* statement is executed which exits from the *repeat-until* loop and the next simulation trial is initiated. If the current system time is still smaller than the specified design life *a*, a system failure is registered by incrementing the system failures counter system_failures_counter. The expected number of failures during the design life *a* is obtained in the variable Expected_number_of_failures. For the special case $\lambda_A = \lambda_B = 2$ year^{-1}, and for a design life $a = 20$ years, the algorithm yields approximately 26 average number of system failures per 20 years which is much smaller compared to 40 expected number of failures if the system consisted of a single component only.

Now let us come back to the *k-out-of-n* systems introduced in Chapter 2. Since the number of components *n* is larger than the value of *k*, redundancy is built into the *k-out-of-n* system. The use of factor of safety in engineering designs provides in effect a form of redundancy. If a structure containing *n* load-carrying components requires only half of them to carry the maximum design load, essentially a factor of safety of 2 is used (e.g. wires in cables, columns supporting a building, etc.).

In cases where high reliability is required, a cold standby redundancy can be used. Suppose that the switch *S* in Fig. 2.7 is perfect (never fails). While the time to failure of the full active redundant system is equal to the largest among the times to failure of its components, the time to failure of the cold standby system is the sum of the times to failure of all components. In other words, in case of perfect switching, the time to failure of a standby system including *n* components is larger than the time to failure of the corresponding full active redundant system (also based on *n* components) by the sum of the times to failure of $n-1$ components. The larger the number of components, the larger the difference in the times to failure. Theoretically, by providing a sufficiently large number of standby components, the reliability of a standby system with perfect switching can be made arbitrarily close to 1. Indeed, for the special case of *n* cold standby components with a constant hazard rates λ, the reliability associated with time *t* of a standby system with perfect switching is (Tuckwell, 1988):

$$R(t) = \exp(-\lambda t)\left[1 + \frac{(\lambda t)^1}{1!} + \frac{(\lambda t)^2}{2!} + \cdots + \frac{(\lambda t)^{n-1}}{(n-1)!}\right] \qquad (11.1)$$

With increasing the number of components n,

$$\lim_{n\to\infty}\left[1 + \frac{(\lambda t)^1}{1!} + \frac{(\lambda t)^2}{2!} + \cdots\right] = \exp(\lambda t)$$

and, as a result, $\lim_{n\to\infty}[R(t)] = 1$. The number of standby components however is limited by constraints such as size, weight and cost. Standby units may not necessarily be identical. An electrical device for example can have a hydraulic device for backup.

11.2 REDUCING THE RISK OF FAILURE BY INCREASING THE CONNECTIVITY OF THE RELIABILITY NETWORKS

Let us consider the active redundant system in Fig. 11.2(a). A redundancy has been introduced at a system level. In other words, to the existing system composed of a single branch with n components logically arranged in series, with reliabilities $r_1, r_2, \ldots, r_n$, an identical redundant branch has been added.

The reliability of the arrangement in Fig. 11.2(a) can be increased significantly if cross-links are introduced such as in Fig. 11.2(b). This alteration of the reliability network effectively transfers the redundancy from a system level to a component level. The resultant topology is characterised by a significantly greater connectivity compared to the arrangement in Fig. 11.2(a), which makes it less sensitive to failures in both branches. While failures of two components, each in a separate branch always fail the arrangement in Fig. 11.2(a), the arrangement in Fig. 11.2(b) fails only if the two failed components have the same indices. There are n^2 different possible ways of having a single failure in each branch. While all of these possible ways invariably mean failure for the system in Fig. 11.2(a), only n of them fail the system in Fig. 11.2(b).

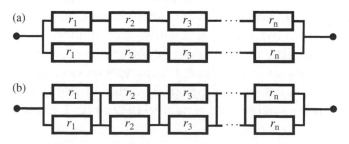

Figure 11.2 Redundant system with redundancy at (a) system level and (b) component level.

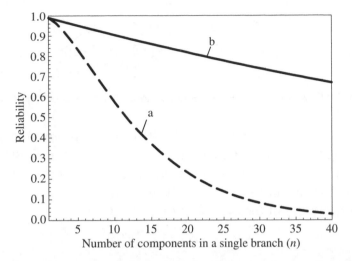

Figure 11.3 Improving the reliability of an active redundant section by transforming the redundancy at (a) a system level into (b) a component level.

These conclusions are confirmed if the reliabilities of both arrangements are determined. Assume for the sake of simplicity that all components have the same reliability $r_1 = r_2 = \cdots = r_n = r = 0.9$. The reliability of arrangement 'a' with redundancy at a system level is given by

$$R_a = 1 - (1 - r^n)^2 \qquad (11.2)$$

while the reliability of arrangement 'b' with redundancy at a component level is

$$R_b = [1 - (1 - r)^2]^n \qquad (11.3)$$

These two dependencies have been plotted in Fig. 11.3. As can be verified, the arrangement with a redundancy at a component level (Fig. 11.2(b)) is significantly more reliable compared to the arrangement with redundancy at a system level (Fig. 11.2(a)), particularly for a large number of components.

Improving the reliability by increasing the connectivity of the system can be illustrated by the system for transmitting messages in Fig. 11.4. Assuming for simplicity that all transmitters are characterised by the same reliability r, the reliability of a configuration consisting of n pairs of transmitters (Fig. 11.4(a)) is

$$R_a = 1 - (1 - r^2)^n \qquad (11.4)$$

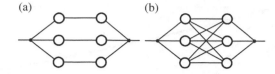

Figure 11.4 Two systems with different connectivity for transmitting messages.

The reliability of the system in Fig. 11.4(a) can be improved significantly by increasing the number of connections between the transmitters. For the system in Fig. 11.4(b), only a single working transmitter is required in each of the two columns for the system to be working.

For the reliability of the system in Fig. 11.4(b) we get

$$R_b = [1 - (1 - r)^n]^2 \tag{11.5}$$

The graphs of equations (11.4) and (11.5) corresponding to reliability $r = 0.1$ of a single transmitter are presented in Fig. 11.5. Clearly, for a transmitter characterised by a small reliability, the benefit from increasing the connectivity of the system is significant.

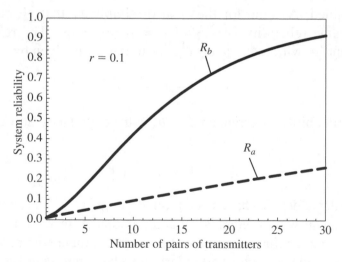

Figure 11.5 Reliability variation with the number of pairs of transmitters for arrangements (a) and (b) in case of a small reliability of a single transmitter.

If the reliability of a single transmitter is increased to $r = 0.6$ however, the benefit from having a system with larger connectivity is reduced significantly. For transmitters with high reliability (e.g. $r = 0.9$), the difference between the two curves is negligible and the effect from the extra connectivity is insignificant (Fig. 11.6).

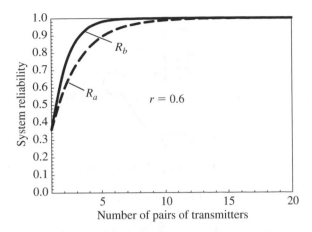

Figure 11.6 Variation of reliability with the number of pairs of transmitters for arrangements (a) and (b), the reliability of a single transmitter is $r = 0.6$.

These examples indicate that reliability investment towards increasing connectivity is most efficient in cases where the reliability of components is small. For each particular system configuration, before a reliability improvement, an assessment should be made regarding the benefits from the improvement. If the impact is insignificant, the cost towards increasing the complexity of the system topology may not be justified by the small risk reduction.

11.3 DECREASING THE PROBABILITY OF AN ERROR OUTPUT BY USING VOTING SYSTEMS

Suppose that a component A receiving a particular input produces an error with probability p (Fig. 11.7). The probability of an error can be reduced by creating a voting system. Voting is based on replicating the initial component A to n identical components, each of which receives the same input as the original component.

Each component operates independently from the others and with probability p, each component produces an error in the output. All outputs from the separate components are collected by a voter device V (Fig. 11.7). Suppose that the output of the voter device is determined by the majority vote of the components' outputs. In other words, in order for the voter to produce an error output, more than half of the components must produce an error output. Thus, for $n = 2k + 1$ identical components, at least $k + 1$ outputs must be error outputs. Since the distribution of the number of error outputs

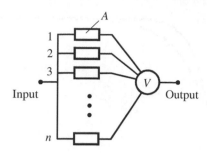

Figure 11.7 A voting system reduces significantly the probability of an error output.

X is described by the binomial distribution:

$$P(X = x) = \frac{n!}{x!(n-x)!} p^x (1-p)^{n-x} \tag{11.6}$$

the probability that the number or error outputs will be at least $k+1$ is given by the cumulative binomial distribution:

$$P(X \geq k+1) = 1 - P(X \leq k) = 1 - \sum_{x=0}^{k} \frac{n!}{x!(n-x)!} p^x (1-p)^{n-x} \tag{11.7}$$

For $n = 11$ and $p = 0.1$ for example, the probability of an erroneous output is

$$P(X \geq 6) = 1 - \sum_{x=0}^{5} \frac{n!}{x!(n-x)!} 0.1^x (1-0.1)^{n-x} \approx 0.0003 \tag{11.8}$$

Thus, the relatively high probability of an error output of $p = 0.1$ characterising a single component has been decreased 333 times by using a voting system!

11.4   REDUCING THE RISK OF FAILURE BY REDUCING THE SENSITIVITY TO FAILURE OF SINGLE COMPONENTS

In many cases, a catastrophic failure associated with large losses is triggered by a single component which sets a chain of failures and ultimately causes failure of the whole structure/system. There have been known cases where the collapse of large structures has been triggered by failure of a single, often a small structural component. A typical example of a sensitive (vulnerable) design can be given with the explosion of the space shuttle *Challenger* in January 1986 caused by the failure of an O-ring seal. Another

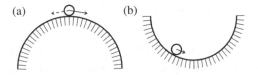

Figure 11.8 Vulnerable and non-vulnerable designs can be compared to (a) unstable and (b) stable state of equilibrium.

example is the collapse of a structure caused by failure of a pin-joint in supporting chain, or a single tension rope or a single supporting column. Such designs can be compared to the unstable state of equilibrium in Fig. 11.8, where a minor failure or a mistake disturbs the balance and causes the whole structure/system to collapse.

In a broader sense, the sensitivity of a design is measured by its susceptibility to faults due to imperfections in the material, design, manufacturing, assembly, maintenance, and variations in the operating conditions. These faults act as weak links originating catastrophic failures under particular circumstances.

Sensitive designs often do not have built-in redundancy, which means that they fail whenever a particular component fails. A way of counteracting this is to make the designs insensitive to failures of separate components and even to failure of several components. This is commonly achieved by building in redundancy, especially in supporting structures. Another way of counteracting the sensitivity to failures of key components is to increase significantly the reliability of all key components.

A common design measure to decrease the sensitivity of designs to failures of single components is to avoid domino-effects where failure of a single load-carrying component entails overloading of other load-carrying components. Such is for example a construction design where all horizontal concrete plates are assembled first on supporting columns. A shear failure of the horizontal top plate has been known to cause the top plate to fall and by impacting and overloading the next horizontal plate below, to cause another failure and so on until the whole structure collapsed. Another common example of a domino-type failure exists in cases where a single component failure causes damage to other components and triggers multiple secondary failures (e.g. an overheating failure of a cooled device because of failure of the cooling system, or failure of a fixture supporting a suspended component).

Sensitivity of a design to a single component failure is also present in cases where a particular component is overloaded with too many functions and demands. The component is 'over-stretched' and its strength can be easily exceeded by one of the multiple demands. This usually means that the component can no longer perform some of the required functions. A common design error of this type, which has caused a number of failures, is combining the critical functions of load carrying and sealing in the design of a single joint. Such an example related to the *Challenger* booster's O-ring, which also took the pressure of combustion has been discussed by Ullman (2003). Additional functions required from a component make the design sensitive because the number of different modes in which the component can fail is essentially increased. With adding more failure modes, the overall hazard rate of the component always increases.

A typical example of sensitivity to a single failure is present in cases where preventing the release of toxic substances in the environment depends on the reliable operation of a single control detector. An accidental damage of the detector could entail failure with grave consequences.

Particularly sensitive to a single failure are systems concentrating a large amount of energy, whose safe containment depends on the safe operation of one or several components or on a safe sequence of operations. Such systems can be compared to loaded springs accumulating a large amount of potential energy controlled by a single lock. Failure of the lock releases a large amount of stored energy with huge destructive power.

Such are for example the dams built from non-compacted material containing large amount of water. Their strength depends on the reliable operation of the draining system. Such dams are vulnerable because if the draining system fails (which can easily happen if the draining pipes are blocked by silt or debris), the strength of the dam can be eroded quickly.

Another example of sensitivity to single failures are the systems whose safe operation overly depends on the absence of human error. In this case, a human error during performing a critical operation can trigger a major failure associated with grave consequences. A way of counteracting this type of sensitivity is to build in fail-safe devices or failure prevention systems that make conducting the dangerous operation impossible. Such are the various failure prevention interlocks which do not permit conducting an operation until particular safety conditions are guaranteed.

11.5 REDUCING THE RISK OF FAILURE BY DERATING

Derating is one of the most powerful tools available to the designer for reducing the likelihood of failures. It is commonly done by reducing operating stresses, temperatures and flow rates below their rated levels. Life of many components and systems increases dramatically if the stress level is decreased. This makes the components and systems robust against the inevitable variations of the load and strength. The intensity of the wearout and damage accumulation also decreases significantly with reducing the stress magnitude.

Fatigue damage, corrosion or any other type of deterioration is a function of the time and a particular controlling (wearout) factor p (Fig. 11.9). During fatigue for example, the wearout factor can be the stress or strain amplitude. During corrosion, the wearout factor can be the density of the corrosion current. As can be seen from Fig. 11.9, reducing the stress-intensity level from p_1 to p_2 enhances the component's life because of the increased time to attain the critical level of damage, after which the component is considered to have failed.

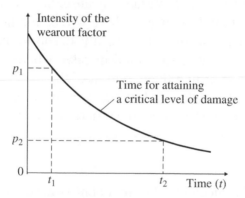

Figure 11.9 Time to failure for different intensity levels of the wearout factor.

Reducing the stress levels also reduces the volumes of the regions subjected to high stresses. As a result, the probability of having a critical flaw in such zones is reduced and from this – the probability of failure.

Derating essentially 'overdesigns' components by separating the strength distribution from the load distribution thereby reducing the interaction between the distribution tails. The smaller the interaction of the distribution tails, the smaller the probability that the load will exceed strength, the smaller the probability of failure. Derating however is associated with inefficient use of the components' strength capacity (Fig. 11.10).

In general, the greater the derating, the longer the life of the device. Voltage and temperature are common derating stresses for electrical and electronic components. The life of a light bulb designed for 220 V for example, can be enhanced enormously, simply by operating it at a voltage

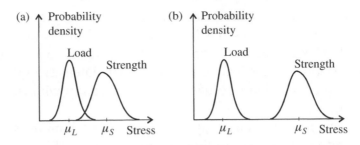

Figure 11.10 Stress and strength distribution (a) before and (b) after derating.

below the rated level (e.g. at 110 V). For mechanical components, common derating stresses are the operating speed, load, temperature, pressure, etc.

Underrating however can sometimes result in shorter life. Smith (1976) reports that resistors of carbon composition, for example, greatly increase their life at 50% of their rated wattage because they run cooler. If derating is further increased to 90% however, there will be too little heat generated to drive out absorbed moisture which results in shorter life.

Derating can be used as a safeguard in cases where there exists a great deal of uncertainty regarding the variation of the load and strength. It can also be used to limit the risk of failure below a specified level. This import-ant application of derating will be illustrated by the following engineering problem.

Example:

The strength of a mechanical component is described by the three-parameter Weibull model

$$F(\sigma) = 1 - \exp\left[-\left(\frac{\sigma - \sigma_0}{\eta}\right)^m\right]$$

with parameters $\sigma_0 = 200$ MPa, $\eta = 297.7$ MPa and $m = 3.9$. The cost of failure of the component is \$950,000. What should be the maximum working stress specified by the engineer-designer, which guarantees that the risk of failure will not be greater than $K_{max} = \$75,000$?

Solution:

Since the maximum acceptable risk is $K_{max} = \$75,000$ and the cost of failure is $C = 950,000$, the maximum acceptable probability of failure is $p_{f_{max}} = K_{max}/C = 75,000/950,000 = 0.0789$.

Solving $F(\sigma) \leq p_{f_{max}} = K_{max}/C$ with respect to σ yields

$$\sigma \leq \sigma_{max} = \sigma_0 + \eta(-\ln[1 - K_{max}/C])^{1/m} \qquad (11.9)$$

for the upper bound of the working stress of the component. A working stress in the range $0 \leq \sigma \leq \sigma_{max}$ limits the risk of failure below the maximum acceptable level K_{max}. Substituting the parameter values results in

$$\sigma_{max} = 200 + 297.7 \times (-\ln[1 - 0.0789])^{1/3.9} \approx 356.86 \, \text{MPa}$$

for the maximum working stress of the component. Working stress in the range $(0, 356.86 \, \text{MPa})$ limits the risk of failure below the maximum tolerable level of £75,000.

11.6 IMPROVING RELIABILITY BY SIMPLIFYING COMPONENTS AND SYSTEMS

Simplifying systems and components can be done in various ways: reducing the number of components, simplifying their shape, simplifying the function, reducing the number of functions carried out, etc.:

1. Simplifying the system can be achieved simply by reducing the number of components and blocks arranged in series.

    Complex designs are often associated with difficult maintenance and small reliability due to the large number of interactions between components which are a source of faults and failures. The larger the number of blocks in the system, the more possibilities for failures, the lower the reliability of the arrangement. Indeed, the reliability of a system composed of n components arranged in series is $R = R_1 \times R_2 \times \cdots \times R_n$. Without loss of generality, suppose that the last m components are removed. The reliability of the simplified system then becomes $R' = R_1 \times R_2 \times \cdots \times R_{n-m}$ and since $R_r = R_{n-m+1} \times R_{n-m+2} \times \cdots \times R_n < 1$ then $R = R' \times R_r < R'$. In other words, the simpler system is associated with larger reliability.
2. The shape of components and interfaces can also be simplified. This aides manufacturing, creates fewer possibilities for manufacturing faults, reduces the number of regions with stress intensification, improves the load-carrying capacity of the components by a better distribution of the stresses in the volume of the components.

3. Simplifying components' functions and reducing their number improves significantly their reliability. Reducing the number of functions reduces the number of failure modes. Failure modes characterising a particular component are logically arranged in series (activating any failure mode causes the component to fail) and the effect from reducing the number of failure modes is similar to the effect from reducing the number of components in a system in series. During the design of electro-mechanic devices, where possible, *the complexity should be transferred to the software*. Design should be oriented towards simpler but more refined mechanical components combined with powerful software to guarantee both performance and flexibility (French, 1999).

Every function should be designed in a clear and simple way. The *principle of uniformity* should be used if there is no reason for departures (French, 1999). For example, the components should be uniformly stressed, the stress distribution should be smooth and uniform, uniform rotating motion should be preferred to a non-uniform, alternating motion, etc.

11.7   IMPROVING RELIABILITY BY ELIMINATING WEAK LINKS IN THE DESIGN

Consider again a common example of a system with n components, logically arranged in series, with reliabilities $R_1, \ldots, R_n$. The system also contains a weak link with reliability r, logically arranged in series with the rest of the components. In other words, $r < R_1$, $r < R_2, \ldots, r < R_n$ are fulfilled and the reliability of this arrangement $R = R_1 \times R_2 \times \cdots \times R_n \times r$ is smaller than the reliability of the weakest link. (Indeed, since $R' = R_1 \times R_2 \times \cdots \times R_n < 1$ then $R = R' \times r < r$.)

Interfaces often appear as weak links in the chain, thereby limiting the overall reliability of the assembly. Consider a common practical example related to two very reliable components with high reliabilities R_1 and R_2 connected with an interface with relatively low reliability $r \ll R_c$ (Fig.11.11).

The reliability of the arrangement in Fig. 11.11 is smaller than the reliability r of the interface and in order to improve the reliability of the arrangement, the reliability of the interface must be increased. One of the reasons why so many failures occur at interfaces, despite the fact

Figure 11.11 Two very reliable components with reliabilities R_1 and R_2 connected with an unreliable interface with reliability r.

that the interfaced components are usually very reliable is the fact that *often interfaces are not manufactured to match the reliability of the corresponding components*. Seals, for example, commonly appear as weak links. A high quality of the seal and fine surface finish in the area of the seal is a necessary measure preventing external leakage. Internal leakage in valves for example is prevented by tapered joints which provide a tighter contact between the gate and the seat.

Figure 11.12 features an example (Altshuller, 1974) where the weakness of the joint (Fig. 11.12(a)) has been eliminated by redesigning the joint (Fig.11.12(b)).

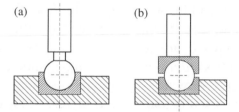

Figure 11.12 Reducing the vulnerability of designs by removing a weak link.

A weak link can be counteracted by including redundancy. A redundancy in a system in series is usually allocated on the component with the smallest reliability. Indeed, suppose that a system contains n components logically arranged in series, with reliabilities $r_1 < r_2 < \cdots < r_n$. The reliability of the system is given by $R = r_1 \times r_2 \times \cdots \times r_n$. Allocating active redundancy for the ith component will increase this reliability to

$$R' = r_1 \times r_2 \times \cdots \times [1 - (1 - r_i)^2] \times \cdots \times r_n \qquad (11.10)$$

Since $1 - (1 - r_i)^2 = r_i(2 - r_i)$, the reliability R' becomes

$$R' = R \times (2 - r_i) \qquad (11.11)$$

As can be verified R' is maximum when r_i is minimum (Elsayed, 1996). In other words, allocating redundancy for the weakest link yields the largest reliability increase. This principle for redundancy allocation however, as already demonstrated in Chapter 9, usually minimises the losses from multiple failures of a repairable system but does not necessarily minimise the risk of failure for a non-repairable system. If, for example, the cost of the redundant component is significant, the risk reduction from including redundancy may not be able to outweigh the cost of the redundant component. In other words, *maximising reliability and minimising the total cost are not equivalent.*

Useful principles in designing interfaces is to make the joints near the maximum section or diameter of the assembled parts. A large bending moment is easily resisted in the largest section.

11.8   REDUCING THE RISK OF FAILURE BY A PROPER DESIGN OF MOVING PARTS AND REDUCING THEIR NUMBER

Moving parts fail more frequently compared to stationary parts. This is usually due to the increased kinetics energy, vibration, fatigue loading, wear, corrosion, erosion and heat generation associated with them. The increased kinetics energy of moving parts (e.g. impellers, fans, turbines , etc.) makes them prone to overstress failures if their motion is suddenly restricted due to lodged foreign objects. Moving parts are also associated with large inertia forces which cause pulsating loading and increased fatigue. Uniform rotation is associated with smaller out-of-balance forces and should be preferred to an alternating motion which is associated with larger out-of-balance forces.

One of the design principles is to prefer small components working fast or at a high frequency. Compared to a larger part moving slowly, a small part moving fast can develop as much power and will weigh and cost less (French, 1999). However, if out-of-balance forces are present in the rotating parts and excitation frequencies are reached, the resonance amplitudes are a frequent cause of failures. A rotating shaft, for example, both in flexure and torsion can be regarded as an elastic/spring component with attached mass elements (e.g. gears, flywheels) and the mass of the shaft itself. Damping due to friction also exists. Although an accurate modelling of its vibration response may become a complicated task, a simple preliminary estimate of the *fundamental natural frequency* and a comparison with the system's *forcing frequency* is often recommended to make sure that *resonance* is avoided. This comparison is necessary in order to establish the critical rotational speed which is the lowest shaft speed that excites a resonance in the shaft assembly. It is also important to assure that the shaft stiffness is sufficient to keep the fundamental natural frequency well above the forcing frequency (Collins, 2003).

Vibration is always associated with moving parts and promotes fast wearout and fretting fatigue. Moving parts are sensitive to tolerance faults because they require more precise alignment. The friction and heath generated by moving parts requires lubrication and cooling which make moving parts very sensitive to failures or faults associated with the lubrication or cooling system. In this respect, designing pivots should be preferred to slides. Pivots can be better protected from dirt and wear, do not suffer from jamming and can be made to perform accurately (French, 1999). Motion should be transferred at the optimum ratio involving the combination of travel and force which minimises the wear and the possibility of an overstress failure.

11.9 REDUCING THE RISK OF FAILURE BY MAINTAINING THE CONTINUITY OF ACTION

Maintaining the continuity of action avoids high resistance forces and dynamic transient stresses from start–stop regimes. Thus, the resistance of pressure vessels to thermal fatigue is significantly enhanced if the start–stop regimes which induce high thermal stresses are avoided. The resistance to jamming of sliding surfaces (e.g. stems in valves) is enhanced by maintaining continuity of motion which prevents the formation of build-ups of corrosion products. Maintaining a continuous flow of granular substances and fluids improves the continuity of the flow.

11.10 REDUCING THE RISK OF FAILURE BY INTRODUCING CHANGES WITH OPPOSITE SIGN TO UNFAVOURABLE CHANGES DURING SERVICE

Typical examples where the risk of failure is reduced by introducing changes with opposite sign to the changes the component experiences during service are the allowances for lost wall thickness. The *corrosion, erosion* and *wear allowances* added to the computed sections are based on estimating the total loss of wall thickness.

By deliberately creating residual stresses which oppose the operational stresses, the negative impact of unfavourable stress states can be reduced.

This principle is often used in the construction where the tensile stresses from bending of concrete beams can be reduced if preloaded in tension steel ropes or rods are inserted in the beam. Consequently, after the concrete sets and the tension load is released, the beam is loaded in compression. Since during bending, the tensile stresses need first to overcome the compressive residual stresses, the effective tensile stress during service is reduced significantly.

The same principle can be applied to shafts transmitting torque where residual stresses are created with opposite sign to the actual working stresses induced by the torque. Components working in close contact (e.g. piston-cylinder) and moving relative to each other generate heat which, if not dissipated, causes intensive wear, reduced strength and deformations. The risk of failure of such an assembly can be reduced if one of the parts (e.g. the cylinder) is cooled which reduces friction and wear and dissipates the released heat.

In order to compensate the tensile stresses at the surface and improve fatigue resistance, *shot-peening* has been used as an important element of the manufacturing technology (Niku-Lari, 1981; Bird and Saynor, 1984). Introduction of residual compressive stresses is most effective for materials with high yield strength. Compressive residual stresses at the surface, compensating the service stresses, can also be created by a special heat- and thermochemical treatment such as case-hardening, gas-carburising and gas-nitriding. During case-hardening of steels for example, a relatively thin surface layer is heated above the critical temperature marking the start of the austenitic phase transformation. When the part is quenched, the austenite at the surface transforms into martensite which has a higher specific volume. Since the core remains unchanged, it pulls the case into compression. As a result, compressive residual stresses are induced at the surface, which are counter-balanced by tensile residual stresses in the core.

The risk of failure can also be reduced by introducing load-carrying elements in unfavourably stressed regions. Such is the reason behind reinforcing a brittle matrix with fibres possessing large tensile strength. This permits the component to endure both tensile and compressive stresses.

Preloading of assemblies is often carried out by clamping components in such a way that the tension in one part is counterbalanced by compression in other parts. Preloading has many advantages: elimination of unwanted clearance gaps between parts, increased stiffness and improved fatigue resistance. Preloading is frequently applied to bolted joints and flange-and-gasket assemblies. Tensile preloading increases the fatigue life of a part subjected to a completely reversed zero-mean alternating stresses. The mean stress is indeed increased, but the *equivalent completely reversed cyclic stress* is reduced significantly and as a result, the fatigue life is increased substantially (Collins, 2003).

Cold forming can also be used to create favourable residual stresses at the component's surface. In order to protect against excessive operational stresses in a particular direction, the material is overstressed and yielded in the same direction. In this respect, *pre-stressing* is a useful way of creating beneficial residual stresses. It is an overloading which causes local yielding producing a residual stress field favourable to loads acting in the same direction. For compression springs for example, pre-stressing, also known as '*pre-setting*' consists of compressing the spring and yielding the material so that beneficial residual stresses are introduced. As a result, the *fatigue resistance* and the *load-carrying capacity* of the spring are increased.

Axial tension pre-stressing creates local yielding of the material at the roots of surface notches. As a result, residual compressive stresses are created in these locations and the resistance against yielding caused by loading in the same direction is improved.

11.11 REDUCING THE RISK OF FAILURE BY REDUCING THE FREQUENCY OF LOAD APPLICATIONS

Suppose that a random load, characterised by a cumulative distribution function $F_L(x)$, is applied a number of times during a finite time interval with length t and the times of load application follow a homogeneous Poisson process with intensity ρ. Suppose that the strength is characterised by a probability density distribution $f_S(x)$. It is also assumed that the load and strength are statistically independent random variables.

The probability of no failure (the reliability) associated with the finite time interval $(0, t)$ can be calculated from the overstress reliability integral derived in Todinov (2004e):

$$R(t) = \int_{S_{\min}}^{S_{\max}} \exp\left[-\rho t(1 - F_L(x))\right] f_S(x)\, dx \qquad (11.12)$$

The term $\exp\left[-\rho t(1 - F_L(x))\right]$ in the *overstress reliability integral* (11.12) gives the probability that none of the random loads in the time interval $(0, t)$ will exceed strength with magnitude x.

In the case of a constant strength ($S =$ constant), the integral yields

$$R(t) = \exp[-\rho t(1 - F_L(S))] \tag{11.13}$$

for the reliability associated with the time interval $(0, t)$, which decreases with increasing the number density ρ of the load applications.

If equation (11.13) is presented as

$$R(t) = \exp[-\lambda t] \tag{11.14}$$

where $\lambda = \rho[1 - F_L(S)]$, an expression for the reliability associated with the time interval $(0, t)$ is obtained, where λ can be interpreted as a constant hazard rate. In this respect, the constant hazard rate has a fundamental significance. In cases where the strength is constant and the load applications follow a homogeneous Poisson process in the time interval $(0, t)$, the failure rate of the component is constant, equal to the product of the number density of the load applications and the probability of failure during a single application. Equation (11.14) also provides the opportunity for calculating the hazard rate from the number density of the load applications and the probability of failure associated with a single load application.

It is interesting to investigate the effect on reliability of a smooth and rough loading. For a rough loading, where there exists a non-zero probability that load will be greater than any possible value of strength, with increasing the number density of the load applications, reliability approaches asymptotically zero (Fig. 11.13(a)).

In order to guarantee perfectly smooth loading (the variance of the load is zero), assume that the load is constant, equal to L. The cumulative distribution function of the load then becomes $F_L(x) = 0$, if $x < L$ and $F_L(x) = 1$,

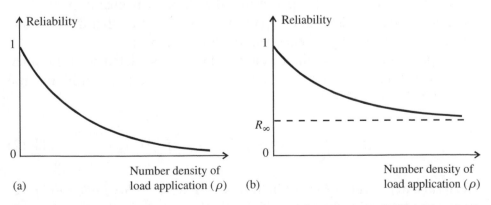

Figure 11.13 Influence of the number density of load applications on reliability (a) perfectly rough and (b) perfectly smooth loading.

if $x \geq L$. The integral from (11.12) can then be presented as

$$R(t) = \int_{S_{\min}}^{L-} \exp\left[-\rho t(1-0)\right] f_S(x)\,dx + \int_{L}^{S_{\max}} \exp\left[-\rho t(1-1)\right] f_S(x)\,dx$$

$$= \exp\left(-\rho t\right) \int_{S_{\min}}^{L-} f_S(x)\,dx + \int_{L}^{S_{\max}} f_S(x)\,dx \tag{11.15}$$

This result shows that, during ideally smooth loading, with increasing the number of load applications, reliability decreases monotonically, approaching the value $R_{\infty} = \int_{L}^{S_{\max}} f_S(x)\,dx$, which is the probability that strength will be larger than the load. In other words, in case of a small variation of the load, increasing the number of load applications beyond a particular value has no practical effect on reliability. Reliability values tend to the probability $R_{\infty} = \int_{L}^{S_{\max}} f_S(x)\,dx$ that strength will be larger than load. Now suppose that the constant load is so large, that strength always remains smaller than load. In this case, $\int_{S_{\min}}^{L} f_S(x)\,dx = 1$ and $\int_{L}^{S_{\max}} f_S(x)\,dx = 0$ in equation (11.15) and reliability becomes

$$R(t) = \exp\left(-\rho t\right)$$

This is yet another reason for the origin of the exponential distribution of the time to failure and the flat region of the bathtub curve. Even if all components undergo wearout and deterioration during the time interval $(0, t)$, the time to failure is still described by the exponential distribution if failure is controlled by random load applications whose times follow a homogeneous Poisson process.

11.12 RISK REDUCTION BY MODIFYING THE SHAPE OF COMPONENTS AND CHANGING THE AGGREGATE STATE

11.12.1 Risk Reduction by Modifying the Shape of Components

A good example where the risk of failure is reduced by modifying the shape of components is the case where in order to increase heat dissipation, the components' surface is increased by flattening, or by introducing cooling ribs which increase the surface-to-volume ratio. Conversely, in cases where heat conduction is unwanted, the shape is made spheroidal which decreases the surface-to-volume ratio. Thus, in order to reduce erosion of the cladding in blast furnaces due to interaction with molten metal, the cladding components are often made spheroidal.

In another example, discharging wet sand through a symmetrical funnel is often associated with an arch formed by the sand above the opening and which interrupts the flow. This failure is eliminated completely by a funnel with asymmetrical shape (Ullman, 2003).

Now consider the pressure vessel in Fig. 11.14(a) with diameter D, length L and thickness of the shell s. The vessel contains fluid exerting pressure p on the inside of the shell (Fig. 11.14(a)). Changing the shape of the pressure vessel from that in Fig. 11.14(a) to the one in Fig. 11.14(b) by keeping the same volume

$$V_1 = \pi D^2 \left(\frac{D}{6} + \frac{L}{4} \right) = V_2 = \pi d^2 \left(\frac{d}{6} + \frac{l}{4} \right)$$

reduces significantly the hoop stress, which is the largest principal tensile stress acting on an element from the shell. The axial principal tensile stress is also reduced.

Thus, by modifying the shape, the hoop stress decreases from $\sigma_{H_1} = pD/(2s)$, (Fig. 11.14(a)), to $\sigma_{H_2} = pd/(2s)$ (Fig. 11.14(b)), while the axial stress is decreased from $\sigma_{A_1} = pD/(4s)$ (Fig. 11.14(a)) to $\sigma_{H_2} = pd/(4s)$, ($D > d$), (Fig. 11.14(b)).

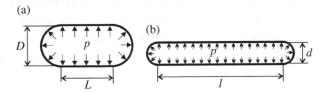

Figure 11.14 Modifying the shape of the pressure vessel by keeping the same volume significantly reduces the stresses.

11.12.2 Reducing the Risk by Changing the Aggregate State

Changing the aggregate state reduces the risk during transportation and storage of hazardous materials. Solidifying nuclear waste and other hazardous materials for example reduces significantly their hazard potential and the likelihood of pollution during an accident.

Changing the aggregate state is also used to protect the measuring equipment in rockets from overheating. They are put in a foam shell which evaporates after launching the rocket. Altshuller (1974) gives an interesting example of the principle of changing the aggregate state, to guarantee reliable separation of oil products transported in the same pipeline. The technical dilemma is that the solid separators cannot pass through the pumps maintaining the necessary working pressure for transportation. Gaseous separators worsen the properties of the flaw, collect in the upper part of the pipeline and essentially lose their capability to be separators. Liquid separators, on the other hand, tend to mix with the oil products and there is a problem related to their removal after transportation.

The problem with reliable transportation of several oil products had been solved by changing the aggregate state of the separator. Sections of liquid ammonia were used as separators. Ammonia does not dissolve in the oil products, does not interact with them and is relatively cheap. At the end of the transportation where the pressure drops, the liquid ammonia evaporates and there is no need to clean the oil products from it.

11.13 REDUCING THE RISK OF FAILURE CAUSED BY HUMAN ERRORS

Human errors account for a significant number of failures. They are an inevitable part of each stage of the product development and operation: design, manufacturing, installation and operation. Following Dhillon and Singh (1981), human errors can be categorised as (i) errors in design; (ii) operator errors (failure to follow the correct procedures); (iii) errors during manufacturing; (iv) errors during maintenance; (v) errors during inspection and (vi) errors during assembly and handling.

A thorough analysis of the root causes, conditions and factors promoting human errors is an important step towards reducing their number. Some of the most important error-promoting conditions and factors are listed below:

(i) Factors related to the work environment
- Poor organisation of the work place
- High noise levels
- Poor layout
- Many distractions
- Crowded space and poor accessibility
- Identically looking devices or information panels
- Difficult to operate controls or too sensitive controls
- Inadequate tools and faulty equipment
- High humidity, low or high temperature.

(ii) *Factors related to the individuals*
   - Time pressure and stress
   - Poor discipline and safety culture
   - Forgetting and omitting essential operations
   - Inattention and lack of concentration
   - Unfamiliarity with the equipment and the necessary procedures
   - Poor work skills and lack of experience
   - Poor health
   - Low confidence
   - Negative emotional states and disempowering beliefs
   - Overload and fatigue
   - Poor relationships with the management or with other members of the team
   - Poor motivation and work attitude.

(iii) *Factors related to the management and the organisation*
   - Inadequate command management style promoting low confidence and motivation in the workforce
   - Lack of ownership of management errors and blame culture
   - Conflicting requirements, directives and priorities
   - Inadequate information, specifications and documentation
   - Poor safety culture
   - Lack of enforcement of safety policies and practises
   - Suppressing the initiative and the ideas generated by the workforce
   - Too complex tasks and operations
   - Poorly defined responsibilities
   - Inadequate operating procedures
   - Poor training
   - Poor communication between designers, manufacturers and installers
   - Lack of feedback and interaction between teams and individuals
   - Inadequate management structure impeding the interaction between teams working on the same project.

Instructions and procedures must be clearly written, easy to follow and well justified. The procedures must also reflect and incorporate the input from people who are expected to follow them. It must always be kept in mind that human beings are prone to forgetting, misjudgement, lack of attention, creating false pictures of the real situation, etc. – conditions which are difficult to manage. Hardware systems and procedures are much easier to manage and change than human behaviour; therefore, the efforts should concentrate on devising hardware and procedures significantly reducing the possibility for human errors.

Learning from past failures and making available the information about past human errors which have caused failures is a powerful preventive tool. In this respect, compiling formal databases containing descriptions of failures and lessons learnt, and making them available to designers, manufacturers and operators are of significant value.

Frequent reviews, checks and tests of designs, software codes, calculations, written documents, operations or other products heavily involving people is a major tool for preventing human errors. In this respect, *double checking* of the validity of calculations, derivations or a software code are invaluable in preventing human errors. To eliminate common cause errors associated with selected approaches and methods, double checking based on two different methods is particularly helpful. In the area of probabilistic modelling in particular, where there are few props for the intuition, we found that testing the theoretical derivations by a Monte Carlo simulation is particularly useful and this practice has been followed in the book.

Since human errors are inevitable, protective barriers should be put in place, in order to contain and mitigate the consequences. Collaborative management style as opposed to a command management style increases confidence and motivation levels and reduces human errors. Typical features of a poor management is ignoring ideas for improvement generated by the workforce; not taking the ownership for mistakes and errors and blaming the workforce or external factors instead. Poor management often tacitly approves the violation of safety procedures and practices if this could speed up the accomplishment of a task.

A number of human errors arise in situations where a successful operation or assembly is overly dependent on human judgement. For example, over-tightening of bolted flanges may damage the seals while under-tightening promotes leakage and erosion of the seals. Human errors of this type can be avoided by using tools/devices which rely less on a correct human judgement. Wrenches with controlled torque are an example of devices which could be used to eliminate the problem related to incorrect tightening of flanges. Design features simplifying the assembly, Poka Yoke design features, and special recording and marking techniques could be used to prevent assembling parts incorrectly. Blocking against common cause maintenance errors could be achieved by avoiding situations where a single person is responsible for all pieces of equipment. Splitting the maintenance of redundant components between several operators prevents creating a common cause maintenance fault which could induce failures in all redundant components.

Most of the human errors are associated with performing operations and tasks. Many handling errors occur because storage and transport are not in accordance with the manufacturer's recommendations. Proper storage of components is important in order to avoid damage and deterioration by ageing. Components could be damaged as a result of storage under heavy parts, not being separated from other components and contaminated by absorbing moisture, oil and dirt.

Many early-life failures could be prevented if the number of inspection errors is reduced. These can broadly be divided into type-I errors where a substandard component is accepted and type-II errors where a quality component is rejected. Type-I inspection errors are particularly dangerous because they promote early-life failures associated with significant losses. In this respect, a special care must be exercised during assembly, which is accountable for a great number of early-life failures. Dents, scratches, tool marks, cuts and chips from anticorrosion coatings must be avoided, because they promote fast corrosion and fatigue deterioration. A tool mark or a small scratch on a heavily loaded component (e.g. a spring) reduces its fatigue life dramatically.

A thorough task analysis reveals weaknesses in the timing and the sequences of the separate operations and is a key factor for improving their reliability. Additional training and reducing the number of operations have a great impact on the probability of successfully accomplishing a task. The influence of additional training can be illustrated by the following generic example. Suppose that a task consisting of 50 identical operations needs to be accomplished and the probability of successfully accomplishing an operation without additional training is 0.98. Suppose also that additional training increases this probability to 0.99. Comparing the probability $0.98^{50} \approx 0.36$ of accomplishing the task without training and the probability $0.99^{50} \approx 0.60$ of accomplishing the task with additional training, clearly demonstrates the impact of training.

In another example illustrating the effect from reducing the number of operations, a task consisting of 50 identical operations needs to be accomplished, where the probability of successfully accomplishing an operation is 0.98. The probability of successfully accomplishing all of the operations is $p_{50} = 0.98^{50} \approx 0.36$. If the number of operations is halved, the probability of successfully accomplishing the task is increased dramatically to $p_{25} = 0.98^{25} \approx 0.60$.

11.14 REDUCING THE RISK OF FAILURE BY REDUCING THE PROBABILITY OF CLUSTERING OF EVENTS

Reliable and smooth operation often depends on not having two or more events clustered within a critical time interval. In other words, the failure-free operation depends on the existence of minimum critical distances between the time occurrences of the events. Consider a finite time interval of length $a = 1$ year, during which $n = 10$ large consumers connect to a supply system independently and randomly. Suppose that the supply system needs a minimum time interval of 1 week ($s = 7$ days) to recover and stabilise after a demand from a consumer. The supply system is overloaded if two or more demands follow (cluster) within the critical time interval $s = 7$ days. For a fixed number of consumers, the probability of failure of the supply system can be calculated by using the equation (Todinov, 2004a)

$$p_c = 1 - \left(1 - \frac{(n-1)s}{a}\right)^n = 1 - \left(1 - \frac{9 \times 7}{365}\right)^{10} \approx 0.85 \qquad (11.16)$$

where 1 year $\approx$ 365 days has been assumed. There exists 85% chance that there will be clustering of two or more demands within a week! Equation (11.16) yields an unexpected result which has also been confirmed by a Monte Carlo simulation.

Suppose now that the times of the demands follow a homogeneous Poisson process. In other words, the number of demands in the time interval is a random variable. According to an equation rigorously derived in (Todinov, 2004e),the probability of clustering p_c of two or more random demands within the critical distance s is

$$p_c = 1 - \exp\left(-\lambda a\right)\left(1 + \lambda a + \frac{\lambda^2 (a-s)^2}{2!} + \cdots + \frac{\lambda^r [a - (r-1)s]^r}{r!}\right)$$
$$(11.17)$$

where r denotes the maximum number of demands, with demand-free gaps of length s between them which can be accommodated into the finite time interval with length a ($r = [a/s] + 1$), where $[a/s]$ is the greatest integer which does not exceed the ratio a/s.

Now let us solve the problem related to clustering of random demands within a week, by assuming that the random demands follow a homogeneous

Poisson process with an average of 10 demands per year ($\lambda = 10$ year^{-1}). Substituting the values $a = 365$ days, $s = 7$ days and $\lambda = 10/365$ day^{-1} in equation (11.17) results in $p_c \approx 0.775$ for the probability of clustering of two or more demands within a week. Although in this case, the number of random demands is a random variable itself, the probability of clustering is still very large: 77.5%!

These examples demonstrate, how easy it is, without a proper calculation, to underestimate the probability of clustering of demands which can result in poor risk management decisions regarding the resources necessary to meet all demands.

Let us consider another example. A single spare component is kept as a standby redundant component. In case of failure of the working component it is replaced by the spare component, a new spare component is ordered immediately and the production continues. The time for delivery of the spare component is 1 week. Suppose that failures follow a homogeneous Poisson process with expected number of 10 failures a year (any of the failures is equally likely to occur on any day of the year). The probability that production will be lost because no spare component will be available at the time of failure is 77.5%, which is a very large probability.

Here are some common examples where reliability depends on the existence of minimum critical distances *between* the occurrences of random events (Todinov, 2005a):

- Users using the same equipment for a fixed time s. Collisions between user demands occurs if two or more users arrive within the time interval s, allocated per each user.
- Forces acting on a loaded component which fails if two or more forces cluster within a critically small distance.
- Limited available resources for repairs. In this case, it is important to guarantee that failures will be apart, at distances greater than a specified minimum distance, so that there will be no shortage of resources for repair.
- Clustering of two or more random flaws over a small critical distance decreases dangerously the load-carrying capacity of thin fibres and wires. As a result, a configuration where two or more flaws are within the critical distance cannot be tolerated during loading.

Decreasing the number of events in the time interval is an efficient way of decreasing the probability of clustering of events. Thus, for the discussed example related to random demands to a supply system, if the number of users is reduced from 10 to 5, the probability of clustering of demands is reduced significantly, from 85% to 33%:

$$p_c = 1 - \left(1 - \frac{(n-1)s}{a}\right)^n = 1 - \left(1 - \frac{4 \times 7}{365}\right)^5 \approx 0.33 \qquad (11.18)$$

Similarly, if the number of demands is a random variable, reducing the number density of the demands from an average of 10 demands per year to 5 demands per year, reduces the probability of clustering from 77.5% to 34%. We must point out that the alternative way of decreasing the probability of clustering by increasing the length of the time interval during which these demands occur is not as efficient (Todinov, 2004e).Indeed, according to equation (11.16), doubling the time interval from 1 to 2 years by keeping the same number of demands ($n = 10$) still yields a high probability of clustering (59%):

$$p_c = 1 - \left(1 - \frac{9 \times 7}{2 \times 365}\right)^{10} \approx 0.59$$

Similarly, if the demands follow a homogeneous Poisson process, doubling the time interval to two years by keeping the expected number of demands the same, results in a probability of clustering 56.7% (see equation (11.17)), which is still large.

12

REDUCING THE RISK OF FAILURE BY REDUCING THE NEGATIVE IMPACT FROM THE VARIABILITY OF DESIGN PARAMETERS

12.1 IMPROVING RELIABILITY BY REDUCING THE VARIABILITY OF DESIGN PARAMETERS

Variability of reliability-critical parameters can broadly be divided into the following categories: (i) variability associated with material and physical properties, manufacturing and assembly; (ii) variability caused by the product deterioration (iii) variability associated with the loads the product experiences in service; (iv) variability associated with the operating environment.

Strength variability caused by production variability and variability of properties is one of the major reasons for an increased interference of the strength distribution and the load distribution, resulting in overstress early-life failures. A heavy lower tail of the distribution of properties usually yields a heavy lower tail of the strength distribution, thereby promoting early-life failures. Low values of the material properties exert stronger influence on reliability than do high or intermediate values.

Reducing the losses from failures caused by variability of design parameters can be done in two main ways:

(i) By *reducing the variability of design parameters*.
(ii) By *making the design robust*, in other words, by decreasing its sensitivity to the variability of design parameters.

Variability of critical design parameters (e.g. material properties and dimensions) caused by processing, manufacturing and assembly is an important factor promoting early-life failures. Material properties such as (i) yield stress; (ii) static fracture toughness; (iii) fatigue resistance; (iv) modulus of elasticity and elastic limit; (v) shear modulus; (vi) percentage elongation; and (vii) density, often vary significantly. Defects and unwanted inhomogeneity are also sources of variability. Residual stress magnitudes are typically associated with large variation.

An important way of reducing the lower tail of the material properties distribution is the high-stress burn-in. The result is a substantial decrease of the strength variability and increased reliability on demand due to a reduced interference of the strength distribution and the load distribution.

Defects like shrinkage pores, sand particles and entrained oxides from casting, micro-cracks from heat treatment and oxide inclusions from material processing are preferred sites for early fatigue crack initiation. These flaws are also preferred sites for initiating fracture during an overstress failure. Segregation of impurities along grain boundaries reduces significantly the local fracture toughness and promotes *intergranular brittle fracture*. Impurities like *sulphide stringers* for example reduce the corrosion and fatigue resistance.

Since variability is a source of unreliability (Carter, 1997), reducing it is a particularly important factor for reducing early-life failures. Due to the inherent variability of the manufacturing process however, even items produced by the same manufacturer can be characterised by different properties. Production variability during manufacturing, not guaranteeing the specified tolerances or introducing flaws in the manufactured product, leads to a significant number of failures. Depending on the supplier, the same component, of the same material, manufactured to the same specifications is usually characterised by different properties. Between-suppliers variation exists even if the variation of the property values characterising the individual suppliers is small. A possible way of reducing the 'between-suppliers variation' is to use only the supplier producing items with the smallest variation of properties.

Suppose that three different suppliers, with market shares p_1, p_2 and p_3 ($p_1 + p_2 + p_3 = 1$) produce spring rods with yield strength characterised by means μ_1, μ_2 and μ_3 and standard deviations σ_1, σ_2 and σ_3, respectively (Fig. 12.1). Without loss of generality, suppose that $\sigma_1 < \sigma_2 < \sigma_3$. The variance σ^2 of the whole batch of rods is then given by (Todinov, 2002).

$$\sigma^2 = p_1\sigma_1^2 + p_2\sigma_2^2 + p_3\sigma_3^2 + p_1p_2(\mu_1 - \mu_2)^2$$
$$+ p_2p_3(\mu_2 - \mu_3)^2 + p_3p_1(\mu_3 - \mu_1)^2 \tag{12.1}$$

As can be verified, the variance is smallest ($\sigma^2 = \sigma_1^2$) when the supplier with the smallest variance is selected. If the shares $p_1, p_2, \ldots, p_M$ from the sources are unknown, a conservative assessment of the maximum possible

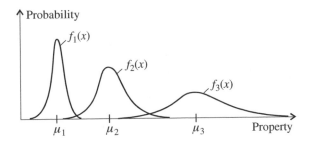

Figure 12.1 Distributions of components properties from three different suppliers.

variance from sampling them can be made by using the *upper bound variance theorem* (Todinov, 2003): *The maximum variance of properties from sampling multiple sources is always attained from sampling a single or at most two sources.* An algorithm for determining the exact upper bound of the variance can be found in (Todinov, 2003).

The 'within-supplier variation' can be reduced significantly by a *statistical process control*, more precise tools, production and control equipment, better specifications, better inspection and quality control procedures. Process control based on computerised manufacturing processes reduces significantly the variation of properties. Process control charts monitoring the variations of output parameters, statistical quality control and statistical techniques are important tools for reducing the amount of defective components (Montgomery et al., 2001).

The manufacturing process, if not tightly controlled can be the largest contributor to early-life failures. Because of the natural variation of critical design parameters, early-life failures are sometimes due to unfavourable combinations of values (e.g. worst-case tolerance stacks) rather than due to particular production defects. Probabilistic methods based on the distributions of critical design parameters can be used to assess the probability of unfavourable combinations of parameter values which cause faults. A comprehensive discussion related to the effect of dimensional variability on the reliability of products can be found in Booker et al. (2001) and Haugen (1980).

12.1.1 Improving Reliability by Reducing the Overall Variability of Load and Strength

Figure 12.2 illustrates a case where low reliability is a result of large variability of strength caused by poor material properties, manufacturing, assembly,

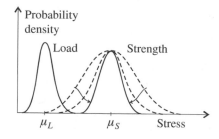

Figure 12.2 Increasing reliability by decreasing the variability of strength.

quality and process control. The large variability of strength leads to a large overlap of the lower tail of the strength distribution and the upper tail of the load distribution. Reliability can be improved by reducing the variance of strength (Fig. 12.2).

If, for example, strength variability is due to sampling from multiple sources, it can be decreased by sampling from a single source – the source characterised by the smallest variance. Low reliability due to increased strength variability is often caused by ageing and the associated with it material degradation. Material degradation can often be induced by the environment, for example from corrosion and irradiation. A typical feature of the strength degradation is an increase of the variance and a decrease of the mean of the strength distribution.

Low reliability is often due to excessive variability of the load. If variability of the load is large (rough loading), the probability of an overstress failure is significant. Mechanical equipment is usually characterised by a rough loading (Fig. 12.3(a)). A common example of smooth loading (Fig. 12.3(b)) is the power supply of electronic equipment through an anti-surge protector.

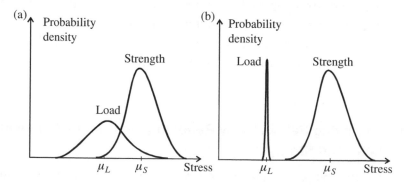

Figure 12.3 (a) Rough and (b) smooth loading.

The most important aspect of the load–strength interaction is the interaction of the upper tail of the load distribution and the lower tail of the strength distribution.

The variability associated with the lower tail of the strength distribution controls this interaction, not the variability of strength associated with the high or central values (Fig. 12.4). Stress screening eliminating substandard items is an efficient way of reducing the variability in the lower-tail region of the strength distribution and increasing reliability.

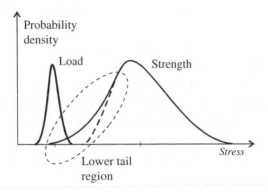

Figure 12.4 Reliability is controlled by the strength variability in the lower-tail region of the strength distribution.

In summary, some of the possible options for increasing the reliability on demand are: (i) decreasing the strength variability; (ii) modifying the lower tail of the strength distribution; (iii) increasing the mean strength; (iv) decreasing the mean load; (v) decreasing the variability of the load and obtaining a smooth loading; and (vi) truncating the upper tail of the load distribution by using stress limiters (See Fig. 13.3 from Chapter 13).

12.2 REDUCING THE VARIABILITY OF STRENGTH BY IMPROVING THE MATERIAL QUALITY

Material quality is positively correlated with the reliability of components. This correlation is particularly strong for highly stressed components, such as helical springs. If for example, a *Si–Mn* spring wire is of poor quality, and contains a large number of sulphide stringers or oxide inclusions, its fatigue life is inferior to the fatigue life of a spring manufactured from cleaner steel. While flaws like oxide inclusions serve as fatigue crack initiation sites, sulphide stringers cause anisotropy of the spring wire and promote longitudinal splitting during fatigue loading.

Sources of material supply must be controlled strictly, without relying on vendor's tradenames or past performance. Changes in the processing and manufacturing procedures often result in materials with poor quality. In case of spring rods for example, the quantity of sulphide and oxide inclusions and manufacturing defects (e.g. seams) varies significantly with the supplier. Spring rods with excessive amount of these affairs result in springs with unsatisfactory fatigue life.

Improper heat treatment such as improper gas carburising, quenching, tempering or case hardening may result in reduced component strength or soft patches on the surface. As a result, overstress loading or large contact stresses cause failure due to yielding, excessive plastic deformation or seizure. Improper processing can also be a source of problems during subsequent manufacturing operations (e.g. welding). A classical case illustrating this point is the improperly processed chromium stainless steel which becomes susceptible to intensive intergranular corrosion. For such a steel, chromium carbides precipitate at the grain boundaries during welding thereby depleting the adjacent matrix of chromium, which reduces the corrosion resistance.

12.3   REDUCING THE VARIABILITY OF GEOMETRICAL PARAMETERS, PREVENTING FITTING FAILURES AND JAMMING

If the tolerances are strictly controlled during manufacturing, there will be less possibility for problems and failures during assembly, less possibility for jamming, seizure, poor lubrication and fast wearout. Precautions must also be taken to prevent degradation or contamination during assembly.

Often variability of geometrical parameters causes *fit failures*, resulting from interference of solid parts containing points with the same space coordinates which makes the assembly impossible. When combined with corrosion deposits or debris build-up, the variability of geometrical parameters also promotes jamming. Jamming does not allow a valve or other control to be operated when required. It can be counteracted by avoiding tolerances smaller than the prescribed ones. Although such tolerances permit the assembly to be made, they do not provide sufficient barrier against jamming. The next example outlines a method based on load-strength interference for calculating the combined probability of a fit failure or a tolerance fault causing jamming.

Example:

In an assembly of a particular equipment, component A in Fig. 12.5 characterised by a normally distributed diameter d with mean $\mu_d = 28$ mm and standard deviation $\sigma_d = 0.5$ mm must fit into component B with normally distributed inside diameter D with mean $\mu_D = 31$ mm and standard deviation $\sigma_D = 1.5$ mm. In order to avoid a fit failure (inability to fit A into B), the inside diameter of component B must be greater than the diameter of component A. In order to avoid jamming, the inside diameter of component B must be greater than the diameter of component A by at least $a = 0.5$ mm. Calculate the percentage of assemblies for which no fit failure or a tolerance fault causing jamming will be present.

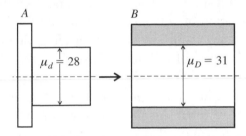

Figure 12.5 Assembly which is affected by a fit failure or a tolerance fault causing jamming.

Solution:

If the minimum tolerance of $a = 0.5$ mm is to exist, the relationship $\Delta = D - (d + a) > 0$ must be fulfilled, where D is the diameter of component B and d is the diameter of component A. Since both diameters are normally distributed, the difference Δ is also normally distributed with mean

$$\mu_\Delta = \mu_D - (\mu_d + a)$$

and standard deviation

$$\sigma_\Delta = \sqrt{\sigma_D^2 + \sigma_d^2}$$

Consequently, the probability $P(\Delta \geq 0)$ can be obtained from

$$p = 1 - \Phi\left(\frac{0 - \mu_\Delta}{\sigma_\Delta}\right) = \Phi\left(\frac{31 - 28 - 0.5}{\sqrt{1.5^2 + 0.5^2}}\right) = \Phi(1.58) \approx 0.94$$

where $\Phi(\bullet)$ is the cumulative distribution of the standard normal distribution. In other words, 94% of the assemblies will be quality assemblies for which a fault causing a fit failure or jamming will not exist.

Experiments have indicated that the time evolution $E(t)$ of the sticking/jamming forces between sliding surfaces can be modelled by the equation

$$E(t) = E_L + (E_U - E_L)\,F(t) \tag{12.2}$$

where t is time, $F(t) = 1 - \exp\left[-k(t - t_0)^m\right]$; k, t_0 and m are parameters depending on the materials, the operating conditions and geometry.

In equation (12.2), E_L corresponds to the minimum level of the jamming force and E_U corresponds to the maximum level of the jamming force (100%) attained after a significant amount of time. Jamming of valve stems can prevent the operation of valves. One of the main failure modes of slab gate valves for example, is failure to open or close due to jamming of the gate, or the stem. The risk of failure to close the valve due to jamming of the stem can be reduced by operating the valve frequently. This helps to dislodge any deposit build-ups and prevents the jamming force from increasing excessively (Fig. 12.6(b)). An alternative way is to use deposit-resistant coatings for the stems, deposit-resistant seals, regular greasing, special design solutions and corrosion control preventing formation of deposits. Filtering the working fluids also helps diminish the likelihood of jamming.

In cases where the valve is closed by a compression spring (e.g. if the hydraulic supply which keeps it open is suddenly lost) it is important to guarantee that the spring force is sufficient to overcome the resistance from deposits and close the valve. In this respect, it is important to prevent

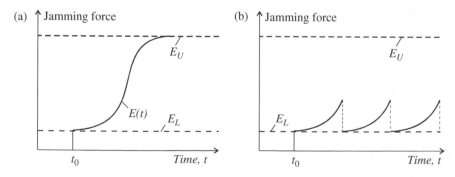

Figure 12.6 Time evolution of the jamming force with time: (a) without operating the sliding surfaces; (b) by frequently operating the sliding surfaces.

the spring from stress relaxation or failure. Stress relaxation is the loss of spring force when the spring is held at load or cycled under load. This can decrease the spring force to such an extent that the fail-safe valve can no longer be closed. Preventing stress relaxation can be achieved by appropriate microalloying and heat-treatment (e.g. increasing the silicon content). Preventing spring failures is achieved by improving the fatigue resistance of the spring steel and preventing stress-corrosion cracking by appropriate corrosion protection.

12.4 REDUCING THE RISK OF FAILURE BY MAKING THE DESIGN ROBUST

12.4.1 Determining the Variation of the Performance Characteristics

In cases where there exists a simple functional relationship, the distribution of the performance characteristics can be determined by transforming the probability density functions of the controlling variables into probability density functions of the performance characteristics (DeGroot, 1989).

Suppose that the functional relationship between the performance characteristic y and the single controlling variable x is specified by

$$y = r(x) \tag{12.3}$$

where $r(x)$ is a continuous, either strictly increasing or strictly decreasing function. Let $f(x)$ be the probability density distribution of the control variable x and $x = r^{-1}(y)$ be the inverse function. The probability density function of the performance characteristic y is then given by

$$g(y) = f(r^{-1}(y)) \left| \frac{dr^{-1}(y)}{dy} \right| \tag{12.4}$$

Similar relationships can also be derived for functions of two or more variables (DeGroot, 1989).

Suppose that the performance characteristic referred to as 'output' is a non-linear function of several statistically independent input parameters x_1, $x_2, \ldots, x_n$: $y = f(x_1, x_2, \ldots, x_n)$. A small variation of the output regarding a particular point $x^0 \equiv \{x_1^0, x_2^0, \ldots, x_n^0\}$ can then be presented as

$$\delta y = \left(\frac{\partial f}{\partial x_1} \Big|_{\mathbf{x}=\mathbf{x}^0} \right) \delta x_1 + \cdots + \left(\frac{\partial f}{\partial x_n} \Big|_{\mathbf{x}=\mathbf{x}^0} \right) \delta x_n$$

Since $\left(\frac{\partial f}{\partial x_i}\big|_{\mathbf{x}=\mathbf{x}^0}\right)$ are constants, from the formula for the variance of a linear combination of statistically independent random variables, the variance s^2 of y is given by

$$s^2 \equiv G(x_1, \ldots, x_n) = \left(\frac{\partial f}{\partial x_1}\right)^2 s_1^2 + \cdots + \left(\frac{\partial f}{\partial x_n}\right)^2 s_n^2 \qquad (12.5)$$

where s_i^2 are the variances of the input parameters x_i. In some cases, between the controlling random variables $x_1, \ldots, x_n$ there might exist equality constraints:

$$g_i(x_1, x_2, \ldots, x_n) = 0 \quad (i = 1, 2, \ldots, m) \quad (m < n)$$

In the general case, determining the mean values of the control variables which minimise the variation of the performance characteristic requires efficient algorithms for constrained non-linear optimisation. In the simplest cases, the mean values of the parameters $x_1, \ldots, x_n$ minimising the variance $G(x_1, \ldots, x_n) = 0$ can be obtained by using *Lagrange multipliers*. Methods based on Lagrange multipliers can be used in cases where the variables are continuous, no inequalities appear in the constraints and the objective and constraint functions possess second-order partial derivatives.

The necessary conditions for extremum of the variance $G(x_1, \ldots, x_n)$ are

$$\frac{\partial G}{\partial x_k} + \lambda_1 \frac{\partial g_1}{\partial x_k} + \lambda_2 \frac{\partial g_2}{\partial x_k} + \cdots + \lambda_m \frac{\partial g_m}{\partial x_k} = 0 \quad (k = 1, 2, \ldots, n) \qquad (12.6)$$

where $\lambda_1, \ldots, \lambda_m$ are Lagrange multipliers. These n equations are solved together with the m constraints $g_i(x_1, x_2, \ldots, x_n) = 0$, from which the $n + m$ unknowns $(x_1, x_2, \ldots, x_n, \lambda_1, \lambda_2, \ldots, \lambda_m)$ are determined.

The parameter values $x_1^*, \ldots, x_n^*$ which minimise the variance define a robust design. For these optimum mean values of the input parameters, the output performance characteristics will be least sensitive to variations in the input parameters.

If the standard deviation of the performance characteristic y is still large, the next step is to reduce the variation of parameters x_i.

The solution of the constrained optimisation problem can also be obtained by solving a sequence of unconstrained optimisation problems whose objective functions include penalty for violating the constraints. The unconstrained problems can be solved by some of the well known methods: for example the *Hooke–Jeeves pattern search method*, the *flexible polygon search of Nelder–Mead* or the *conjugate direction search of Powell* (Press et al., 1992). The *simulated annealing method* based on the *Metropolis criterion* can be used as a probabilistic method for solving the constrained optimisation problem. In case of inequality constraints

$$g_i(x_1, x_2, \ldots, x_n) \leq 0 \quad (i = 1, 2, \ldots, m)$$

the corresponding necessary conditions (*Kuhn–Tucker conditions*) are

$$\frac{\partial G}{\partial x_k} - \lambda_1 \frac{\partial g_1}{\partial x_k} - \lambda_2 \frac{\partial g_2}{\partial x_k} - \cdots - \lambda_m \frac{\partial g_m}{\partial x_k} = 0 \quad (k = 1, 2, \ldots, n) \quad (12.7)$$

$$\lambda_i \, g_i(x_1, x_2, \ldots, x_n) = 0, \quad \lambda_i \geq 0, \quad g_i \leq 0, \quad i = 1, \ldots, m \quad (12.8)$$

In many cases, the relationship between the performance characteristic and the controlling variables cannot be presented in a closed form or if it does exist, it is too complex. Furthermore, the controlling variables may be interdependent, subjected to complex constraints. In all these cases, the Monte Carlo simulation is a powerful alternative to other methods. It is universal, handles complex constraints and interdependencies between controlling variables, and does not require a closed form function related to the performance characteristics. Furthermore, its algorithm and implementation is simple and straightforward. Here is a generic, Monte Carlo simulation algorithm for determining the variation of a performance characteristic as a function of n statistically independent controlling random variables.

Algorithm 12.1
x[n] /* Global array containing the current values of the n controlling variables */
Output[Number_of_trials] /* Global array containing the distribution of
values for the performance characteristic */

procedure **Generate_control_variable (j)**
{
   /* *Generates a realisation (value) of the j-th controlling random variable x[j]* */
}

function **Performance_characteristic()**
{
   /* *For a particular combination of values of the controlling variables*
      x[1], ... , x[n], calculates the value of the performance characteristic */
}

/* Main algorithm */

```
 S1=0;  S2= 0;
For  i = 1 to  Number_of_trials do
 {
    /* Generate the i-th set of n controlling random variables */
    For  j=1 to  n do
       Generate_control_variable (j);

    Output[i] = Performance_characteristic(); /* Calculates the value of the
    performance characteristic in the i-th simulation trial and stores the result
    in Output[i] */

   // Finds the variance of the performance characteristic
      temp = Output[i];
      S1 = S1 + temp;
      S2 = S2 + temp x temp;
}
Output_variance = S2/Number_of_trials + (S1/Number_of_trials)²;
Sort array Output [] in ascending order and plot the cumulative distribution function.
```

In the simulation loop controlled by variable i, a second nested loop has been defined, controlled by variable j, whose purpose is to generate instances of all controlling variables. After obtaining a set of values for the random variables, the function **Performance_characteristic()** is called to calculate the value of the performance characteristic in the ith trial and store it in the ith element of the array Output[]. Simultaneously, the calculated values and their squares are accumulated in variables $S1$ and $S2$, introduced to calculate the variance of the performance characteristic. At the end, the array Output[Number_of_trials] contains the sorted in ascending order values of the performance characteristic while the value Output_variance contains the variance of the performance characteristic. Plotting the ordered values Output[i] against i/ (Number_of_trials + 1) yields the empirical distribution of the performance characteristic.

12.4.2 Reducing the Variation of the Performance Characteristics Through Robust Designs

Robustness is an important property of components and systems. Robust designs are insensitive to variations of the manufacturing quality, drifts in the parameter values, operating and maintenance conditions, environmental loads and deterioration with age (Lewis, 1996). In many cases, the reliable work of components and systems occurs under too narrowly specified conditions. Slight variations in the material quality, the quality of manufacturing, the external load or the values of the design parameters are sufficient to promote failures. In this sense, the non-robust and robust design could be compared to an unstable and stable state.

Design solutions requiring fewer parts, with simple geometry, reduce the susceptibility to manufacturing variability. Designs incorporating appropriate geometry, static determinacy, tolerances and materials with high conductivity reduce the susceptibility to temperature variations and large thermal stresses which are a common cause of failure. Components made of material free from inclusions and other flaws reduce the susceptibility to failures initiated by flaws. A simple example of robust design (Kalpakjian and Schmid, 2001) is presented in Fig. 12.7.

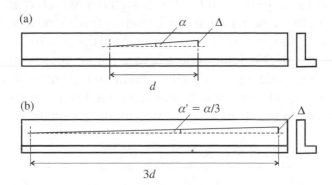

Figure 12.7 A simple example of non-robust (a) and robust design (b).

The angular bracket in Fig. 12.7 should be fixed on the wall so that the bracket is horizontal. For the design in Fig. 12.7(a), a small relative error Δ in the vertical position of the mounting bolts causes a misalignment angle $\alpha \approx \Delta/d$. By increasing the distance between the bolts 3 times the misalignment angle $\alpha' = \Delta/(3d)$ is reduced 3 times. For the same relative error Δ in the vertical position of the bolts, the design in Fig. 12.7(b) is less sensitive compared to the design in Fig. 12.7(a). The more robust design has been achieved by increasing the distance between the mounting bolts, without reducing the variation Δ associated with their vertical position.

The concept 'robust design' can also be illustrated by the two designs in Fig. 12.8. They are characterised by the same loading force F and different design angles $\alpha(\alpha = 85°$ for design A and $\alpha' = 20°$ for design B). For a

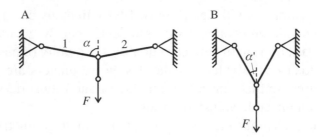

Figure 12.8 Design *A* can be made more robust by reducing angle α to α' (B).

loading force with magnitude F, the force acting in the two struts '1' and '2' is $\frac{F}{2\cos\alpha}$. Its values, at different values of the loading force F, have been plotted in Fig. 12.9.

As can be verified, for design *A* characterised by angle $\alpha = 85°$, a variation of the loading force F within the range 0, 10000 N results in a variation of the strut forces in the range 0, 57369 N. For design *B* however, characterised by angle $\alpha' = 20°$, the same variation of the loading force F results in more than 10 times smaller range (0, 5321 N) of variation of the strut forces. Compared to design *A*, design *B* is more robust because it is characterised by small strut forces whereas design A magnifies the amplitude of the loading force.

Suppose that a minimal design angle α_0 has been determined, which limits the forces in the struts below a maximum acceptable value T_0. It is

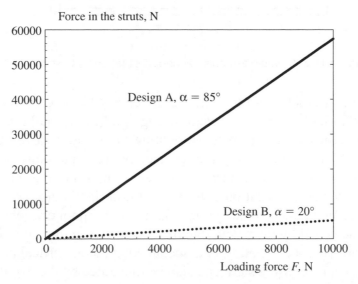

Figure 12.9 Variation of the force acting in struts '1' and '2' as a function of the variation of the loading force *F*.

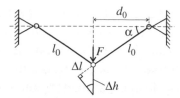

Figure 12.10 The tensile strut forces $T = F/(2 \sin \alpha)$ induced by the loading force F cause elastic deformation of the struts which results in a vertical displacement Δh.

of interest to determine the angle α^* which guarantees the smallest vertical displacement Δh of the pin joint where struts '1' and '2' are connected (Fig. 12.10).

In the domain $\alpha > \alpha_0$ which limits the strut forces T below the critical value $T_0(T < T_0 = F/(2 \sin \alpha_0))$, we assume that the vertical displacement Δh can be approximated with a reasonable accuracy by

$$\Delta h = \frac{\Delta l}{\sin \alpha} \tag{12.9}$$

The absolute deformation Δl of the struts however, caused by the strut forces T, is given by

$$\Delta l = \frac{T l_0}{ES} \tag{12.10}$$

where E is the elastic modulus of the material, l_0 is the non-deformed length of the strut corresponding to a zero loading force ($F = 0$) and S is the cross-sectional area of the strut.

Since $T = F/(2 \sin \alpha)$ and $l_0 = d_0/\cos \alpha$, substituting in equation (12.9) gives

$$\Delta h \approx \frac{F d_0}{2ES \sin^2 \alpha \cos \alpha} \tag{12.11}$$

Clearly, the displacement Δh is smallest when $\sin^2 \alpha \cos \alpha$ has a maximum. Since $\sin^2 \alpha \cos \alpha = y = x - x^3$ where $x = \cos \alpha$, all local extrema can be obtained by solving $y' = 1 - 3x^2 = 0$ ($0 \leq x \leq \cos \alpha_0$) Since $y'' = -6x < 0$ for $0 \leq x \leq \cos \alpha_0$, the obtained local extremum is a maximum. The local maximum is attained for $x = \sqrt{3}/3$ which corresponds to an angle $\alpha^* = \arccos(\sqrt{3}/3)$. If $\alpha_0 \leq \alpha^*$, the absolute maximum of $y = \sin^2 \alpha \cos \alpha$ is obtained at $\alpha = \alpha^*$; if $\alpha_0 > \alpha^*$, the absolute maximum is obtained at $\alpha = \alpha_0$.

The Taguchi robust design methodology focuses on selecting mean values of the design parameters such that the variation of the performance characteristics is minimised. In other words, performance characteristics are made insensitive to variation of the parameter values (Ulman, 2003; Ross, 1988). Compared to reducing the variation of the parameter values, this is a more cost-effective approach. Reducing variability by specifying tighter tolerances for design parameters increases the manufacturing costs and does not decrease the sensitivity to variation associated with the environment (Lewis, 1996).

In conclusion, the methodology for achieving robust designs can be summarised in the following basic steps.

(i) Establishing relationships between the control variables and a performance characteristic;
(ii) From the variation of the control variables, determining the variation of the performance characteristic.
(iii) Determining the mean values of the control variables which minimise the variation of the performance characteristic.

The variation of the performance characteristic can be obtained from the variations of the control variables by a Monte Carlo simulation. In trivial cases, where a simple analytical solution exists, the distribution of the performance characteristic can be determined analytically.

In cases where it is impossible to build a model of the performance characteristic, a robust design through testing can be achieved. This is essentially, the well-documented Taguchi's experimental method for robust design (Phadke, 1989).

13

GENERIC SOLUTIONS FOR REDUCING THE LIKELIHOOD OF OVERSTRESS AND WEAROUT FAILURES

Following Dasgupta and Pecht (1991), the mechanisms of failure of components can be divided broadly into overstress failures (*brittle fracture, ductile fracture, yield, buckling*, etc.) and wearout failures (*fatigue, corrosion, stress-corrosion cracking, wear, creep*, etc.) Overstress failures occur when load exceeds strength. If load is smaller than strength, the load has no permanent effect on the component. Conversely, wearout failures are characterised by a damage which accumulates irreversibly and does not disappear when load is removed. Once the damage tolerance limit is reached, the component fails (Blischke and Murthy, 2000).

13.1 IMPROVING RELIABILITY BY A RELATIVE SEPARATION OF THE UPPER TAIL OF THE LOAD DISTRIBUTION AND THE LOWER TAIL OF THE STRENGTH DISTRIBUTION

Figure 13.1 illustrates clearly one significant drawback of the reliability improvement based on reducing the variance of the load and strength. In cases where reliability is controlled by an interaction of the load and strength, reducing the variance of the load and strength in the area of the interaction of their tails is important, not reducing the total variance. This can be demonstrated by using simple counterexamples. In Fig. 13.1(b), the variances of the load and strength have been decreased significantly, yet reliability on demand is lower compared to the configuration in Fig. 13.1(a). This is because of the larger overlap between the lower tail of the strength distribution and the upper tail of the load distribution characterising the configuration in Fig. 13.1(b). Reducing the variation of the load and strength in the way shown in Fig. 13.1(b) *will result in a less reliable design*!

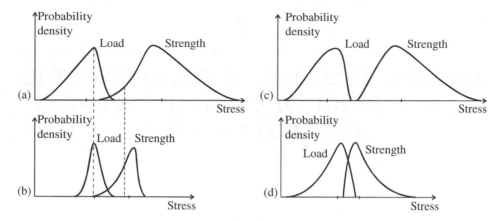

Figure 13.1 Counterexamples demonstrating the importance of the region of the upper tail of the load distribution and the lower tail of the strength distribution.

Conversely, in Fig. 13.1(c), despite the large variation associated with the load and strength, reliability is high because there is no interaction between the tails of the load and strength. It now seems that a reliable design can be achieved by minimising the variation of the load in the region occupied by the upper tail of the load distribution and the lower tail of the strength distribution. Figure 13.1(d) however demonstrates a case where despite the small variation of the lower tail of the strength distribution and the upper tail of the load distribution, the reliability is still low.

The problem related to the criterion for a reliable design can be resolved if we quantify the interaction between the lower tail of the strength distribution and the upper tail of the load distribution. The probability of failure p_f for a single load application is given by (Freudental, 1954):

$$p_f = \int_{S_{\min}}^{S_{\max}} [1 - F_L(x)] f_S(x) \, dx \qquad (13.1)$$

where $F_L(x)$ is the cumulative distribution of the load, $f_S(x)$ is the probability density of the strength, $S_{\min}$ and $S_{\max}$ are the lower and the upper bound of strength for which $f_S(x) \approx 0$ if $x \le S_{\min}$ or $x \ge S_{\max}$.

If the upper bound of the load is smaller than the upper bound of the strength $(L_{\max} < S_{\max})$, the integral in equation (13.1) can also be presented as

$$p_f = \int_{S_{\min}}^{L_{\max}} [1 - F_L(x)] f_S(x) \, dx + \int_{L_{\max}}^{S_{\max}} [1 - F_L(x)] f_S(x) \, dx \qquad (13.2)$$

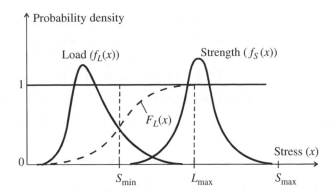

Figure 13.2 Deriving the probability of failure, by integrating within the lower bound of the strength distribution and the upper bound of the load distribution (Todinov, 2005a).

For $x > L_{max}$, $F_L(x) \approx 1$ holds for the cumulative distribution of the load (Fig. 13.2) and the second integral in equation (13.2) becomes zero ($\int_{L_{max}}^{S_{max}} [1 - F_L(x)] f_S(x) \, dx \approx 0$). In other words, there is no need to integrate between L_{max} and S_{max} (Fig. 13.2), because in this interval, load is always smaller than strength and the probability of failure is zero. For similar reasons, there is no need to integrate below S_{min} because $f_S(x) \approx 0$ in this region.

Consequently, the probability of failure becomes

$$p_f = \int_{S_{min}}^{L_{max}} [1 - F_L(x)] f_S(x) \, dx \tag{13.3}$$

The shape of the distributions in the domains ($x < S_{min}$ and $x > L_{max}$) has no influence on the probability of failure! Achieving reliable designs therefore is about selecting mean values of the parameters which minimise the integral in equation (13.3), not minimising *the variances of the load and strength*.

The value of the integral (the probability of failure) can be decreased by narrowing its integration limits. This can be done by increasing the limit S_{min} or decreasing the limit L_{max} or both. In other words, by decreasing the difference $L_{max} - S_{min}$. Other ways of minimising the integral (13.3) can be revealed if it is presented in the form

$$p_f = \int_{S_{min}}^{L_{max}} f_S(x) \, dx - \int_{S_{min}}^{L_{max}} F_L(x) f_S(x) \, dx \tag{13.4}$$

Since $f_S(x) \geq F_L(x) f_S(x)$, this difference is always non-negative. An alternative way of decreasing this difference is to reduce the integral

$\int_{S_{\min}}^{L_{\max}} f_S(x)\,\mathrm{d}x$ by reducing $f_S(x)$ between $S_{\min}$ and $L_{\max}$ (e.g. through stress screening). This integral has an immediate interpretation. It gives the probability mass beneath the lower tail of the strength distribution bounded by the minimum strength and the upper bound of the load variation. Making $f_S(x)$ very small in this region, makes both terms $\int_{S_{\min}}^{L_{\max}} f_S(x)\,\mathrm{d}x$ and $\int_{S_{\min}}^{L_{\max}} F_L(x)f_S(x)\,\mathrm{d}x$ very small and, as a result, the probability of failure p_f is very small.

An alternative way of making p_f small is to make $\int_{S_{\min}}^{L_{\max}} F_L(x)f_S(x)\,\mathrm{d}x$ very close to $\int_{S_{\min}}^{L_{\max}} f_S(x)\,\mathrm{d}x$. Then the difference p_f will be very small. This can be achieved by separating the load and strength. Such a separation can be achieved by increasing the mean strength or decreasing the mean load, by decreasing the variance of the load and strength while not reducing the distance between their mean values, or by altering the shape of the tails.

13.2   INCREASING THE RESISTANCE AGAINST FAILURES CAUSED BY EXCESSIVE STRESSES

Often, poor knowledge and underestimation of dynamic environmental loads result in designs prone to early-life failures. These excessive loads, not normally accounted for during design, are usually due to large uncertainty regarding the loads the component is likely to experience during service, imprecision of the assembly, deformation of components with insufficient stiffness, residual stresses, increased friction, thermal stresses and excessive stresses during transportation and handling. Underestimating the working pressure and temperature, for example, may cause an early-life failure of a seal and a release of harmful chemicals into the environment. An early-life failure of a critical component may also be caused by unanticipated eccentric loads due to weight of additional components or by high-magnitude thermal stresses during start-up regimes. Early-life failures often result from failure to account for the extra loads during assembly. Installation loads are often the largest loads a component is ever likely to experience. Such are for example the loads during installation of pipelines on the sea bed for deep-water oil and gas production, which may easily induce cracks, plastic deformation or other damage. A faulty assembly often introduces additional stresses not considered during design. Misalignment of components creates extra loads, susceptibility to vibrations, excessive centrifugal forces on rotating shafts, and larger stress amplitudes leading to early-fatigue failures. A faulty

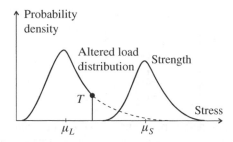

Figure 13.3 High reliability achieved by altering the upper tail of the load distribution by using stress limiters.

assembly and mishandling often causes excessive yield and deterioration of protective coatings, which promotes a rapid corrosion.

The nominal loading stress can be reduced by using *stress limiters* or by redesigning the component in order to endure larger stresses (e.g. using better material, increasing the thickness of critical sections, using fillets with large radii, etc.). Possible design solutions against shock loading, which often causes overstress failures, are the *shock absorbers* and *alloyed steels*.

Altering the upper tail of the load distribution by stress limiters (Fig. 13.3) is an efficient safeguard against excessive loads. The result is concentrating the probability mass beneath the upper tail of the load distribution (the area beneath the dashed tail in Fig. 13.3) into the truncation point T (Todinov, 2005a). Typical examples of stress limiters are the safety pressure valves, the fuses and switches, activated when pressure or current reaches critical values.

In cases where pressure build-up could cause a pressure vessel to explode, failure is avoided by designing safety pressure valves for releasing the excessive pressure. Friction clutches which slip if the torque of the driving shaft exceeds the maximum acceptable torque of the rotating machine is another example of a stress limiter.

13.3   REDUCING THE RISK OF FAILURE BY OPTIMISING LOADING AND AVOIDING UNFAVOURABLE STRESS STATES

A general principle to designing the loading is to ensure short, direct force paths. The lines of the *force flow* should be kept as direct as possible. Figure 13.4 illustrates this principle. Configuration (a) should be preferred to configuration (b) in design (Collins, 2003).

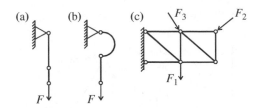

Figure 13.4 (a and b) A loading configuration illustrating the direct load path principle. (c) A loading configuration replacing bending with tension/compression.

The material should be used efficiently during loading. In case of torsion and bending, the stress distribution is non-uniform over the cross section. Since the highest stresses are carried by a relatively small part of the cross-section, the limit state occurs when the stresses in these parts reach unacceptable magnitudes. A hollow cylinder provides the most efficient use of material during torque, because it provides an almost constant stress throughout its cross-section.

During compression/tension, the stresses are uniform over the cross-section, the limit state occurs approximately simultaneously for all parts of the cross-section and the material is used more efficiently. A rod in tension or compression is the simplest shape which guarantees the most efficient use of material and also a uniform distribution of the stress. A *truss* carries all the loads by creating only compression and tension in its links. This principle is illustrated by Fig. 13.4(c) where any combination of loads applied solely in the joints, will result in uniformly distributed tensile and compressive stress in each of the links. If the friction in the pinned joints is negligible, no bending stresses will be introduced.

Triaxial tensile stress tensors, for which the three principal stresses σ_1, σ_2 and σ_3 are compressive, will impede initiation of fracture during overstress or fatigue cycling. One of the reasons is that if such a stress tensor is present, for any flaw orientation, the crack initiation is impeded. Even if a crack already exists, the triaxial compressive stress state will cause crack closure and its propagation rate will be small or zero. Less favourable would be a stress state where two of the principal stresses are compressive and one is tensile.

Triaxial tensile stress tensors, for which the three principal stresses σ_1, σ_2 and σ_3 are tensile, promote initiation of fracture during overstress or fatigue cycling. One of the reasons is that the triaxial tensile stress tensor is associated with a large normal opening tensile stress on the crack flanks, whose magnitude is high irrespective of the crack orientation. Once a crack has been initiated, it propagates easily in a stress field characterised by a tensile triaxial stress tensor. Loading leading to stress tensors with three principle tensile stresses should be avoided where possible or the volumes where such stress tensors are present should be reduced in order to reduce the probability that a flaw will reside inside.

For loading associated with one or two principal compressive stresses, for example, the crack will be unstable only for particular orientations. A stress tensor with one tensile and two compressive principal stresses is preferable to a stress tensor characterised by two tensile and one compressive principal stress. A stress state where two of the principal stresses are tensile and one of them is compressive will promote crack propagation for a larger number of crack orientations.

In order to avoid excessive stresses, the *principle of least constraint* should be followed where possible. Static determinacy should be maintained. Using self-alignment, particularly in the design of bearings, is a well-documented approach (French, 1999).

13.4   REDUCING THE RISK OF FAILURE DUE TO EXCESSIVE DEFORMATION

Increased deformation of components usually causes failures long before the critical stresses develop. Excessive deformation violates the prescribed tolerances and location of the components and leads to loss of function, increased friction, overheating, non-uniform loading, local high stresses and fast wearout. Stiffness is a property which reflects *the capability to resist elastic or viscoelastic deformation* (Larousse, 1995). In relation to reliability, *stiffness* is the capability of a component to withstand loads with minimum deformations which do not impair the function and the reliability of the system. Formally, stiffness is the ratio of force to deflection and is the reciprocal of *compliance*. Thus, for a bar with length l, elastic modulus E and constant cross-section with area S subjected to tension/compression in the elastic region, stiffness K is determined from $K = F/\delta l$, where δl is the displacement caused by the force with magnitude F. Since, according to the Hooke's law: $\sigma = E\delta l/l$ and $F = \sigma S$, stiffness is determined from $K = ES/l$. The link between force, stiffness and displacement is given by $K\delta = F$ and it can be shown (see Astley, 1992) that a similar relationship is valid for a complex pin-jointed frame where all loads are applied at the joints so that no bending occurs in the individual members (Fig. 13.4(c)).

The deflection of the entire structure is defined by the displacements of the joints, determined from

$$\mathbf{Kd} = \mathbf{F} \tag{13.5}$$

where $\mathbf{d}$ is the *nodal displacement vector*, $\mathbf{K}$ is the *stiffness matrix* and $\mathbf{F}$ is the *nodal force vector*.

In case of a torsion loading of a cylindrical bar with length l, the stiffness is determined as a ratio of the acting moment M and the angular displacement φ: $K = M/\varphi = GI_p/l$, where I_p is the polar moment of the cross-section and G is the shear modulus. In case of bending of a bar with constant cross-section, stiffness is determined from the ratio $K = F/f = a\,EI/l^3$, where I is the second moment of area and a is a coefficient depending on the loading geometry. While during tension and torsion, the deformation is proportional to the bar length l, during bending the deformation is proportional to the third power of the bar length.

Stiffness depends on the elastic modulus, the component geometry, the location of the applied forces and the type and location of the supports. In case of tension/compression, the possibilities of increasing the stiffness are limited because it is not affected by the shape of the cross-section. If the material with its elastic modulus has already been specified, the only way of increasing the stiffness is by reducing the length of the bar. In case of torsion, apart from diminishing the length of the region subjected to torsion, a particularly efficient way of increasing the stiffness is to increase the diameter of the bar because the polar moment increases in proportion to the fourth power of the diameter. Efficient ways of increasing the stiffness during bending are reducing the length of the region subjected to bending, selecting a cross-section with large second moment of area and changing the loading scheme.

Figure 13.5 features increasing the stiffness by modifying the shape consisting of moving material away from the neutral bending axis x. This increases the second moment of area of the cross-section and increases the resistance of the sheath against deflection due to bending. Similar is the purpose of the T- and I-shaped beam cross-sections.

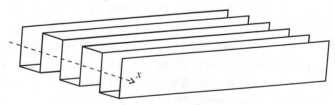

Figure 13.5 Increasing the stiffness of a sheath of metal by modifying its shape.

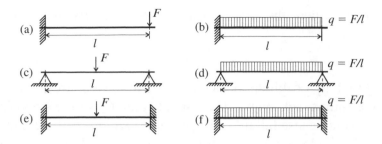

Figure 13.6 Increasing the stiffness of a bar subjected to bending (from (*a*) to (*f*)) with modifying the loading scheme and the support (Orlov, 1988).

Figure 13.6 features increasing the stiffness with modifying the loading scheme and the support.

An important design method for increasing the stiffness of components is to reduce bending by partly replacing it with compression (Fig. 13.7).

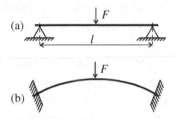

Figure 13.7 Reducing bending by partly replacing it with compression.

Stiffness can also be increased by triangulating the design of shapes where possible by providing shear webs and ribs which resist the various forces applied to the structures.

In some cases, excessive deformation is desirable. Many components are mounted on cantilever arms rather than on high-stiffness structures (e.g. trusses and triangles). In these cases, the cantilever arms act as spring suspension systems. The drawbacks are the higher stresses at the fixed end of the cantilever.

13.5 REDUCING THE RISK OF FAILURE BY IMPROVING THE RESISTANCE TO FRACTURE

The *stress-intensity factor* is an important measure of the magnitude of the crack-tip stress field. It depends on the geometry of the cracked component and the load configuration. For a tensile mode of loading (mode I) of a surface crack, where the crack surfaces move directly apart, the stress-intensity factor is

$$K_I = Y \sigma \sqrt{\pi a} \qquad (13.6)$$

where $Y \approx 1.1$ is a calibration factor, σ is the tensile stress far from the crack and a is the crack length. The critical value K_{Ic} of the stress-intensity factor that would cause failure is the *fracture toughness* of the material.

The crack with length a becomes unstable at applied stress σ_c, for which the intensity factor K_I in equation (13.6) becomes equal to the critical stress-intensity factor

$$K_{Ic} = Y \sigma_c \sqrt{\pi a} \qquad (13.7)$$

Equation (13.7) summarises the roles of the design stress and geometry characterised by σ_c and the factor Y, the material processing characterised by the size of the defect a, and the material properties characterised by the fracture toughness K_{Ic}. Given that a surface crack of length a is present, in order to prevent fracture, the design stress σ_d must be smaller than the critical stress ($\sigma_d < \sigma_c = K_{Ic}/[Y \sqrt{\pi a}]$).

Consequently, a material with higher fracture toughness K_{Ic} is characterised by higher critical stress σ_c and higher fracture resistance. Some of the routes to attain a higher fracture toughness are: alloying, heat treatment (e.g. quenching, tempering, ageing, recrystallization), cold working, solid solution strengthening, precipitation hardening, reinforcing by particles and fibres, grain size control, transformation and surface toughening, etc.

13.6   REDUCING THE RISK OF OVERSTRESS FAILURE BY MODIFYING THE COMPONENT GEOMETRY

This technique will be illustrated by an example where a modification of the design geometry eliminates the risk of overstress failure due to thermal expansion.

Let α designate the coefficient of linear thermal expansion of a metal pipe. During a temperature change ΔT, the absolute strain of the pipe is $\varepsilon = \alpha \Delta T$. If the ends of the pipe are fully constrained, the pipe cannot expand longitudinally and compressive stress with magnitude $\sigma = \alpha E \Delta T$ is generated in the pipe, where E is the elastic modulus. The thermal stress σ can cause yielding, if the yield strength is exceeded or buckling if the critical buckling stress is exceeded. Expansion offsets are a common generic solution

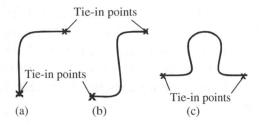

Figure 13.8 (a) 90°; (b) S-shaped and (c) Ω–shaped expansion offset.

to accommodate thermal expansion and prevent developing excessive thermal stresses. The shape and the size of the offsets depend on the amount of thermal expansion that needs to be accommodated (Fig. 13.8).

At a partially constrained end, the expansion strain is consumed to move the end and the thermal stress magnitude is significantly reduced.

The total input energy lost to heat in the system must be smaller than the capacity of the components to dissipate heat energy. Otherwise, the result is overheating of particular components which leads to increased thermal stresses, decreased stiffness and strength, decreased fatigue resistance and increased rate of material degradation.

13.7 GENERIC METHODS FOR REDUCING WEAROUT FAILURES

Fatigue failures are often associated with components experiencing cyclic stresses or strains resulting in permanent damage. This accumulates until it develops into a crack which propagates and causes failure. The process of damage accumulation and failure caused by cyclic loading is called fatigue. Fatigue failures can be reduced significantly if the development of a fatigue crack is delayed by introducing compressive residual stresses at the surface. Such compressive stresses delay the fatigue crack initiation and by causing crack closure also decrease the rate of crack propagation. One of the reasons is that the compressive stresses subtract from the loading stresses thereby producing smaller effective stresses. Eliminating low-strength surfaces can significantly reduce early-life failures due to rapid fatigue or wear. This can be achieved by:

- eliminating soft decarburised surface after austenitisation;
- eliminating surface discontinuities, folds and pores;
- eliminating coarse microstructure at the surface;

- strengthening the surface layers by surface hardening, carburising, nitriding and deposition of hard coatings. For example, TiC, TiN and Al_2O_3 coatings delay substantially tool wear. Failures due to rapid wear can substantially be reduced by specifying appropriate lubricants. These create interfacial incompressible films that keep the surfaces from contacting.

Delaying the onset of fatigue failure for hot-coiled compression springs for example requires (Todinov, 2000b): (i) a small susceptibility to surface decarburisation; (ii) improved quenching to remove the tensile residual stresses at the spring surface; (iii) improved tempering to achieve optimal hardness corresponding to a maximum fatigue resistance; (iv) special surface treatment (e.g. shot peening) resulting in compressive residual stresses at the spring surface; (v) selecting a cleaner steel with small number of oxide inclusions which serve as ready fatigue crack initiation sites and (vi) smaller number density of sulphide inclusions, which promote anisotropy and reduce the spring wire toughness.

These measures increase the number of cycles needed for fatigue crack initiation, slow down the rate of fatigue crack propagation and delay significantly the onset of fatigue failure.

13.7.1 Increasing Fatigue Resistance by Limiting the Size of the Flaws

Usually, the fatigue life of machine components is a sum of a fatigue crack initiation life and life for fatigue crack propagation. The fatigue life of cast aluminium components and powder metallurgy alloys for example is strongly dependent on the initiating defects (e.g. pores, pits, cavities, inclusions, oxide films, etc.).

The crack growth rate da/dN is commonly estimated from the Paris–Erdogan power law (Hertzberg, 1996):

$$\frac{da}{dN} = C \, \Delta K^m \qquad (13.8)$$

From equation (13.8), the fatigue life N (number of loading cycles) can be estimated from the integral:

$$N = \int_{a_i}^{a_f} \frac{da}{C \, (\Delta K)^m} \qquad (13.9)$$

where C and m are material constants and ΔK is the stress-intensity factor range. The integration limits a_i and a_f are the initial defect size and the final fatigue crack length. Most of the loading cycles are expended on the early stage of crack extension when the crack is small. During the late stages of fatigue crack propagation, a relatively small number of cycles is sufficient to extend the crack until failure. This is the reason why fatigue life is so sensitive to the size a_i of the flaws. Consequently, limiting the size of the flaws in the material increases the fatigue life.

It has been reported (Ting and Lawrence, 1993) that in cast aluminium alloys, the dominant fatigue cracks (the cracks which caused fatigue failure) initiated from near-surface casting pores in polished specimens or from cast surface texture discontinuities in as-cast specimens. During fatigue life predictions for such alloys, the distribution of the initial lengths of the fatigue cracks is commonly assumed to be the size distribution of the surface (subsurface) discontinuities and pores. It is also implicitly assumed that the probability of fatigue crack initiation on a particular defect is equal to the probability of its existence in the stressed volume. This assumption however is too conservative, because it does not account for the circumstance that not all pores/defects will initiate fatigue cracks. Fatigue crack initiation is associated with certain probability because it depends on the type of the defect, its size and orientation, the mechanical properties and the microstructure of the matrix, the residual stresses and the stress state in the vicinity of the defect. Further discussion related to these points is provided in Chapter 16.

13.7.2 Increasing Fatigue Life by Avoiding Stress Concentrators

Sharp notches in components result in high-stress concentration which reduces fatigue resistance and promotes early-fatigue failures. Such are the sharp corners, keyways, holes, abrupt changes in cross-sections, etc. Fatigue cracks on rotating shafts often originate on badly machined fillet radii which act as stress intensifiers. Because of this, they are reliability critical elements (Thompson, 1999) and their appropriate design and manufacturing quality should be guaranteed. Reducing the stress magnitude in the vicinity of a fillet can be achieved by increasing its radius. In this way, the resistance to an overstress failure or fatigue failure is enhanced. In the vicinity of a notch, the stress is characterised by a sharp gradient. The smaller the curvature of the notch, the larger the stress magnitude, the lower the resistance to overstress and fatigue failures.

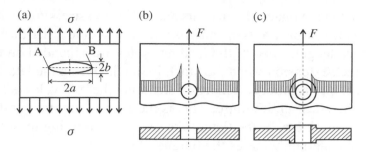

Figure 13.9 (a) Elliptical hole in a plate; (b) stress concentration around a circular (bolt) hole in a plate and (c) reducing the stress concentration by appropriate design.

Fatigue crack initiation is also promoted at the grooves and the micro-crevices of *rough surfaces*. These can be removed if appropriate treatment (grinding, honing and polishing) is prescribed.

Suppose that an elliptical through hole exists in a plate with practically infinite size relative to the hole. If the major axis of the hole is of length $2a$ and the minor axis is of length $2b$ (Fig. 13.9(a)), the stress at the tips of the major axis can be determined from (Inglis, 1913):

$$\sigma_A = \sigma_B = \sigma \left(1 + 2\frac{a}{b} \right) \tag{13.10}$$

The stress concentration factor is described by the ratio $k_t = \sigma_A/\sigma$. With increasing the eccentricity of the hole, the stress concentration factor increases. For the special case of a circular hole (e.g. a bolt hole, Fig. 13.9(b)), according to equation (13.10), the stress concentration factor becomes $k_t = (1 + 2a/a) = 3$. Reducing the maximum stress in the vicinity of a hole could be achieved by appropriate design as shown in the example from Fig. 13.9(c) (Orlov, 1988).

Excessive bending of flexible pipes in dynamic applications for example can also be associated with excessive localised stresses and strains. Bend restrictors and bend stiffeners (a bend stiffener is shown in Fig. 13.10) are common design measures to allow some degree of bending and at the same time to restrict excessive bending. The bend restrictors consist of interlocking rings around the pipe which do not restrict decreasing the curvature until a particular critical value is reached. Bending beyond this critical value causes the rings to lock and no further decrease of the curvature is possible.

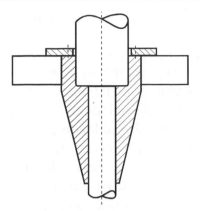

Figure 13.10 Bend stiffener.

13.7.3 Improving Fatigue Resistance by Improving the Condition of the Surface Layers

The fatigue life of components depends strongly on the condition of the surface. Numerous observations confirmed that fatigue failures usually start from surface imperfections. One of the reasons is that the surface layers usually carry the largest stresses. Furthermore, the surface layer is usually saturated with discontinuities and defects, and is exposed directly to the negative influence of the processing and working environment. High-strength steels and alloys are particularly sensitive to surface defects. Some of the surface imperfections are a direct result from the manufacturing process. The surface roughness for example is a function of many parameters: geometry of the cutting tool, type of machined material, homogeneity, cutting speed and feed, rigidity of the fixtures, vibration resistance of the cutting machine, degree of wearout of the cutting blade, presence of lubricants and coolants, etc. The machined surface contains a large number of grooves of different depth and sharpness, causing local stress concentrations and reduced fatigue strength. The greater the material strength, the more detrimental the effect of these stress concentrators. Surface roughness is decreased and fatigue life is improved if the machined materials have a homogeneous microstructure, characterised by a small grain size. Surface roughness is decreased by using sharp cutting blades, increasing the cutting speed, applying lubricants and coolants, eliminating vibrations by using damping devices and fixtures of high rigidity. The size of the surface irregularities and their direction has a profound effect on fatigue. Consequently, surface roughness from machining can be reduced significantly and fatigue life further increased by *grinding*, *polishing honing* and *superfinish*.

Strain-hardening operations such as burnishing, rolling and *shot peening* increase fatigue life because the strain-hardened surface layers resist the formation and propagation of fatigue cracks. As a consequence, in strain-hardened components, the initiation of fatigue cracks occurs at higher stresses and after a greater number of loading cycles compared to components which have not been strain hardened. During burnishing for example, the surface roughness is decreased, surface layers are strain hardened and residual compressive stresses are generated. Burnishing also raises the fatigue limit at high temperatures (Zahavi and Torbilo, 1996). As a result, burnishing applied as a finishing operation to shafts, bars, pistons and cylinders ensures high reliability.

Even insignificant decarburisation of steels with martensitic structure causes a significant reduction of their fatigue strength. Decarburisation diminishes the fatigue resistance of steel components by: diminishing the local fatigue strength due to the decreased density of the surface layer, increased grain size and diminished fracture toughness and yield strength (Chernykh 1991; Todinov, 2000b). These factors create low cycle fatigue conditions for the surface and promote early-fatigue crack initiation and premature fatigue failure. Consequently, in order to delay the onset of fatigue failure, during austenitisation of steel components, decarburisation and excessive grain growth should be avoided.

13.8   IMPROVING RELIABILITY BY ELIMINATING TENSILE RESIDUAL STRESSES AT THE SURFACE OF COMPONENTS

Tensile residual stresses at the component surfaces after quenching, increase the effective net stress range and the mean stress during fatigue loading. This accelerates the fatigue crack initiation and increases the fatigue crack propagation rate. Unlike the compressive residual stresses at the surface, tensile residual stresses increase the negative effect from the damaged surface. A comprehensive discussion related to the genesis of thermal and transformation residual stresses is provided in (Todinov, 1998a).

By using computer simulations (Todinov, 1999a), it has been shown that the tensile residual stress at the surface from quenching steel bars can be eliminated by special quenching conditions. These must guarantee a maximum cooling rate shifted towards high temperatures. Although the quench oil produces small thermal gradients at martensitic temperatures, at high

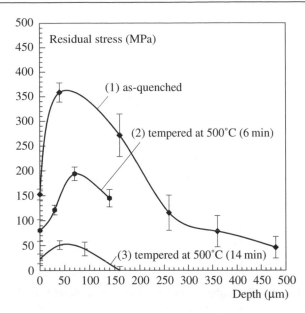

Figure 13.11 Decreasing the residual stresses at the surface by tempering (Todinov, 2000b).

temperatures, during the intensive thermal contraction, it does not generate sufficient plastic strain and corresponding compressive residual stresses to compensate the tensile residual stresses due to decarburisation. To eliminate the tensile residual stress at the surface, the cooling rate at high temperatures should result in sufficiently large plastic strains, guaranteeing large compressive residual stresses from thermal contraction. At the same time, to minimise the tensile transformation stresses at the surface, the thermal gradient in the steel bar, during the phase transformation stage, should be small.

If the heat treatment ends with tensile residual stresses at the surface, stress relief annealing or tempering can be used to decrease their magnitude as shown in Fig. 13.11, depicting the residual stresses of as-quenched and tempered Si–Mn steel bar with diameter 12 mm. The longer the exposure time, the larger the decrease of the residual stress magnitude.

In order to compensate the tensile residual stresses at the surface and improve fatigue resistance, shot peening has been used as an important element of the manufacturing technology (Niku-Lari, 1981; Bird and Saynor, 1984). Figure 13.12 shows the net stress distribution $\sigma_r(x)$ near the surface of a loaded, shot-peened helical spring, obtained from the superposition of the load-imposed principal tensile stress $\sigma_t(x)$ and the residual stress $\sigma_{rs}(x)$ from shot peening. Shot peening decreases the effective net tensile

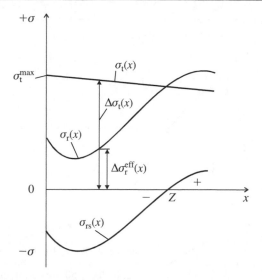

Figure 13.12 Net stress distribution near the surface of a loaded, shot-peened compression spring.

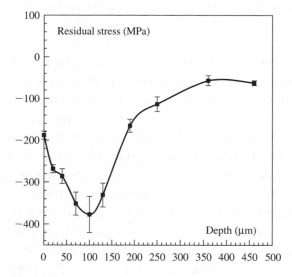

Figure 13.13 Residual stress variation in a shot-peened Si–Mn spring steel bar with diameter 12 mm (Todinov, 2000b).

stress at the surface during service, delays the fatigue crack initiation and impedes the fatigue crack propagation.

Figure 13.13 depicts the residual stress variation measured in a shot peened Si–Mn steel bar with diameter 12 mm. The stress distribution with depth was produced by X-ray measurements followed by dissolving the surface layers (Todinov, 2000b).

Residual stresses may also develop in the absence of thermal processing. During cold rolling of metals for example, the surface fibres are stretched more than the inner material. After the cold rolling of a section, the requirement for compatibility of the deformations results in surface layers loaded in compression and core loaded in tension. Compressive residual stresses at the surface, compensating the service stresses can also be created by a special heat- and thermochemical treatment such as case-hardening, gas-carburising and gas-nitriding.

13.9   REDUCING THE RISK OF FAILURE BY MITIGATING THE HARMFUL EFFECT OF THE ENVIRONMENT

Reducing the harmful effect of the environment can be done by (i) replacing it with inert environment, (ii) modifying it in order to reduce its harmful effect and (iii) improving the resistance against its harmful effect.

The corrosion intensity is a function of the environment. Aggressive environments combined with low corrosion resistance cause expensive early-life corrosion failures. The severity of the environment depends on its pH, and on the presence of particular gases (e.g. O_2, CO_2, H_2S), salts, acids, alkalis, solvents, catalysts or other substances.

Oxidation corrosion can occur in dry conditions and at high temperatures while wet corrosion involves aqueous solutions of electrolytes. Corrosion can proceed as uniform attack but in many cases *pitting* occurs which is a highly localised corrosion attack resulting in isolated pits or holes while the surrounding area is unaffected. This is a particularly insidious form of corrosion because it is difficult to detect corrosion pits on the surface. Consequently, pitting attack often remains undiscovered until the component is lost due to a through-thickness perforation. Pitting usually occurs in the presence of neutral or acidic solutions containing chlorides and is accelerated by small imperfections on the metal surface (e.g. sulphide inclusions, voids, dents, scratches, marks, etc.). Austenitic stainless steels for example, are particularly prone to pitting attack in salt water. Also, horizontal surfaces are more susceptible to pitting than vertical surfaces.

Arc welding, shielded by an inert gas such as argon or carbon dioxide is an example of reducing the harmful effect of the environment and improving the reliability of welds. This principle is used for example in the MIG (metal inert gas) and TIG (tungsten inert gas) welding techniques. Another example is using inert environments in hermetic or plastic encapsulated integrated electronic circuits to protect them from the harmful action of humidity and air.

Corrosion inhibitors are compounds which modify the corrosive environment thereby reducing the rate of corrosion. Corrosion inhibitors are often injected into a pipeline where they mix with the product to reduce corrosion. Corrosion retarding inhibitors are added to the corrosive medium to reduce the intensity of the anode/cathode processes. They can also reduce the corrosion rate by forming barrier films separating the protected surface and the corrosive environment. There exist also inhibitors which passivate the protected surface by reacting with it and forming compounds which serve as anti-corrosion barrier. Inorganic inhibitors usually passivate the protected surface, while organic inhibitors usually form a protective film on the surface. There exist also polar inhibitors for which one part of the molecule binds to the protected surface while the other part is hydrophobic and serves as a barrier.

Removal of corrosive agents from production liquids is another example of a measure reducing the rate of corrosion by modifying the corrosive environment.

13.9.1   Improving the Resistance Against Failures Due to Corrosion

Corrosion is a name for the degradation of mechanical, microstructural and physical properties of materials due to the harmful effect of the environment. Material degradation due to corrosion is often the root cause of failures entailing loss of life, damage to the environment and huge financial losses. Corrosion also incurs large capital, operation and maintenance costs due to the need for constant corrosion protection. Materials differ significantly according to their resistance to corrosion. The corrosion rate is significant and the need for corrosion protection is particularly strong where inexpensive, mostly low-grade carbon steels are used. Such are for example the carbon steel pipelines used for transportation of fluids in the processing industry and the oil and gas industry.

Methods increasing the corrosion resistance include *cathodic protection, corrosion allowance, protective coatings, plastic or cement liners*, use of *corrosion resistant special alloys*. Corrosion control can also be ensured by condition monitoring.

Corrosion rate can be significantly reduced by selecting *corrosion-resistant alloys* (e.g. martensitic, austenitic or duplex stainless steels, titanium alloys, etc.). Such materials are particularly useful in cases of highly corrosive working fluids where using chemical inhibitors for corrosion protection is inefficient or uneconomical. A common drawback of the corrosion-resistant alloys is their high cost and difficulty in welding which contributes to the overall cost. In each specific case, a careful cost–benefit analysis should be made to make sure that the investment into a more expensive alloy will be outweighed by the reduction of the expected losses from failures because of the increased corrosion resistance.

Often, material processing during manufacturing and assembly decreases the corrosion resistance. For example, the corrosion resistance of welded duplex or super duplex steels decreases significantly in the area of the heat-affected zone of the weld. As a rule, a non-homogeneous microstructure has a lower corrosion resistance compared to a homogeneous microstructure. A typical non-homogeneous structure with increased susceptibility to corrosion attack is the welded structure consisting of a central zone, a zone of columnar crystal growth and reheated zone.

For metals with insufficient corrosion resistance, protective coatings and paint can be used to ensure a barrier against corrosion. Since no coating can be free from defects, coatings often also guarantee cathodic protection. For example, steel structures are often covered with zinc coatings that acts as a sacrificial anode. In the atmosphere and most aqueous environments, zinc is anodic to steel and protects it. In order to prevent the corrosion of the zinc layer, an organic coating often protects the zinc coating. Proper protective coatings, heat treatment, diffusion treatment and surface finish can substantially decrease the corrosion rate. A number of examples of design solutions preventing corrosion are discussed in Budinski (1996).

Generic ways to delay corrosion failures is the *cathodic protection* and *corrosion allowance*. *Cathodic protection* is used in cases of galvanic corrosion. This can be cased by dissimilar metals immersed in conductive fluid. The metal lower in the galvanic series acts as cathode and its corrosion rate is small or non-existent. The less noble metal (the metal higher in the galvanic series) acts as anode and its corrosion rate will be substantial. The increased corrosion of the anode material produced by coupling to a cathodic material is known as galvanic corrosion (Sedricks, 1979). Interfaces including incompatible metals operating in sea water for example can promote a rapid galvanic corrosion.

Since corrosion and material loss occurs at the anode, the metal part acting as cathode is protected. Cathodic protection can be based on a sacrificial anode or impressed current. Sacrificial anodes are pieces of metal less noble than the protected metal. With the protected metal they form a galvanic corrosion cell and are preferentially corroded.

An estimation of the total anode material dependent on the required life, current density and the existing anode consumption rate is an important part of the design calculations. A design based on lower than the actual temperature may promote rapid anode wastage and decreased cathodic protection. A generic rule is that the materials of the interfacing parts working in aggressive environments should not be far apart in the galvanic series. Consulting corrosion databases in order to avoid environment–material combinations promoting rapid galvanic corrosion is an efficient prevention technique.

Underground steel pipes, underwater structures and ship hulls are all protected in this way. In a corrosion system, the same total current flows through anodic and cathodic parts. Accordingly, the current densities and the corrosion rates vary inversely with the anode size. Consequently, small anodes in contact with large cathodes will corrode more severely than large anodes in contact with small cathodes. Indeed, small steel rivets used to clamp large copper plates will corrode much faster than copper rivets clamping large steel plates (Ohring, 1995). The reason is that in the first case, the current density at the steel rivets will be large due to the small surface area and they will corrode faster. In the second case, the current density of the steel plates acting as anode will be small and their corrosion rate will be small.

Cathodic protection based on impressed current uses an external source of electric current and is applied in cases where the current required to guarantee protection is relatively large and a source of electrical energy is readily available.

Crevice corrosion is a local corrosion occurring within crevices and other shielded areas containing small volumes of stagnant solutions. Similar to pitting corrosion, it is particularly severe in neutral or acidic chloride solutions. Crevice corrosion often occurs at flange joints, threaded connections, in poorly gasketed pipe flanges and under bolt heads. Similar to pitting, crevice corrosion can also be very destructive because of the localised damage. Fresh supply of oxidants is restricted in the crevice and as a result, oxygen and other oxidants are quickly consumed. Since the oxidants can no longer maintain a passive surface layer which would normally serve as a barrier, the corrosion rate increases substantially and the crevice corrosion attack often causes failure in a very short time.

This type of corrosion can be minimised by avoiding design geometry or assembly associated with crevices that can accommodate corrodents. This is of particular importance to equipment which is likely to be exposed to an environment containing chlorides. Examples of design modifications which prevent crevice corrosion are the use of welded joints instead of bolted or riveted joints to avoid formation of crevices. Components should be designed to avoid collecting stagnant solutions (Mattson, 1989). The susceptibility to crevice corrosion of steels can be reduced by increasing the chromium, molybdenum and nitrogen content.

Stress-corrosion cracking, *corrosion fatigue* and *hydrogen embrittlement* collectively referred to as *environment-assisted cracking* are responsible for a substantial portion of field failures. Stress-corrosion cracking is a spontaneous corrosion-induced cracking under static stress. The required static stress is usually significantly below the yield stress and may not necessarily originate from external loading. It can be due to residual or thermal stresses. Designs associated with high static tensile stresses should be avoided because they promote stress-corrosion cracking. Some materials, inert in particular corrosive environments, become susceptible to corrosion cracking if stress is applied. Salts, particularly chloride, intensify the stress-corrosion cracking. Aluminium alloys, for example, widely used in the aerospace industry, automotive industry and in the process industries are prone to corrosion fatigue and stress-corrosion cracking in chloride solutions. Consequently, non-destructive inspection techniques such as ultrasonic and eddy current techniques are used to detect stress-corrosion cracks in components and structures made of high-strength aluminium alloys. Such cracks often appear within the bore of fastener holes of rolled plates from high-strength aluminium alloys and propagate along the grain boundaries.

Hydrogen embrittlement refers to the phenomenon where certain metal alloys experience a significant reduction in ductility when atomic hydrogen penetrates into the material. During welding, hydrogen can diffuse into the base plate while it is hot. Subsequently, embrittlement may occur upon cooling, by a process referred to as cold cracking in the heat-affected zone of the weld (Hertzberg, 1996). Hydrogen can also be generated from localised galvanic cells where the sacrificial coating corrodes: for example, during pickling of steels in sulphuric acid. The atomic hydrogen diffuses into the metal substrate and causes embrittlement. Increasing strength tends to enhance the susceptibility to hydrogen embrittlement. High-strength steels, for example, are particularly susceptible to hydrogen embrittlement.

13.9.2 Improving the Resistance Against Failures Due to Erosion and Cavitation

Poor design of the flow paths of fluids containing abrasive material promotes rapid erosion which can be minimised by a proper material selection and design. Structural design features promoting rapid erosion (Mattson, 1989) should be avoided. Such are for example the bends with small radii in pipelines or other obstacles promoting turbulent flow. Increasing the pipeline curvature and removing the obstacles results in less turbulent flaw and reduced erosion. Erosion is significantly reduced by appropriate *heat treatment* increasing the surface hardness.

If the possibility for increasing the curvature of the flow paths is limited, *internal coatings* resistant to erosion may be considered at the vulnerable spots. Polyurethane coatings and ceramic modified coatings for example have excellent *abrasion* resistance. These coatings however are costly and to be economical, their use is often restricted to vulnerable locations like bends. A careful cost–benefit analysis is necessary to make sure that the investment is outweighed by the losses from prevented failures. Erosion is often present in leaking seals and gaskets. A combination of working fluids containing erosion particles (e.g. sand) and inappropriately tightened flanges with small leaks can cause erosion of the seals. Metal coatings or leather jackets are examples of simple solutions protecting the threads from erosion particles. Filters for separating corrosion products, contaminating solid particles and maintaining the cleanliness of the working liquids can significantly reduce the erosion intensity.

Erosion is often combined with corrosion which is known as *erosion–corrosion*. Presence of solid particles increases the corrosion rate in corrosive environments. Because of the presence of erosion particles, the passive surface layer is consistently destroyed and new metal surface is exposed. As a result, the passive surface layer no longer acts as anti-corrosion barrier and the corrosion rate is high.

Cavitation is generation of cavity bubbles in liquids by rapid pressure changes. When the cavity bubbles implode, close to a metal surface they cause pitting erosion. Typical spots of cavitation damage are: (i) suction pipes of pumps and impellers, narrow flow spaces, sudden changes in the flow direction (bends, pipe tees) which cause turbulence.

This type of damage can be avoided by designing the flow paths in such a way that sharp pressure drops are avoided (especially below the atmospheric pressure). This can be achieved by designs guaranteeing a multistage pressure drop. Avoidance of turbulence by streamlining the flaw is an important measure decreasing cavitation.

Alternatively, the flow paths can be designed in such a way that the cavitation bubbles implode in the fluid but not next to the metal surface. As a result, cavitation is still present but the metal surfaces are not affected. The susceptibility to cavitation damage can be reduced by using cavitation-resistant materials, welded overlay of metals, sprayed metal coatings or elastomeric coatings.

14

REDUCING THE RISK OF FAILURE BY REMOVING LATENT FAULTS, AND AVOIDING COMMON CAUSE FAILURES

14.1 FAULTS AND FAILURES

An efficient way of reliability improvement is the removal of latent faults from products, systems and operations. A *fault* is an incorrect state, or a defect resulting from errors during material processing, design, manufacturing, assembly or operation, with the potential to cause failure under particular conditions. Such is for example a large size flaw existing in the material of a component subjected to cycling stresses which is a primary cause of fatigue failure. Faults also result from an unfavourable combination of values of controlling parameters. A software fault is synonymous with bug and is in effect a defect in the code that can cause software failure.

Fault is not the same as failure. A system with faults, may continue to provide its service, that is, not fail until some triggering condition is encountered. For example, a material flaw is a fault which may not cause failure during normal operation but if the component is overstressed it may cause a catastrophic failure. Similarly, the presence of a software fault does not necessarily result in immediate software failure. Failure will be present only when the branch containing the faulty piece of code is executed for the first time. A system is said to have failed, if the service it delivers to the user deviates from compliance with the specified system function, for the specified operating conditions.

The link between faults and failures is cause. A fault can lead to other faults, or to failure, or neither. In practice, some faults are likely to remain in a complex design after development. Consider a flexible pipe transporting hydrocarbons under water (Fig. 14.1).

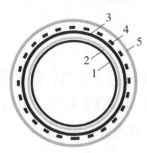

Figure 14.1 Components of a flexible pipe transporting hydrocarbons under water.

The pipe is composed of a stainless steel *internal carcass* (1); an *internal sheath* (2) which is extruded polymer barrier; *a pressure armour* (3) – a carbon-steel interlocked circumferential layer and *a tensile armour* (4) – helically wound carbon-steel layers for axial strength. Externally, the pipe is protected by *extruded sheath* (5). The internal carcass prevents collapse of the internal sheath and also ensures mechanical protection. The internal sheath ensures the integrity of the transported fluid while the function of the pressure armour and the tensile armour is to provide resistance against radial and tensile loads. The external sheath is a mechanical barrier shielding the pipe's internal structural elements from the environment. A damaged external sheath during assembly is a fault. Despite that the pipeline will still operate (no immediate failure will be present), the contact with the environment (e.g. sea water) will induce material degradation of the armours due to corrosion and corrosion fatigue. Soon, material degradation will erode the load carrying capacity of the armours and the pipe will fail.

Similar effect will have a scratch on the painted layer protecting the surface of a compression spring made of steel which is not corrosion resistant. The spring will still operate but the exposed surface from the scratch will start to corrode at a fast rate which will eventually lead to the nucleation of a fatigue crack. As a result, the fatigue life of the spring will be reduced significantly.

For packages containing harmful waste materials for example, the expected behaviour is not to release toxic or radioactive substance during impact or fire. Even the most strictly controlled process is subjected to faults due to the natural variation of process parameters. A particularly important issue is the situation where two or more parameters have values in certain ranges thereby promoting critical faults. If a low strength of a container packed with hazardous materials is combined with large dynamic loading stresses from an impact during handling, the container's integrity could be compromised and toxic substances released. Another example is the combination of failure of a measuring instrument and inappropriate content in a waste package.

Given that a critical fault is present, the probability p_f of missing the fault after n identical, independent quality control (QC) checks is $p_f = p_{mf}^n$, where p_{mf} is the probability of missing the fault after a single quality check. Clearly, in case of a large probability of missing the fault during a single check ($p_{mf} \rightarrow 1$), increasing the number of checks n cannot reduce significantly the probability p_f of missing the fault after n checks. Conversely, in case of a small probability of missing the fault, increasing the number of checks n reduces significantly the probability of missing the fault after n checks.

14.2 ASSESSING THE LIKELIHOOD OF LATENT CRITICAL FAULTS

Let us take as an example packages containing harmful waste material (e.g. radioactive waste). Faults in the waste packages should always be considered in the context of the subsequent interim storage. A waste package can be risk-free if handled shortly after production, and associated with unacceptable risks if handled after 10 or more years of interim storage.

Important objectives during risk assessment of waste packages is to devise strategies for decreasing the likelihood of critical faults.

The likelihood of critical faults in the waste packages can for example be determined by developing a fault-stream simulation model related to the genesis and propagation of critical faults. This model could be based on the functional block diagram of the waste packaging process. Conservative (upper bound) estimates of the fault frequencies associated with the different operations, processes, components and the frequencies of human errors, guarantee a conservative estimate regarding the likelihood of undetected faults in the waste packages.

An important part of such a study are the material deterioration processes during the interim storage which reduce the strength of the waste packages. These, processes could be:

- corrosion of the welds, bolts and the flanges of the waste container;
- fracture (impact) toughness degradation of the welds;
- stress-corrosion cracking of the bolts on the lid.

Welds suffer impact toughness degradation due to corrosion, irradiation and ageing. Even if the entire container is made of stainless steel, its components are still susceptible to a corrosion attack. Thus, the bolts which are subjected to tensile residual stresses are susceptible to stress-corrosion cracking in presence of chlorides. The flanges of the lid are never perfectly flat and form cavities which, when filled with electrolyte, create conditions promoting intensive crevice corrosion. Horizontal surfaces, such as the base and the lid, are susceptible to pitting corrosion, etc.

By using empirical cumulative distribution functions characterising the impact toughness of the deteriorated and non-deteriorated material, the welds of the waste containers can be tested for impact toughness degradation. Such a test is depicted in Fig. 14.2, where the magnitude of the shift in the empirical cumulative distribution of the impact toughness indicates the degree of impact toughness degradation.

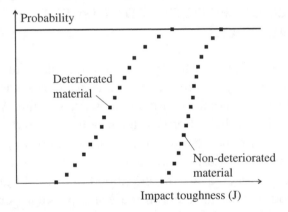

Figure 14.2 Cumulative distributions of the impact toughness of deteriorated and non-deteriorated material. The large separation between the two distribution curves indicates impact toughness degradation.

A necessary prerequisite of a model related to the genesis and propagation of critical faults is determining what constitutes a critical fault in the waste package. A critical fault for example could be any combination of parameter values leading to excessive material degradation of the container. Among the factors affecting the genesis and propagation of critical faults during waste packaging are:

(i) combination of values for the controlling parameters leading to packages with degraded strength or weak resistance to deterioration;

(ii) defects in the material of the container (inclusions, cavities, surface imperfections);

(iii) human errors;
(iv) failures of the equipment for waste packaging and waste treatment;
 (v) failures of the control equipment;
(vi) design errors;
(vii) poor control of the manufacturing process;

On the basis of the functional block-diagram of the waste packaging process, a fault-stream diagram related to the genesis and propagation of critical faults can be developed. Such is the fault-stream diagram in Fig. 14.3 related to a generic waste packaging process.

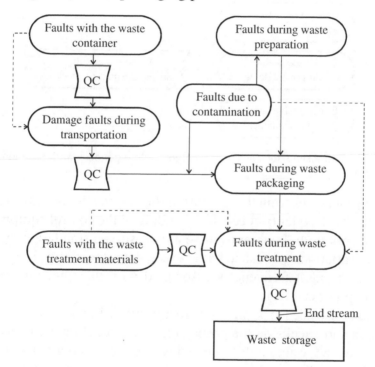

Figure 14.3 A block diagram showing the origin of faults and the faults propagation during a generic waste packaging process.

The fault-stream diagram is a set of blocks connected with arrows. Some of the blocks are sources of faults which increase the number of faults in the end stream, others are fault sinks – they remove faults from the fault streams. Examples of fault sinks are all quality control (QC) stages. The streams of faults are depicted by continuous arrows; the fault sources are depicted by rectangles with convex ends while the fault sinks are depicted by rectangles with concave ends. Fault-neutral processes which neither generate nor remove faults from the fault streams are depicted by regular rectangles. The dashed arrows indicate dependencies between the waste packaging stages. Thus, faults with the waste container after manufacturing will increase the likelihood of damage faults during transportation and this is indicated by a dashed arrow. Faults caused by a contamination and faults associated with the waste treatment materials will induce faults at the waste treatment stage, etc.

Figure 14.4 depicts a simplified logical block related to the genesis of an undetected container fault only.

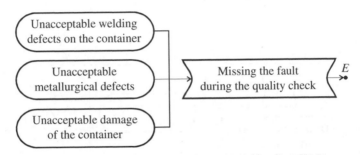

Figure 14.4 A simplified logical block related to the genesis and propagation of a container fault.

A fault propagates from the left part of the diagram to the end stream E if it is missed by the quality check due to failure of the control equipment or a human error or both. In order to solve a fault-stream simulation model, the parameter variations associated with the separate stages must be quantified, as well as the fault frequencies associated with the stages of the waste packaging process.

Estimates also need to be produced regarding the impact of quality control procedures and checks on the propagation of critical faults. Conservative estimates related to the fault frequencies (e.g. upper bounds) and the efficiency of the quality control procedures (e.g. lower bounds of the probability of detecting a fault) yield conservative estimates related to the probability of latent faults in the waste packages.

A number (N) of packages (Monte Carlo trials) are specified in the simulation, for each of which faults are simulated at the fault-generation blocks and removed at the fault-sink blocks in Fig. 14.3. If the estimated fault frequency of a particular fault-generation stage is $x\%$, generating a fault is conducted by generating a random number r in the range 0, 1. The number is compared with the fault frequency $x/100$ and if $r \leq x/100$, a fault is simulated. Similar procedures are applied at the fault-sink stages.

During the simulation trials, the number of faults in the end stream are accumulated in the variable N_f. The total number of packages which have reached the end stream is accumulated in the variable N_{total} and the probability P_f of a fault in a waste package is estimated from $P_f = \dfrac{N_f}{N_{total}}$.

This approach will be illustrated by a simple numerical example. In Fig. 14.5, the values of two parameters are compared during a quality check after an assembly. If the value of the second parameter is greater than the value of the first parameter by more than 1.5, the assembly is fault-free and therefore accepted, otherwise the assembly is rejected.

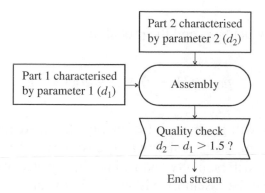

Figure 14.5 A simple production stage where the assembly of two parts is followed by a quality check.

It is known that both parameters follow Gaussian distributions with means and standard deviations $\mu_1 = 16.0$, $\sigma_1 = 5.0$ for the first parameter and $\mu_2 = 20.0$ and $\sigma_2 = 3.5$ for the second parameter, respectively.

The quality check is associated with probability $q = 0.80$ of rejecting an assembly given that it is faulty ($d_2 - d_1 \leq a = 1.5$). This means that in 20% of the cases, a faulty assembly will be classified as fault-free because the fault will be missed by the quality check. The quality check is also associated with an error of type II, which is characterised by the probability $r = 0.10$ that a fault-free assembly will be rejected.

An analytical or a simulation method can be employed to determine the probability of a faulty assembly in the end stream.

Using a demand-capacity model (see Chapter 12) gives

$$p = P(d_2 - d_1 \geq a) = \Phi\left(\frac{\mu_2 - \mu_1 - a}{\sqrt{\sigma_1^2 + \sigma_2^2}}\right) \tag{14.1}$$

for the probability p that the value of parameter d_2 will exceed the value of parameter d_1 by more than a, where $\Phi(\bullet)$ in equation (14.1) is the cumulative distribution of the standard normal distribution. The probability that the assembly will be faulty and the quality check will miss it is $p(1-q)$. The probability p_f of a faulty assembly in the end stream after the quality check stage is given by the ratio

$$p_f = \frac{p(1-q)}{p(1-q) + (1-p)(1-r)} \qquad (14.2)$$

of the probability $p(1-q)$ and the probability

$$p_0 = p(1-q) + (1-p)(1-r) \qquad (14.3)$$

that an assembly will go through the quality check. According to the total probability theorem, the probability p_0 is a sum of the probability $p(1-q)$ that the assembly will be faulty and will not be rejected and the probability $(1-p)(1-r)$ that the assembly will not be faulty and will not be rejected. Substituting the numerical values gives $p_f = 0.1035$ which is confirmed by the value 0.1040 obtained from a direct simulation.

A sensitivity study revealing critical stages and operations of the waste packaging process, accounting for most of the undetected critical faults is an inseparable part of the modelling exercise. Using the results from the sensitivity study, a Pareto chart could be constructed, ranking the different stages and operations according to the associated risk of critical faults. In turn, this will provide a basis for informed decisions regarding the optimal allocation of resources on the few stages and processes associated with the largest contribution to the total risk. A specific parametric study related to the optimal scheduling of the fault-control checks during waste packaging can also be conducted.

Using the model, the impact of different management strategies decreasing the likelihood of undetected critical faults in the waste packages can be assessed. It is necessary to assess two basic types of risk: (i) the risk of accepting waste packages which contain faults and need over-packaging and (ii) the risk of rejecting waste packages as faulty when they are in fact fault-free, which results in unnecessary over-packaging. Measures need to be devised for reducing both types of risk.

14.3 REDUCING THE RISK OF FAILURE BY REMOVING LATENT FAULTS AND DESIGNING FAULT-TOLERANT SYSTEMS

14.3.1 Reducing the Number of Faults in a Piece of Software

Unlike the bathtub curve characterising the rate of occurrence of failures in hardware, the software bathtub curve is usually decreasing, with no wearout region, because software does not degrade with time. Although decreasing, after each re-writing (new release) of the software, new errors are introduced which cause a sharp increase in the rate of appearance of software errors (Fig. 14.6).

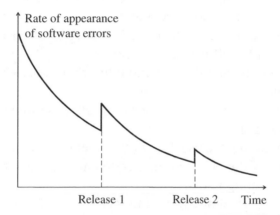

Figure 14.6 Rate of appearance of software errors as a function of time (Beasley, 1991).

Reducing the number of software faults is a necessary condition for reducing the number of software failures. Unlike hardware, where no two pieces of equipment are absolutely identical and therefore there exists a substantial variation in the failure pattern, all copies of a piece of software are identical and there is no variation in the failure pattern. If a software fault exists, it is present in all copies of the software and causes failure if particular conditions and a combination of input data are present. Software is particularly prone to common cause faults if the redundant routines are designed by the same programmer/team. A fault in one of the software modules is likely to be present in the back-up module too.

Reducing the number of faults in a designed piece of software requires:

(i) Well-documented and unambiguous user requirements, functional and design specification. Often the programme output fails to comply with the expected behaviour because of omissions of requirements in the specification. A typical example is the case where a programme for reliability analysis of a large system takes an inordinate amount of time to accomplish the computation because no requirements have been specified regarding the size and complexity of the analysed system and the desired time interval within which to perform the analysis. As a result, no resources are allocated to devise efficient algorithms for system reliability analysis of large systems with complex topology. Algorithms, appropriate for small-size systems with simple topology are used instead and, as a result, the time spent on determining the reliability of a large and complex system is not acceptable.

(ii) Adopting the principle of modular design, based on well-defined modules with well-defined functions:
   – Defining the architecture and interrelation between the software modules.
   – Using the 'top–down' approach in programme planning and development.
   – Discouraging the use of *goto* statements which increase the complexity of the programming code and increase the possibility of programming faults.

(iii) Following a good programming style:
   – Consistent, clear and self-documenting coding. Consistent names for the software components, constants, variables, arrays, structures, functions, procedures, objects, classes, etc.
   – Testing the modules as soon as they have been developed. It is much easier to identify errors in small sections of code than in a large programme.
   – Initialising all pointers immediately after declaration. Storing data at an address pointed by a non-initialised pointer destroys the existing data at this address.
   – A good memory management (e.g. a check before allocating memory and freeing the allocated memory when it is no longer needed). Avoiding shared use of memory locations.
   – Range checking of arrays.
   – Checking stacks for overflow.
   – Using standard names for constants and variables.

(iv) Documenting the developed programming segments in order to maintain them easily.

(v) Testing and debugging the developed procedures and functions:
  - Testing all possible functions of the developed modules.
  - Testing as many as possible branches in the programme by running it with different combinations of input data.
  - Monitoring the values of key variables during controlled execution of the programme.
  - Using hardware emulators to test the software before running it on a real system.
  - Using profilers to reveal bottlenecks during the execution and to optimise the code.
(vi) A large amount of errors are associated with handling combinations of input values from the extremes of the data ranges and with handling exceptional events. Particular care must be exerted in writing correct exception handling routines.
(vii) Using formal mathematical methods to develop and prove the validity of software code, significantly reduces the possibility for ambiguity and logical errors.
(viii) Formal examination of the source code for non-referenced variables, double declarations, use of non-initialised variables, unreachable code, compatibility of different data types, correctness of assignments, etc. Testing the conditions for exiting the loops. Avoiding the programme execution becoming locked in a loop.
(ix) Reducing software complexity by using standard tested software libraries.
(x) Use of modern integrated development environment including a compiler capable of automatic detection of a large number of logical and syntax errors and a debugger capable of step-by-step execution of programming statements, selective execution and execution between specified break-points, monitoring the values and the addresses of variables, pointers and other data structures.
(xi) Writing the programme by using simple rather than complex programming techniques. Avoiding unnecessarily complex constructs which, although marginally more efficient to standard tested and proved solutions, are prone to faults.
(xii) As a rule, reducing the size of the software modules increases reliability.
(xiii) Error-avoidance and error-recovery software capable of avoiding the use of modules if an error has been discovered. Programmes should be written to be fault tolerant. This can be achieved by developing

internal tests and exception handling routines which set up safe conditions in case of errors.

(xiv) Programming to guarantee error confinement – avoiding side effects and spreading errors between modules.

(xv) Avoiding timing errors during programming of real-time engineering systems.

(xvi) Introducing various dynamic checks – check sums, parity checks, dynamic memory and file existence checks, checks for overflow and underflow, checks for loss of precision or attaining the desired precision, iteration counters, check of the state of key variables, etc.

(xvii) Using routines for syntax and semantic checks on input data. Capability of handling data inputs which, although incorrect, are possible.

(xviii) Guaranteeing the reliability of the transmitted data – data corruption due to noise or memory faults can be prevented by using parity checks and various error detection codes.

(xix) Using self-diagnostics routines.

14.3.2 Reducing the Risk of Failure by Reducing the Number of Faults and by Designing Fault-Tolerant Systems

A powerful tool for reducing the number of design faults are the frequent design reviews, particularly after a design change. Without these, design changes frequently result in latent faults causing expensive field failures. At the design stage, information regarding the environmental stresses and the operating conditions must be available in order to reduce the possibility of design faults.

Because a significant number of latent faults are caused by human errors, it is important to identify management strategies aimed at reducing the possibility of human errors (e.g. mislabelling, mishandling) which promote latent faults. In this respect, computer-based systems for selecting materials and manufacturing processes should be used to reduce the possibility of human error.

Designing additional prevention barriers (e.g. quality checks) also reduces the number of faults. Before designing additional barriers, the strength of the existing barriers must be assessed, such as the physical and chemical stability of the waste, the barriers against material deterioration, etc. Furthermore, controls and barriers could be implemented to reduce the hazard potential of the waste, for example solidifying liquid waste or intermediate processing and conversion into a passive safe form. Devices indicating faults (e.g. detecting a build up of flammable or toxic gases or radioactivity) need also to be implemented.

A fault may not cause failure if the system is *fault tolerant*. Differentiation between failures and faults is essential for fault-tolerant systems. An example of a fault-tolerant component is a component made out of tough material which is resistant to cracks, defects and other imperfections. At the other extreme is a component made of material with low toughness (e.g. hardened high-strength steel), sensitive to inclusions and mechanical flaws. A system with built-in redundancy is fault tolerant as opposed to a system with components logically arranged in series.

The *k-out-of-n* redundancy discussed in Chapter 11 is a popular type of redundancy because it makes the system fault tolerant. Such is for example the power supply system based on eight energy sources where only three sources are sufficient to guarantee uninterrupted supply. The system will then be resistant to faults and failures associated with the separate power supply sources. Another common example is an aeroplane that requires two out of three engines for a successful flight.

Reliability networks with large connectivity are another example of fault-tolerant systems. Thus, the topology of the reliability network in Fig. 11.2(b) is characterised by a larger connectivity compared to the reliability network in Fig. 11.2(a), which makes it less sensitive to faults in both branches.

14.4 IMPROVING COMPONENT RELIABILITY BY TESTING TO PRECIPITATE LATENT FAULTS

In reliability testing the emphasis is not on accepting or rejecting a production lot; the emphasis is on determining causes of failure. The natural way of determining the reliability of a product is by *field tests*: testing in the environment and operating conditions experienced during service. Recorded details about failure events include:

   (i) failed component or equipment,
  (ii) failure mode,
 (iii) time of failure,
 (iv) time of introducing the equipment into service,
  (v) operating conditions at the time of failure,
 (vi) location of the equipment and working environment,
(vii) date of manufacture and user characteristics.

The intention is to feed back this information to the manufacturer in order to identify the cause of failure and to implement corresponding corrective measures.

The major drawback of this approach is the significant amount of time and money needed for field tests. This is not only expensive, but causes a significant delay in the introduction of the product. Highly reliable units will require long operational (testing) periods before useful data become available. If the loads during service are relatively smooth, some of these tests can be run for thousands of hours without the service load ever being able to exceed strength and induce failure. Even if the loads encountered during the test are fully representative and cover the whole distribution, because of the relatively small number of test samples, the lower tail of the strength distribution where the interaction with the load distribution occurs is usually not covered at all!

Indeed, suppose that three items are tested. The actual strengths of these three items are likely to be around the expected values of the order statistics of the strength distribution (Fig. 14.7(a)). The expected order statistics of the strength distribution are the means of the distributions of the order statistics if a very large number of tests involving measuring the strength of series including three items were carried out. In this case (Fig. 14.7a), none of the tested items is likely to fail because the lower tail of the strength distribution which interacts with the load is not likely to be covered and for none of the tested items, the load is likely to exceed strength. Such a test,

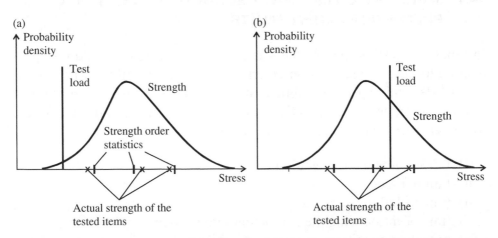

Figure 14.7 (a) The actual strength of three tested items is likely to be close to the expected values of the order statistics and probably none of the tested items will fail. (b) Increasing the test load will probably cause two of the tested items to fail.

despite its high cost, *will not yield useful failure data*, hence no opportunity for reliability improvement will be present.

Furthermore, in field tests, failure modes will be discovered too late to implement timely corrective action during design and manufacturing. As a result, problems discovered during field testing are often accepted as a permanent limitation of the systems/components.

Laboratory and prototype tests are aimed at gaining information concerning failure modes. A typical approach is based on testing the equipment until it fails, a design modification is introduced to remove the failure cause and the procedure is repeated until a desired level of reliability is achieved. Similar to the field tests, if the test stresses are within the range the equipment is expected to experience during service, no failures may be induced by the test loads and no useful information regarding the failure modes may be gained.

The *qualification tests* subject the tested equipment to conditions intended to simulate the extremes expected during service: extreme loads, temperatures, humidity, vibration, shocks, radiation, electric and magnetic fields, etc. Faults revealed during qualification tests must be removed before production is authorised.

The objective of *environmental stress screening* (ESS) is to simulate expected worst-case service environments. The stresses used during the ESS are aimed at eliminating (screening) the part of the population with faults. This part of the population causes a heavy lower tail of the strength distribution which is the primary reason for many early-life failures.

The test has been illustrated in Fig. 14.8(a) where the lower tail of the strength distribution has been altered by stress screening which has removed substandard items with insufficient strength. In Fig. 14.8(b), the strength distribution is a mixture of two distributions: a main distribution reflecting the strength of the strong population of items and a distribution

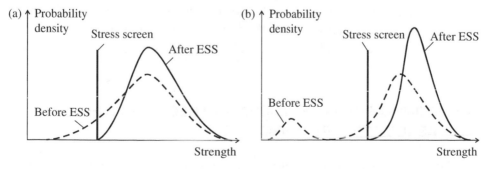

Figure 14.8 Altering the lower tail of the strength distribution by ESS.

characterising the population of items with substandard strength. ESS has improved the strength distribution by removing the weak population (Fig. 14.8(b)). By trapping faults and substandard items before they are released to the customer, this test reduces or eliminates completely early-life failures caused by items with substandard strength. ESS also helps to discover and eliminate sources of faults and weaknesses during design, manufacturing and assembly.

During ESS (*burn-in*), it is important to find operating and environmental test conditions which permit efficient screening without consuming a substantial part of the life of the remaining components which have passed the screening test (Jensen, 1995).

Particularly useful tests which reveal a large number of failure modes and reduce the test time from years to days and hours are the HALT (Highly Accelerated Life Testing) and HASS (Highly Accelerated Stress Screens). The purpose is to expose (precipitate) faults and weaknesses in the design, manufacturing and assembly in order to provide basis for reliability improvement. The purpose *is not* to simulate the service environment. Precipitation of a fault means to change its state from latent/undetected to a detectable state. A poor solder joint is usually undetectable unless it is extremely poor. Applying vibration, thermal or electrical stress helps to precipitate the fault, conduct failure analysis and perform appropriate corrective action (Hobbs, 2000). The precipitated faults and weaknesses are used as opportunities for improvement of the design and manufacturing in order to avoid expensive failures during service. In this respect, HALT and HASS are particularly useful. The stresses used during HALT and HASS testing are extreme stresses applied for a brief period of time. They include all-axis simultaneous vibration, high-rate broad-range temperature cycling, power cycling, voltage, frequency and humidity variation, etc. (Hobbs, 2000). During HALT and HASS, weaknesses are often exposed with a different type of stress or stress level than those which would expose them during service. This is why, the focus is not on the stress and the test conditions which precipitate the weaknesses but on the weaknesses and failure modes themselves.

14.5 COMMON CAUSE FAILURES AND REDUCING THE RISK ASSOCIATED WITH THEM

A common cause failure is usually due to a single cause with multiple failure effects which are not consequences from one another (Billinton and Allan,

1992). Common cause failures are usually associated with significant losses because a common cause can: (i) induce failure of several components; (ii) increase the joint probability of failure for a large number of components or for all components in a system and (iii) destroy a number of redundant paths, at the same time. Even in blocks with a high level of built-in redundancy, in case of a common cause failure, all redundant components in a block can fail at the same time or within a short period of time and the advantage from the built-in redundancy is lost completely.

Failure to account for a common cause usually leads to optimistic reliability predictions: the actual reliability is smaller than the predicted. Considering that most of the common cause effects are very subtle, there is a real possibility that some of the common cause effects will be missed in the analysis.

A significant source of common cause failures are *common design, processing, manufacturing and assembly faults in several components* which reduce their reliability simultaneously. In presence of a common cause, the affected components are more likely to fail, which reduces the overall system reliability. Typical examples of common cause failures are the common design, manufacturing or assembly faults which increase the susceptibility to corrosion of several components in an installation. By simultaneously increasing the hazard rates of the affected components, corrosion increases the probability of system failure. Another example of a common fault is the anisotropy of the spring wire due to excessive sulphide inclusions in the batch of springs rods used for coiling the springs. All manufactured springs from this batch will then fail prematurely due to the common fault. Manufacturing faults like 'scratches' or 'tool-marks' on the spring wire surface will have a similar effect: springs with these faults will fail prematurely.

Maintenance and operating actions common to different components and branches in a system are a major source of common cause failures. An example of this type of common cause failure is the case where applying insufficient torque on flange bolts can induce leaks in a number of assembled flanges, at the same time.

If A and B denote the events 'components a and b will fail within a specified time interval' and $\overline{A}$, $\overline{B}$ denote the events 'components a and b will survive the specified time interval', the common cause failure can be defined as

$$P(A \cap B | C) > P(A \cap B | \overline{C}) \tag{14.4}$$

The joint probability $P(A \cap B|C)$ of failure of components a and b given the common cause C is greater than the joint probability of failure $P(A \cap B|\overline{C})$ of the components without the common cause.

If a common cause C is present, the failures of components A and B can either be statistically independent or statistically dependent. Using a simple example, we will show that we could even have both: statistical independence and statistical dependence, given the same common cause! In other words, for the same common cause C both

$$P(A \cap B|C) = P(A|C) \times P(B|C) \tag{14.5}$$

$$P(A \cap B|C) > P(A|C) \times P(B|C) \tag{14.6}$$

could be fulfilled depending on whether there exists or there is no barrier against the common cause.

Indeed, let A and B be two communication centres located randomly in a space with area S (Fig. 14.9). Suppose that the area in Fig. 14.9 is being shelled by a large number n of identical missiles. Each missile has a radius of destruction r. If a communication centre is within the radius of destruction of a landing missile, the centre is destroyed and stops transmitting signals. Clearly, the missiles are the common cause C of failure of the communication centres.

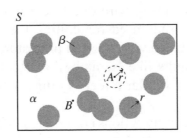

Figure 14.9 Two communication centres located randomly in a space with area S.

The probability $P(A|C)$ that communication centre A will fail given the common cause C is equal to $1 - P(\overline{A}|C)$ where $P(\overline{A}|C)$ is the probability that all n missiles will land outside the circular shield zone with radius r with centre the location of communication centre A (Fig. 14.9):

$$P(\overline{A}|C) = \left(\frac{S - \pi r^2}{S} \right)^n = (1 - \psi)^n = \exp\left[n \ln \left(1 - \psi \right) \right] \tag{14.7}$$

where $\psi = \pi r^2/S$ denotes the areal fraction of the destruction area. If ψ is small, $\ln(1 - \psi) \approx -\psi$ and equation (14.7) can be presented as $P(\overline{A}|C) = \exp(-\lambda s)$ where $\lambda = n/S$ is the number density of the missiles and $s = \pi r^2$ is the area of the destruction zone corresponding to a single missile. Consequently, the probability of failure of communication centre A becomes

$$P(A|C) = 1 - P(\overline{A}|C) \approx 1 - \exp(-n\psi) = 1 - \exp(-\lambda s) \qquad (14.8)$$

Consider the two shield zones with radii 'r', with centres the two communication centres (Fig. 14.10).

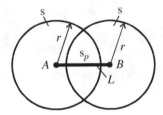

Figure 14.10 Shield zones with radii 'r', around the two communication centres.

Let s_p stand for the overlapped part of the two shield zones in Fig. 14.10. The probability of the event $\overline{A} \cap \overline{B}|C$ that neither of the communication centres will be destroyed given the common cause C:

$$P(\overline{A} \cap \overline{B}|C) = \exp[-\lambda(2s - s_p)] \qquad (14.9)$$

is equal to the probability that no missile will land in the region covered by shield zones of the communication centres (Fig. 14.10). If no overlapped part existed ($s_p = 0$)

$$P(\overline{A} \cap \overline{B}|C) = \exp(-2\lambda s) = [\exp(-\lambda s)][\exp(-\lambda s)]$$
$$= P(\overline{A}|C)P(\overline{B}|C) \qquad (14.10)$$

that is, events $\overline{A}$ and $\overline{B}$ are statistically independent. However, events $\overline{A}$ and $\overline{B}$ are statistically dependent, $P(\overline{A} \cap \overline{B}|C) \neq P(\overline{A}|C)P(\overline{B}|C)$, if an overlapped zone exists (if $s_p > 0$, then $\exp[-\lambda(2s - s_p)] > \exp(-2\lambda s)$). The probability that both communication centres will be destroyed can be determined by subtracting from unity the probability of the complimentary event 'at least one of them will survive':

$$P(A \cap B|C) = 1 - P(\overline{A \cap B}|C) = 1 - P(\overline{A} \cup \overline{B}|C) \qquad (14.11)$$

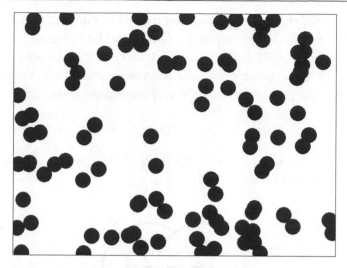

Figure 14.11 Random hits from 100 missiles with a radius of destruction $r = 15$ length units which have landed in the area S.

Considering that $P(\overline{A} \cup \overline{B}|C) = P(\overline{A}|C) + P(\overline{B}|C) - P(\overline{A} \cap \overline{B}|C)$, equation (14.11) results in

$$P(A \cap B|C) = 1 - 2\exp(-\lambda s) + \exp[-\lambda(2s - s_p)] \qquad (14.12)$$

If no overlapped zone existed ($s_p = 0$) equation (14.12) transforms into $P(A \cap B|C) = [1 - \exp(-\lambda s)]^2 = P(A|C)P(B|C)$, that is, the events that communication centres A or B will fail given that the common cause is present, are statistically independent. However, the events A and B are statistically dependent, $P(A \cap B|C) \neq P(A|C)P(B|C)$ given the common cause C, if an overlapped zone is present ($s_p > 0$), because

$$1 - 2\exp(-\lambda s) + \exp[-\lambda(2s - s_p)] > [1 - \exp(-\lambda s)]^2 \qquad (14.13)$$

In other words, the existence of an overlapped zone s_p increases the probability that both communication centres will fail. The area of the overlapped zone s_p is a function of the radius of the destruction zone and the distance between the communication centres. It is given by

$$s_p = 2r^2 \arccos\left(\frac{L}{2r}\right) - L\sqrt{r^2 - (L/2)^2} \qquad (14.14)$$

The area of the section in Fig. 14.11, shows random hits from 100 missiles with a radius of destruction $r = 15$ length units. The probability that both communication centres at a distance greater than $2r = 30$ ($L > 2r = 30$)

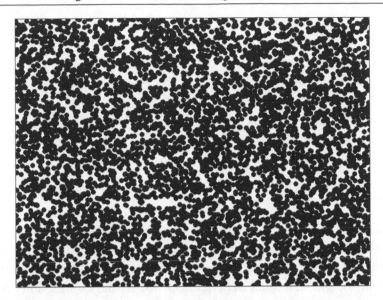

Figure 14.12 Random hits from 6200 missiles with a radius of destruction $r = 4$ length units, which have landed in the area S.

will be destroyed was determined to be: $P(A \cap B|C) \approx 0.04$. The probabilities that each of the communication centres will be destroyed by a missile were calculated by a Monte Carlo simulation: $P(A|C) = P(B|C) = 0.21$. As can be verified, in this case $P(A \cap B|C) = P(A|C)P(B|C)$, that is, the events $A|C$ and $B|C$ are statistically independent. Another Monte Carlo simulation has also been performed, where the distance L between the communication centres was equal to nine length units $L = 9 < 2r$ which guaranteed an overlapped zone. The probability $P(A \cap B|C) \approx 0.14$ that both centres will be destroyed is significantly greater than $P(A|C)P(B|C) = 0.04$ which shows that in this case, the failures of the communication centres are statistically dependent.

The same type of Monte Carlo simulation was also performed by using 6200 missiles with a radius of destruction $r = 4$ units, (Fig. 14.12). A distance between the communication centres $L = 9$ units now guaranteed zero overlapped zone ($s_p = 0$, $L = 9 > 2r = 8$).

The probabilities that each of the communication centres will be destroyed by a missile were calculated by a Monte Carlo simulation and were found to be $P(A|C) = P(B|C) = 0.7$. The probability that both communication centres will be destroyed was found to be $P(A \cap B|C) = 0.49 = P(A|C)P(B|C)$. In other words, despite that a significantly larger area has been destroyed by the missiles compared to the case in Fig. 14.11,

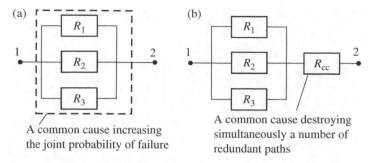

(a) A common cause increasing the joint probability of failure

(b) A common cause destroying simultaneously a number of redundant paths

Figure 14.13 Impact of a common cause failure on a system including three components logically arranged in parallel.

events $A|C$ and $B|C$ are still statistically independent! This is because the distance between the communication centres in this case guarantees absence of an overlapped zone ($L = 9 > 2r = 8$). In short, given the common cause (the missiles), the statistical dependence/independence of events $A|C$ and $B|C$ (destruction of targets A and B) is solely controlled by the distance between the targets. Whether the two events are statistically dependent or independent depends only on whether the distance between their centres is smaller or greater than the destruction diameter of the missiles and *does not depend on the destruction area covered by the missiles*!

Figure 14.13(a) illustrates a case where the common cause only increases the joint probability of failure without actually destroying any of the three redundant paths from node 1 to node 2. Figure 14.13(b) illustrates the case where the common cause destroys all of the redundant paths between node 1 and node 2.

Contamination of the fluid in a hydraulic, cooling or lubrication system affects only the components in contact with the fluid. This increases the joint probability of failure of the affected components. Contamination of a hydraulic system with debris for example, simultaneously affects a number of control valves and filters. Corrosive production fluids have a similar effect: they affect simultaneously a number of production valves which are in contact with the fluid and shorten their life. Contamination of the production fluids with sand has a similar effect. It induces intensive erosion of the valves and pipelines and shortens their life.

Work in dusty atmosphere is a common cause which affects simultaneously all rotating components. The wearout intensity increases and the joint probability of wearout failure associated with the rotating components increases, irrespective of whether they work independently from one another or not.

A faulty maintenance can also be a common cause: using the wrong type of oil or coolant for several rotating machines, can increase the joint probability of failure of all machines. A common cause is present if the main cooling system and the emergency cooling system are located in the same room and fire destroys both systems.

Similar type of failure is present if a common cause (e.g. fire) destroys signals from detectors or to control devices. A damage of the bundle of control cables caused by severing or fire destroys all connections simultaneously. In case of several redundant paths (Fig. 14.13(b)), the reliability of the system is always smaller than the probability R_{cc} ($R < R_{cc}$) of not having a common cause failure, no matter how large the reliabilities of the components are. Indeed, for n identical redundant components, each with reliability R_0, the reliability of the system in Fig. 14.13(b) is $R = [1 - (1 - R_0)^n] \times R_{cc}$. With increasing the reliabilities of the redundant components ($R_0 \rightarrow 1$), system reliability tends to the probability of not having a common cause: $R = [1 - (1 - R_0)^n] \times R_{cc} \rightarrow R_{cc}$. Correspondingly, the probability of failure $p_f = 1 - R \rightarrow 1 - R_{cc} = p_{cc}$ tends to the probability p_{cc} of occurrence of the common cause.

An example of a common cause maintenance failure which destroys two or more redundant paths is the incorrect assembly of series of three valves on a pipeline transporting production fluid. In terms of ability to stop (isolate) the fluid, the valves are logically arranged in parallel (Fig. 14.13(b)), that is, there exists a built-in redundancy. At least one of the valves must be working for the production fluid to be isolated. A faulty assembly, however, can induce a common cause failure where all valves will be leaking internally and none of them will be able to isolate the production fluid.

Typical conditions promoting common cause failures are: common design faults, common manufacturing faults, common installation and assembly faults, common maintenance faults, abnormally high temperature, pressure and humidity, erosion, corrosion, radiation, dust, poor ventilation, frequent start–stop cycles, vibration, electromagnetic interference, impacts and shocks, spikes in the power supply, etc. A common cause can also be due to a shared faulty piece of software. Thus, two programmable controllers produced from different manufacturers, assembled and installed by different people can still suffer from a common cause failure if the same faulty piece of software code has been recorded in the controllers.

It is of particular importance to design against common causes failures in order to reduce losses. This can be done by: (i) identifying and eliminating sources of common faults and common cause failures; (ii) decreasing the likelihood of the common causes and (iii) reducing the consequences from common cause failures.

Due to the nature of the latent faults, common cause failures are difficult to identify and are frequently overlooked if little attention is paid to the working environment and the possibility of latent faults. Common causes should be *designed* out and *their impact reduced* wherever possible. Designing out common causes is not always possible but it should be done if opportunity arises. Corrosion of a cooling or a hydraulic system caused by working fluid for example can be eliminated by selecting non-corrosive working fluids. Erosion caused by production fluids can be reduced if filters eliminating the abrasive particles are installed. A common cause due to insulation catching fire can be eliminated by selecting a fireproof material for the insulation.

An example of reducing the impact of the common cause is the use of corrosion inhibitors which, if mixed with the cooling agent (e.g. in the car radiators), reduce significantly the corrosion rate. The impact of common cause can be reduced by *strengthening the components* against it. Such are for example all corrosion protection measures listed in Chapter 13. Another example are the water-tight designs and couplings in underwater installations which reduce the possibility of contamination due to sea water ingress.

Common cause failures can be reduced by *decreasing the likelihood of occurrence* of common cause events. Frequent design reviews and strict control of the manufacturing process and the assembly reduce the likelihood of latent faults which could be a common cause for expensive failures. A strict control on the maintenance operations reduces the number of maintenance faults. Furthermore, providing maintenance of the redundant paths by two different operators reduces the likelihood of a common cause failure due to faulty maintenance. A common cause due to an incorrect calibration of measuring instruments due to a human error can for example be avoided if the calibration is done by different operators.

Blocking out against a common cause is an efficient technique for reducing common cause failures. The idea is to make it impossible for all of the redundant components to be affected by the same common cause.

Suppose that two pumps (a main pump and an emergency pump) participate in cooling down a nuclear reactor. Failure of both pumps creates an emergency situation. If the two pumps are from different manufactures, the common cause failure due to the same manufacturing fault will be blocked out. If in addition, the two pumps are serviced/maintained by different operators, the 'faulty maintenance' common cause will also be blocked out. In other words, making use of the principle of diversity in processing, manufacturing, maintenance and operation, helps to block out common cause failures. If finally, the pumps are located in different rooms, the common cause failure due to fire will also be blocked out.

Separating the components at distances greater than the radius of influence of the common cause is an efficient way of blocking against the common cause. Thus, separating two or more communication centres at distances greater than the radius of destruction of a missile increases the probability of survival of at least one of the centres. Multiple back-ups of the same vital piece of information kept in different places blocks against the loss of information in case of fire, theft or sabotage.

Another implementation of this principle is the separation of vital control components from a component whose failure could inflict damage. A typical example is separating control lines at safe distances from aeroplane jet engines. In case of explosion of an engine no flight controls will be lost.

Insulating some of the redundant components from contact with the working environment. Blocking against a possible common cause failure due to excessive dust or humidity or temperature for example can be avoided if some of the redundant components are physically insulated.

Avoiding common links which can be affected by a common cause is an efficient way of blocking out common causes. Such are for example the common conduits for cables, common location for components and common cooling and lubricating systems.

Sundararajan (1991) suggests preliminary common cause analysis which consists of identifying all possible common causes to which the system is exposed and their potential effects. The purpose is to alert design engineers to potential problems at an early stage of the design.

15

CONSEQUENCE ANALYSIS AND GENERIC PRINCIPLES FOR REDUCING THE CONSEQUENCES FROM FAILURES

15.1 CONSEQUENCE ANALYSIS AND CONSEQUENCE MODELLING TOOLS

Given that an accident/failure has occurred, for each identified set of initiating events, an assessment of the possible damage is made. In case of loss of containment for example, depending on the *release rate* and the *dispersion rate*, the consequences may vary significantly. In case of a leak to the environment, the consequences are a function of the magnitude of the leak and the dispersion rate. For a leak with large magnitude and substantial dispersion rate, a large amount of toxic substance is released for a short period of time before the failure is isolated. Where possible, the distribution of the conditional losses (consequences given failure) should be determined. This distribution gives the likelihood that the consequences given failure will exceed any specified critical threshold. In case of n mutually exclusive failure scenarios, the conditional cumulative distribution $C(x|f)$ of the loss given failure is described by the equation

$$C(x|f) = p_{1|f}\, C_1(x|f) + p_{2|f} C_2(x|f) + \cdots + p_{n|f} C_n(x|f) \qquad (15.1)$$

where $p_{i|f}$ is the conditional probability that the ith failure scenario will occur first ($\sum_i p_{i|f} = 1$).

Suppose that a leak is initiated, caused by a dropped object penetrating a vessel containing fluid under pressure. The size of the hole made by the dropped object can vary anywhere from $d = d_0$ to $d = d_{max}$. If no information about the distribution of the hole size is available, it can be assumed to be uniformly distributed in the range (d_0, d_{max}). The time to discover and repair the leak and the pressure inside the vessel are also random variables, characterised by particular distributions.

By using simulation, the distributions of the hole size, the pressure inside the vessel and the time to repair can be sampled and subsequently, for each combination of sampled parameters, the amount of released toxic substance can be calculated. Repeating these calculations a large number of times will produce a distribution of the amount of released substance, from which the probability that the released amount will be greater than a critical limit can be estimated easily.

For the expected value $\bar{C}$ of the conditional loss (consequences) given failure, the following equation holds:

$$\bar{C} = p_{1|f}\,\bar{C}_1 + p_{2|f}\bar{C}_2 + \cdots + p_{n|f}\bar{C}_n \qquad (15.2)$$

where $\bar{C}_i$ is the expected loss given the i-th failure scenario.

The full spectrum of possible failure scenarios should be analysed. Event trees are often employed to map all possible failure scenarios. The conditional probabilities $p_{i|f}$ of the separate scenarios are calculated by multiplying the probabilities of the branches composing the corresponding paths. This can be illustrated by the event tree in Fig. 15.1 related to a simplified case where different types of flammable substance could be released. Depending on the type of substance, the released quantity and whether fire is present, the conditional losses vary significantly – from losses associated with pollution and cleaning of the environment in the absence of fire to losses associated with pollution, fatalities and damage to production facilities in case of fire.

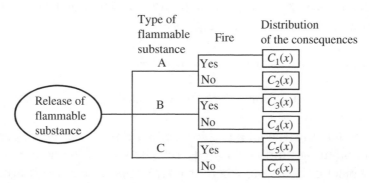

Figure 15.1 Event tree used for mapping possible scenarios of an accident and the distribution of the conditional losses (consequences).

The distribution of the consequences given that a flammable substance has been released is

$$P(X \leq x) \equiv C(x|f) = \sum_{i=1}^{6} p_{i|f} \, C_i(x|f) \qquad (15.3)$$

The conditional probabilities $p_{i|f}$ are determined from the products of the probabilities of the separate branches. Thus, $p_{3|f} = P$(type B flammable material) $\times$ P(fire).

Following the previous example, in case of a release of toxic chemical or contaminant in a confined space for example, depending on the volume released, the concentration in the air will vary and the consequences due to exposure will also vary. Suppose that the released amount varies uniformly in the range (V_0, V_{max}). The number of people exposed to the toxic substance depends on the actual occupancy of the space, which varies during the year. Given that a person has been exposed to the toxic substance, the probability of developing a particular condition is a function of the concentration of the substance (the released volume) and the duration of the exposure. The distribution of the consequences, in other words the distribution of the number of fatalities can then be determined by a simulation. This involves sampling from the distributions related to the released amount, the occupancy of the space and the percentage of people developing the condition. Repeating this calculation for a large number of simulation trials yields the distribution of the number of fatalities. In this way, the conditional losses can be determined as well as the associated uncertainty.

Usually, in case of a large number of different failure scenarios or complex interrelationships among them, special Monte Carlo simulation software is needed to determine the distribution of the conditional losses. Consequence modelling tools help to evaluate the consequences from dispersion of toxic gases, smoke, fires, explosions, etc.

15.2 GENERIC PRINCIPLES AND TECHNIQUES FOR REDUCING THE CONSEQUENCES FROM FAILURES

Implementing appropriate failure protection principles can have a significant impact on the consequences from failures by limiting their *evolution*, *magnitude* and *duration*. It needs to be pointed out however that protective measures are reactive measures. They do not prevent losses from materialising.

15.2.1 Protective Barriers

Implementing appropriate protective barriers can reduce significantly the consequences from failures. Protective barriers control an accident by limiting its extent and duration. They can arrest the evolution of the accident so that subsequent events in the chain never occur. Protective barriers can also prevent particular event sequences and processes which cause damage, by blocking the pathways through which damage propagates.

The defence against a release of toxic substance for example combines (Fig. 15.2):

- Passive physical barriers (protection equipment, clothing, gloves, respiratory masks).
- Active physical barriers (ventilation triggered by a detector).
- Immaterial barriers (handling rules minimising the released quantity in case of an accident, e.g. handling a single container with toxic material at a time).
- *Human actions barriers* and *organisational barriers* (evacuation).
- *Recovery barriers* (first aid, medical treatment).

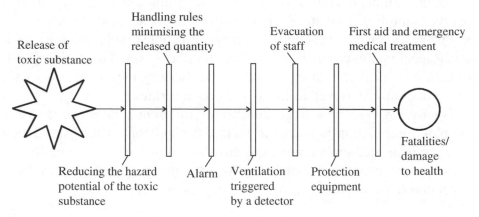

Figure 15.2 Different types of protective barriers mitigating the consequences following the release of a toxic substance.

A number of different types of protective barriers reducing the consequences from failures are given below.

15.2.1.1 *Using Passive Protective Barriers: Separating the Sources of Hazards and the Targets.*   This principle for minimising the damage in case of failure is the reason behind the safety practice of building residential areas beyond the radius of harmful influence of toxic substances from chemical plants, compost production facilities, fuel depots, etc. Separating people from hazards is an important measure for reducing the damage if the control over hazards is lost.

Passive protective barriers physically separate the hazards (the energy sources) from targets. Physical barriers isolate and contain the consequences and prevent the escalation of accidents. They provide passive protection against the spread of fire, radiation, toxic substances or dangerous operating conditions. A blast wall, for example, guards against the effects of a blast wave. Separating physically, or increasing the distance between sources of hazards and targets, minimises the damage in case of an accident.

Examples of passive barriers are: the safeguards protecting workers from flying fragments caused by the disintegration of parts rotating at a high speed; the protective shields around nuclear reactors or containers with radioactive waste; the fireproof partitioning; the double hulls in tankers preventing oil spillage if the integrity of the outer hull is compromised, etc.

15.2.1.2 *Using Active Protective Barriers* The consequences from an accident or failure can be mitigated significantly by activating protective systems. Typical examples of active barriers designed to mitigate the consequences from accidents are: the safety devices such as airbags for protecting passengers during a car accident; activating sprinklers for limiting the spread of fire; activating surge barriers to limit the consequences from floods; automatic brakes in case of a critical failure; automatic circuit breakers in case of a short cut, etc.

15.2.2 Damage Arrestors

A typical example of a damage arrestor can be given with buckling of a pipeline subjected to a high hydrostatic pressure. Buckling could be reduced by increasing the thickness of the pipeline but this option is associated with significant costs. Control of buckling propagation achieved by using buckle arrestors is a cheaper and more beneficial option. Buckle arrestors are thick steel rings welded to or attached at regular intervals to the pipeline, in order to halt the propagating buckle and confine the damage to a relatively small section (Fig. 15.3). In this way, the losses from buckling are limited to the length of the section between two buckle arrestors. In case of failure, only the buckled section will be cut and replaced. The spacing between buckle arrestors can be optimised on the basis of a cost–benefit balance between

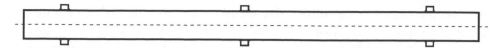

Figure 15.3 A pipeline with buckle arrestors of type stiffened welded rings.

the cost of installation of the arrestors and the expected cost of intervention and repair.

Resistance to buckling caused by a high external hydrostatic pressure on pipelines mounted on the sea bed can also be increased if the degree of ovality is decreased. Resistance to lateral buckling is increased if additional support is introduced. Again, this is associated with investment and detailed cost–benefit analysis is required to justify the design solutions.

Including materials with high toughness, steel crack-arrestor plates or riveted constructions are all measures for reducing the consequences from initiation and propagation of cracks. The purpose is to confine the damage from crack propagation in a relatively small area. Without these measures, the crack could propagate through the entire structure and cause significant damage, especially for wholly welded structures.

15.2.3   Avoiding Concentration of Vulnerable Targets in Close Proximity

An example of this measure is limiting the spread of damage by avoiding large groups of people in small spaces. Another example is avoiding building fuel tanks, pressure vessels, plants or command centres in close proximity. This measure makes the potential targets less vulnerable to a common cause failure from an external source. Secondary failures and domino-type failures are also avoided.

15.2.4   Blocking the Pathways Through Which the Damage Escalates

An efficient way of limiting the consequences from an accident or failure is studying the pathways through which the consequences propagate and where possible, automatically sealing them off in case of an accident. Such are for example the measures taken to prevent the spread of infections, contamination of drinking water, etc.

Activate *protection systems* limit the consequences by blocking automatically the pathways through which the consequences propagate. Such are for example the shut-down systems and fail-safe devices which automatically close key valves in case of a critical failure, thereby isolating toxic or flammable production fluids and reducing the consequences from failures. Various stop buttons interrupting the production cycle in cases of failure are also part of the active protection systems.

15.2.5 Using Fail-Safe Devices

15.2.5.1 *Hardware Fail-Safe Devices* The idea behind these is to eliminate or at least mitigate the consequences should failure occurs. Such is for example the fail-safe slab gate valve in Fig. 15.4 on a pipeline.

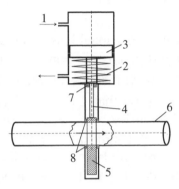

Figure 15.4 A fail-safe gate valve.

The pressure from the hydraulic supply '1' normally maintains the gate valve open by exerting pressure against the compression spring '2' and maintaining the valve bore aligned with the gate bore. In case of failure of the hydraulic power supply '1' associated with loss of pressure, the compressed spring '2' exerts back pressure on the piston 3 and the stem 4 moves the blanked part of gate 5 against the seat assembly 8 and safely isolates the working fluid. Other examples of fail-safe devices are the cut-off switches or fuses which disconnect a circuit if the current exceeds a maximum acceptable value.

15.2.5.2 *Software Fail-Safe Devices* These work by using internal programme tests and exception handling routines which set up safe conditions in case of errors. For example, a control can be set up in 'safe' position and error indicated if an important component or a sensor has failed. Failure of a pressure release valve to release pressure or a thermostat to switch off heating can for example be mitigated by ensuring that the fluid or heath supply will remain switched on for a limited period of time.

15.2.6 Deliberately Introducing Weak Links

Paradoxically, the consequences from failure can be decreased if potential failures are channelled into deliberately designed weak links. Should the unfavourable conditions triggering failure occur, the weak links are the ones to fail and protect the structure or component. In this way, the conditional losses are limited.

In case of M mutually exclusive failure modes, the expected conditional loss $\overline{C}_f$ is given by

$$\overline{C}_f = p_{1|f}\,\overline{C}_{1|f} + p_{2|f}\overline{C}_{2|f} + \cdots + p_{M|f}\overline{C}_{M|f} \tag{15.4}$$

where $\overline{C}_{k|f}$ is the expected conditional loss associated with the kth failure mode $(k = 1, 2, \ldots, M)$ and $p_{k|f}$ is the conditional probability that given failure, it is the kth failure mode that has initiated it $(\sum_{k=1}^{M} p_{k|f} = 1)$. Without loss of generality, suppose that $\overline{C}_{1|f}$ is the failure mode associated with the smallest conditional loss. If the conditional probability $p_{1|f}$ is made significantly larger compared to the other conditional probabilities $(p_{1|f} \gg p_{k|f},\ k = 2, 3, \ldots, M)$ in effect, a weak link has been introduced. Given failure, it is likely that the first failure mode, associated with the smallest conditional losses, has caused it. As a result, the conditional loss $\overline{C}_f$ and the risk of failure $K = p_f\overline{C}_f$ will be limited.

The approach based on deliberately introducing weak links is illustrated in Fig. 15.5 (Altshuller, 1974). In order to protect underground cable lines from cracks caused by freezing of the ground, narrow cracks are deliberately made along the cable line. Thermal stresses from freezing cause only widening of the pre-existing cracks. In this way, the deliberately made cracks prevent the formation of new cracks cutting through the cable lines. Cracks are in general unfavourable, but their negative effect has been deliberately amplified and is no longer negative.

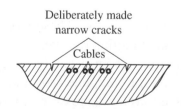

Figure 15.5 Reducing the losses from failures by deliberately introducing weak links (Altshuller, 1974).

Another example given by Altshuller (1974) is associated with installations for liquid helium. These needed lubrication of the seals in order to work properly but at the cryogenic temperatures all lubricants freeze and the reliable work of the seals and lubricants was impossible. The solution found by P.L. Kapitsa (co-recipient of the Nobel Prize for Physics in 1978 for discovering the superfluidity in helium) is an excellent example of deliberately amplifying the negative effect of a weak link. Instead of looking for suitable lubricants and seals, Kapitsa eliminated them; he deliberately increased the gap between the piston and the cylinder so that helium could freely leak through. During the leakage however, the gas expands so quickly that a significant back pressure is created which prevents the leakage of new portions of gas.

15.2.7 Delaying the Rate of Deterioration in Case of Failure

Fireproof coatings of steel-supporting structures for example limit the consequences should fires break out. Without the fireproof protection, in case of fire, the steel will loose quickly its strength causing the entire structure to yield and collapse. Selecting materials which yield as they fail, can significantly limit the damage in comparison with materials which break in a brittle fashion, with no warning.

15.2.8 Reducing the Amount of Released Hazardous Substance in Case of an Accident

Separating hazardous sources (e.g. fuel tanks) at sufficient distances from one another, avoids the domino-effect of multiple explosions and reduces significantly the amount of released toxic gases in case of fire. Dealing with small quantities of hazardous materials in a chemical laboratory also reduces the consequences in case of explosion or fire.

15.2.9 Reducing the Amount of Time Spent in a Hazardous Area

Reducing the amount of time spent in a hazardous area is an important measure limiting the damage to health. In many cases, the extent of the damage (e.g. in case of carbon monoxide poisoning or radiation damage) is strongly correlated with the amount of time spent in the hazardous area.

15.2.10 Reducing the Vulnerability of Targets

Vulnerability of humans is reduced by various barriers, guards, rails and by using personal protective equipment. Examples of personal protective equipment are: protective clothing, harnesses, breathing devices, hats, goggles, boots, gloves, masks, radiation indicators, toxic gas release detectors, lifting and handling equipment, vaccines, etc. Vulnerability of the equipment and systems is reduced by using protection barriers, housing, encapsulation, anti-corrosion and anti-erosion coatings, CCTV surveillance, metal shutters, exclusion zones, security systems for access, removing piles of flammable materials close to buildings, etc.

Vulnerability of data is reduced by using security systems and limiting the access to personal records and confidential data.

15.2.11 Emergency Systems, Equipment and Procedures

A good accidents response management based on well-established rules and training is a major factor mitigating the consequences from accidents. Evacuation procedures, sheltering and rescue operations are designed to reduce the consequences from accidents. Various type of medical emergency equipment also helps to reduce injury, should accidents occur. Adequate first-aid training, washing and decontamination facilities are important mitigating factors.

15.2.12 Using Devices Which Permit Operation in Degraded Conditions

Good examples of such devices are a rigid metal disk in a car tyre or a substance which automatically seals the puncture. These permit the vehicle to continue its travel after puncture so that the driver can maintain control and avoid accidents.

15.2.13 Using Failure Indicators

Using failure indicators is important in cases where the time for discovering the failure is significant. This time can be reduced significantly if failure status monitoring is used.

15.2.14 Using Inexpensive Components with Shorter Life

In the cases where failures are inevitable, losses can be reduced by replacing one or more expensive parts by disposable inexpensive parts with shorter life. By compromising durability to some extent, the losses from failures are reduced significantly. Such is for example the case where cheap plastics instead of expensive glass or ceramic are used if the likelihood of failure during operation is very high.

15.2.15   Reducing the Downtimes for Repair by Keeping Spare Parts

Keeping spare parts as standby components reduces significantly the downtime for repair, and from this, the cost of lost production. If no spare components are kept, each failure of a working component is associated with extra downtime for delivering it. If a single spare component is kept as a standby redundant component, in case of failure of the working component, it is replaced by the spare component, a new spare component is ordered immediately and production continues. In this case, production is disturbed only if clustering of two or more failures occurs within the time for delivery of the failed component.

15.2.16   Risk Planning

The purpose of risk planning is to specify the most appropriate response should failure occurs. Planning guarantees that the optimal course of action will be taken for dealing with the consequences from failure. Usually, in the absence of planning, the quickest and the most obvious actions are taken, which are rarely the optimal ones.

Risk planning prepares for the unexpected. It results in contingency plans for the course of action in case of failure or accident. Planning guarantees proactive rather than reactive attitude to risk and provides more time to react. It is closely linked with the research preparation involving a careful study of the system or process and identifying possible sources of risk. The time invested in such an exercise pays off since the response time and the chances of taking the wrong course of action are reduced significantly. Panic and hasty action are also avoided, which could otherwise promote errors aggravating the consequences from failure.

Planning also provides an answer to the important question 'how much resource to allocate now given the possibility of failure in the future, in order to minimise the total cost'. In this sense, quantifying the risks associated with the different scenarios is at the heart of risk planning.

A particularly important issue for a company is striking the right balance between risk and profitability. Thus, borrowing from banks and investing in projects provides leverage and increases profitability, but also increases the risk. Conversely, an increase in the cash position reduces risk, but also reduces profitability because of the reduced rate of return.

15.2.17   Using Troubleshooting Procedures and Systems

Computer-based expert troubleshooting systems are a powerful tool for reducing the downtimes in case of failure. Expert systems capture and distribute human expertise related to solving common problems and acting appropriately in particular situations. Compared to people, they are more reliable and retain all the time the knowledge about vast number of situations and problems, and the corresponding operating procedures. Furthermore, the troubleshooting prescriptions are objective and not coloured by emotions. Troubleshooting systems can help in training the staff to handle various problems or accidents. They also help counteract the constant loss of expertise as specialists leave or retire (Sutton, 1992).

15.2.18   Using Better Models to Predict the Losses Given Failure

Often, inappropriate model simplifications are made and the normal distribution is used inappropriately to estimate the probability of large losses. Since large losses are often characterised by a significantly heavier tail than the tail of the normal distribution, exceptionally large losses are often underestimated. According to the central limit theorem, the sum $S_n = X_1 + X_2 + \cdots + X_n$ of n consecutive losses if none of them dominates the distribution of the sum is well described by a normal distribution. Often, of particular interest is the distribution of the largest loss:

$$X_{\max} = \max\{X_1, X_2, \ldots, X_n\}$$

Here, the normal distribution is a poor prediction model for $X_{\max}$. The extreme value theory (EVT) offers better tools for predicting exceptionally large losses (King, P. 2001). It uses the *Generalised Pareto Distribution, Fréchet, Weibul* and the *Maximum Extreme Value Distribution*. These models estimate the probability that the conditional loss will exceed a particular limit given that the rare failure event has occurred. A number of applications related to the application of the Maximum extreme value distribution in demand–capacity interference risk models can for example be found in (Todinov, 2005a). Modelling extreme events is associated with difficulties because data in the region of the upper tails of the distributions are rarely available.

A synchronised increase or a decrease of the magnitudes of variables controlling the occurrence of failures (e.g. floods) also occurs. Estimating the probability of such coincidences by simulation or by using theoretical relationships, provides valuable information related to the likelihood of such rare events.

15.3 GENERIC DUAL MEASURES FOR REDUCING BOTH THE LIKELIHOOD OF FAILURES AND THE CONSEQUENCES

In some cases, a risk-reduction measure can be preventive and protective at the same time. Such is the speed restriction barrier (if it is obeyed). It is preventive because it makes an accident less likely by allowing the driver more time for response to a road hazard and at the same time it is protective because it reduces the kinetic energy of the car which provides less challenge to barriers designed to absorb a crash (crumple zones) (Hale et al., 2004). Here are some generic approaches for reducing both the likelihood of failure and the consequences given failure.

15.3.1 Reducing the Risk of Failure by Improving Maintainability

Many production systems are subjected to maintenance of some type. Maintenance actions can be divided broadly into two classes (Smith, 2001): (i) corrective (unscheduled) maintenance and (ii) preventive (scheduled) maintenance.

Corrective (unscheduled) maintenance is initiated only if a critical failure occurs (one or more production units stop production). Failed redundant components are not replaced or repaired until the system fails.

Preventive (scheduled) maintenance is performed at planned intervals with the purpose of keeping the built-in levels of reliability and safety, and prevent the system failure rates from exceeding tolerable levels. This type of maintenance achieves its purpose by the following actions (Smith, 2001):

- Regular service of operating components and sub-system (e.g. lubrication, cleaning, adjustment).
- Checking for and replacement of failed redundant components.
- Replacement of components with excessive wearout.

Frequent inspection and replacement of failed redundant components enhances availability by improving the reliability of the system. Design to improve maintainability is vital to reducing the losses from failures. Improved maintainability reduces significantly the downtimes and the cost of intervention and repair. Availability can be improved significantly by improved accessibility and *modular design* which reduce the cost of intervention and repair and the downtimes. A sufficient number of spares should also be readily available. Using standard items with proven and tested properties also increases reliability.

15.3.1.1 *Guaranteeing Accessibility during Maintenance and Repair* A general principle here is the easy access to failed parts. Maintenance of a failed sub-system should not require the removal of another sub-system. The design must permit access to important components (Thompson, 1999). This design aspect is particularly important in case of maintenance and repair by remotely operated vehicles (ROV) conducted underwater.

15.3.1.2 *Modular Design* Modular design reduces the losses from failures by improving maintainability and reducing the cost of intervention and repair and the downtime. In case of failure of a subsea control module for example, only the failed module needs to be retrieved and repaired instead of retrieving the whole production tree. Compared to retrieving the production tree, replacing a failed control module requires less costly intervention, usually conducted by deploying ROV. Retrieving a production tree to the surface requires a mobilisation of an oil rig which is a significantly more expensive operation compared to deploying ROV. Moreover, while a repair by mobilising an oil rig may require months, repairs involving ROV do not normally require more than several days.

Modules should be connected to other modules as simply as possible, to reduce the downtime and cost associated with the repair/replacement. In this respect, the use of quick-release devices is beneficial (Thompson, 1999).

15.3.1.3 *Efficient Management of the Resources for Maintenance and Repair.* A proper management of the resources for repair reduces significantly the downtimes associated with repair. Optimising the number of spares for critical components or components with high failure frequency is also very important. The optimal number of spares should be estimated by modelling. Condition monitoring can significantly reduce the number of necessary spare components, which reduces the costs and the required storage space.

15.3.2 Reducing the Risk of Unsatisfied Demands by Delaying the Definition of Product/Service

The idea is to keep a product or service only with its common features. When a particular demand arrives which requires a specific feature, this is assembled into the product in order to satisfy the demand. An example can be given with an electronic device sold in places with different standards for the power source (e.g. USA and Europe). If the power source is built into the power cord and not into the electronic device itself, electronic devices appropriate for any destination can be assembled and delivered very quickly if an order arrives.

15.3.3 Reducing the Risk by Abandoning Failing Projects, Products or Services

Abandoning projects, products or services showing clear signs of failure can have a significant impact on the potential losses. Particularly damaging is the unwillingness to admit a mistake and adhering to doomed projects, products or services. These only consume valuable resources which could otherwise be made available to new projects, products or services with much more potential.

A thorough review could be conducted, where projects, products or services are reviewed and assessed against the possibility of meeting particular targets. It is then decided whether they should be continued or stopped. Such a risk-reduction measure is common to the pharmaceutical industry where the development of a new drug is terminated as soon as dangerous side effects are discovered (Pickford, 2001).

15.3.4 Reducing the Hazard Potential

The purpose is to limit the amount of energy possessed by hazards which limits their potential to cause damage. Thus, preventing the formation of large build-ups of snow reduces both the likelihood of an avalanche and its destructive power.

Instead of investing in safety devices and passive barriers, often, it is much more cost efficient to passivate hazardous wastes or spilled hazardous substances. This eliminates or reduces significantly their hazard potential and with it, the associated risk. There are various methods by which this can be achieved:

- Treatment with chemicals which reduce the chemical activity and toxicity of the hazardous substance.
- Reducing the inherent tendency to ignite or burn (e.g. chemicals which cover spilled fuel and prevent it from catching fire).
- Reducing the capability to evaporate.
- Reducing the possibility of auto-ignition (e.g. by avoiding piles of flammable materials).
- Changing the aggregate state. Solidifying liquid toxic waste for example reduces significantly its potential to penetrate through the soil and contaminate underground water.
- Dilution.

15.3.5   Reducing the Values of the Operating Parameters

A typical example is reducing the operating pressure of a process fluid. Decreasing the pressure decreases both the probability of failure associated with loss of containment and the amount of fluid released in case of loss of containment. Limiting the maximum weight which a single worker can handle is another example of a dual measure – it prevents back injury and at the same time reduces its extent should such injury occurs.

15.3.6   Procedures and Devices Providing an Early Warning of Incipient Failures

Procedures or pieces of equipment giving early warning and preventing the development of dangerous operating conditions or monitoring the values of critical parameters act both towards reducing the likelihood of failure and reducing the consequences given failure. Examples of such measures are the inspection procedures, operating, assembly or maintenance procedures, devices for detecting smoke, increased temperature, pressure, vibration, release of toxic substances, humidity and water level gages, radiation detectors and all devices for monitoring deviations from the normal ranges of parameters related to working equipment and processes.

By giving an early warning of incipient failures these devices and procedures help reduce the likelihood of failure and avoid grave consequences.

Thus, a measured increasing temperature and vibrations from a bearing indicates intensive wearout and incipient failure, which can be prevented by a timely replacement of the worn-out bearing. The early warning signals also reduce the downtime for troubleshooting and permit preparations to be made which reduce the damage should failure occurs (e.g. provision of spare parts, evacuation of people, building additional barriers, etc.). Inspection and monitoring of key process parameters such as pressure, concentration and temperature help to control them in safe ranges so that the damage in case of failure is minimised.

15.3.7 Statutory Inspection and a Risk-Based Inspection

Statutory inspections are an important tool for reducing the likelihood of failure and the losses given failure. Such is for example the inspection of critical parts of engines (e.g. the propellers or blades) for fatigue cracks which, if unnoticed, may cause serious accidents. Planned inspection and maintenance minimise significantly the losses from failures for production systems where the cost of failure is very high and unscheduled maintenance is associated with heavy penalties. Lack of inspection or insufficient number of inspections or too long inspection intervals means increased risk of failure. Inspections do reduce the risk of failure but they are also associated with cost overheads. Too frequent inspections and planned maintenance means excessive expenditure and downtime which for the plant operators entails reduced production and profits. There exists an optimum number of inspections which minimises the total cost: the sum of the cost of inspections and the risk of failure. Inspection and quality control techniques are important means for identifying substandard components before they can initiate an early-life failure. Examples of inspection and quality control activities which help reduce early-life failures are:

- Inspection for failed components, and excessive elastic or plastic deformations.
- Using non-destructive inspection techniques (e.g. *Ultrasonic Inspection Technique*) to test for the presence of cracks and other flaws.
- Checking the integrity of protective coatings and whether the corrosion protection provided by the cathodic potential is adequate.
- Inspecting the integrity of interfaces; inspection for leaks from seals.

The risk-based inspection is an alternative to the prescriptive inspection. It attempts to balance the cost of inspection and the benefits from the associated risk reduction. A simple example has been discussed in Chapter 4. In this respect, a central problem is to optimise the scheduling of inspection intervals which minimises the sum of inspection cost and risk of failure. Attempts are also made to prioritise and rationalise the scheduling of inspections and planned maintenance in such a way that the efforts are concentrated in few areas associated with most of the total risk.

15.3.8 Condition Monitoring

Condition monitoring is used for detecting changes or trends in controlling parameters or in the normal operating conditions which indicate the onset of failure. By providing an early problem diagnosis, the underlying idea is to organise in advance the intervention for replacement of components whose failure is imminent, thereby avoiding heavy consequences. Condition monitoring is particularly important in cases where the time for mobilisation of resources for repair is significant. The early problem diagnosis it provides helps to reduce significantly downtime associated with unplanned intervention for repair. A planned or opportune intervention is considerably less expensive than unplanned intervention initiated when a critical failure occurs. Ordering spare components immediately after the onset of failure has been indicated, helps to reduce significantly the downtime for repair and the associated cost of lost production. The earlier the warning, the larger the response time, the more valuable the condition monitoring technique. Early identification of an incipient failure reduces significantly:

   (i)   The risk of environmental pollution.
  (ii)   The number of fatalities.
 (iii)   The loss of production assets.
  (iv)   The cost of repair (damaged components require the mobilisation of specialised repair resources).
   (v)   The losses caused by dependent failures.
  (vi)   The loss of production associated with uncontrolled shutdown.
 (vii)   The loss of production due to the time spent on troubleshooting.

   By using the probability laws, information obtained from condition monitoring can be used to obtain new information.

   Indeed, suppose that the relative frequency/probability of event B obtained *via* condition monitoring is $P(B) = 0.7$. Suppose also that the monitored relative frequency/probability of event A in the absence of event

B is $P(A|\overline{B}) = 0.5$. This information is sufficient to derive the probability $P(A \cup B)$ of the union of events A and B. Indeed, by using the probability laws:

$$P(A \cap \overline{B}) = P(A|\overline{B}) \times P(\overline{B})$$

and since $P(\overline{B}) = 1 - P(B) = 0.3$, $P(A \cap \overline{B}) = 0.3 \times 0.5 = 0.15$.

Since $P(\overline{B}) = P(A \cap \overline{B}) + P(\overline{A} \cap \overline{B})$, for the probability $P(\overline{A} \cap \overline{B})$ we get

$$P(\overline{A} \cap \overline{B}) = 0.3 - 0.15 = 0.15$$

from which:

$$P(A \cup B) = P(\overline{\overline{A} \cap \overline{B}}) = 1 - 0.15 = 0.85.$$

Conditional (Bayesian) updating on the basis of the information provided by the condition monitoring is also an important source of more precise information related to the distribution of the monitored values.

A key feature related to condition monitoring is that the reliability of the monitoring system must exceed the reliability of the monitored equipment (Heron, 1998). Furthermore, the cost of the condition monitoring should be considerably lower than the prevented losses from failure. The infrastructure making it possible to react upon the data stream delivered from the condition monitoring devices must also be in place if the condition monitoring technique is to be of any utility. There is little use of a condition monitoring system giving an early warning of an incipient failure, if the infrastructure for intervening with preventive actions is missing.

Condition monitoring is based on measuring values of specific parameters, critical to the failure-free operation of the monitored equipment. Here are some examples of measured parameters:

- *Temperature measurements* to detect increased heat generation (which is usually an indication of intensive wear, poor lubrication, failure of the cooling system, overloading or inappropriate tolerances).
- *Measurements of pressure and pressure differences* to detect leaks in a hydraulic control system, dangerous pressure levels, etc. Increased pressure difference before and after a filter for example is an indication of clogging or blockage of the filter, while a pressure drop in a hydraulic line is an indication of leakage.
- *Measuring the displacement* of components and parts indicates excessive deformations, unstable fixtures, material degradation, design faults.

- *Vibration monitoring* detects incipient failures in bearings, increased wear, out-of-balance rotating components, etc.
- *Monitoring the cleanliness* of lubricants and hydraulic fluids. Periodic sampling of lubricants and hydraulic fluids for debris is often used to determine the extent of wear undergone by the components. Another useful outcome of sampling hydraulic fluids for debris and their timely replacement is reducing the possibility for jamming of control valves. Such failures could render inoperative large sections of a control system and could be very expensive if the cost of intervention is high. This type of condition monitoring also reveals the degree of deterioration of the lubricants and hydraulic fluids because of poor cooling, lubrication or increased heat generation due to wear.
- *Ultrasonic inspection, radiographic examination, magnetic particles* and *penetrating liquids* are used to detect various flaws (cracks, pores, voids, inclusions) in components, welded joints and castings.
- *Measuring the degree of wear, erosion, and corrosion*. Since most of the mechanical systems undergo some form of degradation caused by wear, corrosion and erosion they benefit from periodic monitoring of the extent of degradation. Deteriorated components can be replaced in time by opportunity maintenance.
- *Measuring the degree of fracture toughness deterioration* of materials due to ageing, corrosion and irradiation. The fracture toughness of steels exposed to deterioration is determined and the ductile-to-brittle transition curve is built. The degree of deterioration is indicated by the shift of the ductile-to-brittle transition region towards high temperatures (Todinov, 2004d).
- *Monitoring the manufacturing and process parameters*. Controlling the tolerances during manufacturing by monitoring the wear rate of cutting tools, reduces the possibility of failures caused by misfit during assembly, less possibility for jamming, poor lubrication and accelerated wearout.
- *Monitoring electrical parameters such as current, voltage and resistance*. Increased current in electric motors for example indicates increased resistance from the powered equipment. This is an indication of jamming due to a build-up of debris or corrosion products, misalignment, increased viscosity of the lubricant or lack of lubrication, damage of the bearings, clogged or blocked filters (in pumps), etc. Decreased conductivity of working fluids (hydraulic fluid) often indicates

contamination which may induce poor heat dissipation, increased energy losses, and accelerated erosion and corrosion.

Condition monitoring is different from status monitoring which determines whether a component is in working or failed state. Condition monitoring however, even used solely as a status monitoring tool, has a high value and can improve the availability of the system immensely if combined with immediate intervention for repair/replacement of failed components. This can be illustrated on a simple system with active redundancy (Fig. 15.6(a)) which consists of two identical components logically arranged in parallel. Without status monitoring, repair is initiated only if the system stops production (no path between nodes 1 and 2). The result from this breakdown-induced intervention will be a sequence of uptime corresponding to the case where at least one of the component works, followed by a downtime when both components have failed and a mobilisation of resources for repair has been initiated (Fig. 15.6(a)). Status monitoring changes this availability pattern dramatically. If the status of both components is constantly monitored and repair is initiated whenever any component fails, the downtime associated with a critical (system) failure could be avoided almost completely (Fig. 15.6(b)). Therefore, in case of large system downtimes status monitoring can help increase availability.

Another example, related to dependent failures, is the common coupling fan-cooled device, where failure of the fan causes an overheating failure of the cooled device. Such a failure can be prevented if the status of the fan is

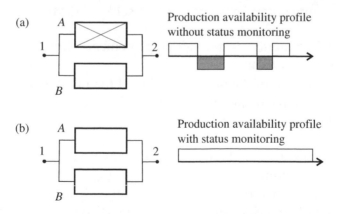

Figure 15.6 Production availability profile if (a) a breakdown policy is adopted and (b) constant status monitoring and immediate replacement of a failed redundant component is adopted.

monitored and, in case of failure, operation is discontinued until the fan is replaced.

We must point out that while status monitoring benefits mainly systems with built-in redundancy, condition monitoring on the extent of degradation of components and structures benefits also systems without any built-in redundancy.

Condition monitoring can be used to predict the approximate time of failure which provides a basis for better planning of the necessary resources for repair. The number of required spare parts and the cost of their preservation and storage can also be reduced significantly. Orders for components whose failure is imminent can be placed for manufacture, thereby reducing the delays associated with their delivery. Condition monitoring also provides the basis for improved designs by feeding back information related to actual times to failure, vulnerable components, root causes of failures, rate of material degradation and the impact of the operating conditions.

15.3.9 Reducing the Financial Impact from Failure by Warranties, Hedging and Insurance

An important way of mitigating the financial consequences from failure are the warranties. They are an integral part of almost all transactions which involve purchase of a product. Broadly, the warranties are contracts establishing liability between the manufacturer and buyer in the event of failure of the purchased product during agreed warranty period from the initial purchase. Manufacturers may agree to (Blischke and Murthy, 2000):

- repair failed items or provide replacements free of charge (*free replacement policy*);
- refund a fraction of the purchase price of the failed item;
- guarantee that the mean time to failure of the purchased equipment will be greater than a specified level.

Hedging limits the risk of a loss due to fluctuations of the market prices. Companies use different hedging methods (e.g. forwards, options and swaps) to manage operational risks associated with variations in exchange rates, cost of raw materials, interest rates, etc. (Crouhy et al., 2001).

Insuring against a risk is a contract, binding a party to indemnify another against specified loss, in return for premiums paid. In a narrow technical context, insurance is an efficient protective measure against losses from failures. However, insurance is no longer a cheap option. Before buying an insurance, it is important to weigh up the cost of insurance against the cost of failure. Another drawback is that because of the numerous exclusion clauses, insurance may not recoup the full amount of losses. Insurance cannot substitute a proper risk management. While some risks are insurable (e.g. security risks) many other risks are less likely to be (financial risks, competitive risks) and these risks require proper control measures.

15.3.10   Reducing the Financial Impact from Failure by Diversification and Portfolio Optimisation

Risk of failure can be reduced by investing in many unrelated sectors whose returns are not correlated. In this way, the variance (volatility) of the return from the whole portfolio of non-correlated stocks is reduced significantly. Investment funds reduce the risk by buying the shares of many companies. Even small investors can obtain a diversified portfolio with low transaction costs by investing, for example, in a unit trust.

The principle of diversification can be illustrated by an example involving a portfolio with n securities. For such a portfolio, the variance of returns σ_p^2 is given by (Teall and Hasan, 2002):

$$\sigma_p^2 = \sum_{i=1}^{n} w_i^2 \sigma_i^2 + 2 \sum_{i<j} w_i w_j \sigma_i \sigma_j \rho_{ij} \tag{15.5}$$

where σ_i is the standard deviation of the returns from security 'i', and ρ_{ij} is the linear correlation coefficient between the ith and jth security. The weight w_i shows how much money is invested in security i relative to the total amount invested in the entire portfolio ($\sum_{i=1}^{n} w_i = 1$). If the portfolio is based on three securities only, equation (15.5) becomes

$$\sigma_p^2 = w_1^2 \sigma_1^2 + w_2^2 \sigma_2^2 + w_3^2 \sigma_3^2 + 2w_1 w_2 \sigma_1 \sigma_2 \rho_{12}$$
$$+ 2w_2 w_3 \sigma_2 \sigma_3 \rho_{23} + 2w_1 w_3 \sigma_1 \sigma_3 \rho_{13} \tag{15.6}$$

For the purposes of the illustration, let us assume equal weights $w_1 = w_2 = \cdots = w_n = 1/n$, variances $\sigma_1 = \sigma_2 = \cdots = \sigma_n = \sigma$ and correlation coefficients $\rho_{ij} = \rho$, $i = 1, \ldots n; j = 1, \ldots n; i \neq j$. Equation (15.5) then becomes

$$\sigma_p^2 = \sigma^2/n + \rho \sigma^2 (1 - 1/n) \tag{15.7}$$

In case of a portfolio based on two securities only, whose returns are perfectly negatively correlated ($\rho = -1$), substituting $n = 2$ in equation (15.7) gives $\sigma_p^2 = 0$. In other words, such a portfolio is associated with no risk.

In case of a portfolio based on a large number n of securities, from equation (15.7), for the portfolio variance,

$$\sigma_p^2 \approx \rho \sigma^2 \qquad (15.8)$$

is obtained. In other words, with increasing the number of securities, the volatility of the portfolio, which is a measure of the risk associated with the portfolio returns, has been reduced to $\rho \sigma^2$. If the returns from the different securities are not correlated ($\rho_{ij} = \rho = 0$), the portfolio volatility (the risk) becomes zero ($\sigma_p^2 = 0$).

By using optimisation techniques, the weights w_i which yield the minimum volatility (risk) of the portfolio specified by equation (15.5) can be determined. A number of additional constraints must also be satisfied. Such is the constraint

$$\overline{R}_{min} = \overline{R}_1 w_1 + \overline{R}_2 w_2 + \cdots + \overline{R}_n w_n \qquad (15.9)$$

where $\overline{R}_1, \ldots, \overline{R}_n$ are the expected returns from the individual securities and $\overline{R}_{min}$ is the minimum expected return from the portfolio required by the investor. Additional constraints are specified by equations (15.10) and (15.11):

$$w_1 + w_2 + \cdots + w_n = 1 \qquad (15.10)$$

$$0 \leq w_i \leq 1, \quad i = 1, n \qquad (15.11)$$

16

LOCALLY INITIATED FAILURE AND
RISK REDUCTION

16.1   A GENERIC EQUATION RELATED TO THE PROBABILITY OF FAILURE OF A STRESSED COMPONENT WITH COMPLEX SHAPE

An important factor affecting failure of components and structures is the presence of flaws due to processing, manufacturing or mechanical damage during service. Presence of flaws in the materials leads to failures at much lower applied stresses. In the presence of flaws, fracture toughness replaces strength as the relevant material property. Failure is controlled by the interaction of the load and strength and occurs when load exceed strength (Fig. 16.1).

Assuming that the weakest-link principle holds, failure occurs if for at least a single flaw, the local load exceeds the local strength. In other words, failure is present if for the particular loading at least one of the flaws initiates it.

The deterministic fracture mechanics approach has been applied successfully to develop designs resistant to failures caused by flaws. Predictions based on the deterministic fracture mechanics approach however are inherently conservative for the following reasons:

(i) It is assumed that failure is initiated in the highest-stressed region of the component.

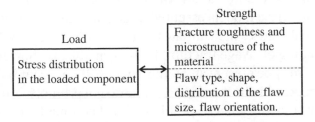

Figure 16.1 Basic parameters controlling failure in material with flaws.

(ii) It is assumed that a flaw *does exist* in the highest-stressed region of the component.

(iii) The flaw used in the analysis is usually taken with its *largest size*.

(iv) The flaw used in the analysis is taken with its *most unfavourable orientation* regarding the local stress tensor and the local texture of the material.

(v) The flaw used in the analysis is taken with its *most unfavourable shape*.

The first two assumptions are fulfilled only if the number density of the existing flaws is so large that it is practically certain that a flaw will reside in the highest-stressed area. In this case, failure will almost certainly be initiated there. If the number density of the existing flaws however is relatively small, there is a high chance that the flaws will lie outside the highest-stressed region. In this case, failure could be initiated outside the highest-stressed region because the presence of flaws lowers significantly the local stress necessary to make the flaw unstable. In other words, the conditions for initiating failure (the failure criterion) may be fulfilled in the low-stress regions containing flaws and not fulfilled in the highest-stressed regions.

Furthermore, presence of flaws in the highest-stressed region is not a sufficient condition for failure. Even if a flaw is present there, its type, size, shape or orientation may be such that no failure will be initiated. In other words, a flaw may exists in this region but it may not be *critical*. A critical combination of values for the controlling factors must be present so that the flaw becomes unstable.

In order to avoid the predicaments associated with the deterministic theories, statistical approaches have been proposed which acknowledged that flaws can reside in the stressed volume with certain probability (Curry and Knott, 1979; Wallin et al., 1984). The predictions from these models however are still conservative because they equate the probability of failure initiated by a flaw in a stressed region with the probability of existence of the flaw in that region. If we denote by F_{d*} the conditional probability of initiating failure, given that a flaw of size $d > d^*$ is present in the highest-stressed region, these models essentially assume $F_{d*} = 1$. This approach is suitable for weak flaws which initiate fracture easily, for example for carbide particles initiating cleavage fracture. In the general case however, a flaw of particular size may be present and failure may still not occur if the flaw type, orientation, bond with the matrix and shape do not promote failure initiation.

An attempt to address these problems has been made in another group of classical models (Batdorf and Crose, 1974; Evans, 1978; Danzer and

Lube, 1996) which take into consideration the flaw size, shape and orientation, and the variation of the stress tensor in the loaded components. These models, however, determine the probability of failure in terms of the expected number $\overline{N}_c$ of critical flaws (flaws which initiate failure):

$$P(\text{failure}) = 1 - \exp(-\overline{N}_c) \qquad (16.1)$$

The expected number of critical flaws $\overline{N}_c$ however is not a measurable quantity. What can be determined by using ultrasonic inspection, radiography or quantitative metallography is the number density of the flaws and their spatial distribution in the material.

In order to avoid the drawbacks of the existing models, in (Todinov, 2005b, 2006a) powerful equations and efficient algorithms have been proposed for determining the probability of failure of loaded components with complex shape, containing multiple types of flaws. The equations are based on the concept '*conditional individual probability of initiating failure*' introduced in (Todinov, 2000a). This is the probability that a single defect/flaw will initiate failure *given* that it resides somewhere in the stressed component. This concept permits to relate in a simple fashion the conditional individual probability of failure characterising a single flaw and the probability of failure characterising a population of flaws. The conditional individual probability characterising a single flaw is estimated by using a Monte Carlo simulation and a failure criterion.

The derived equations have important applications in optimising designs by decreasing their vulnerability to failure initiated by flaws during overloading or fatigue cycling. The classic Weibull distribution (Weibull, 1951) is also a special case from these equations (Todinov, 2006a). Methods have also been developed for specifying the maximum acceptable level of the flaw number density and the maximum size of the stressed volume which guarantee that the probability of failure initiated by flaws remains below a maximum acceptable level.

Suppose that a structure/component with complex shape, with inhomogeneous material containing non-interacting flaws is loaded in an arbitrary fashion. It is assumed that the flaws locations in the volume V follow a non-homogeneous Poisson process. The variation of the flaw number density in the volume of the structure is described by the function $\lambda(x, y, z)$. It gives the flaw number density in the infinitesimal volume dv at a location with coordinates x, y, z (Fig. 16.2).

Suppose that a single flaw is characterised by the conditional individual probability F_c of initiating failure *given* that the flaw resides with certainty

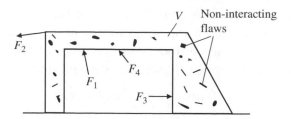

Figure 16.2 A structure with complex shape, loaded with arbitrary forces.

in the stressed structure/component. The index 'c' in F_c means that the individual probability of initiating failure has been conditioned on the existence of a flaw in the component. This probability is different from the probability p_f of failure of the component associated with a population of flaws. The probability p_f is related to the whole population of flaws and, in addition, it is not conditioned on the existence of flaws in the component. In other words, p_f is still a valid concept even if flaws are not present at all in the component while F_c is defined on the basis that a single flaw is already present somewhere in the component. The probability p_f of failure associated with a population of flaws has been derived in (Todinov, 2006a). It is given by

$$p_f = 1 - \exp(-F_c \int_V \lambda(x, y, z) \, dv) \qquad (16.2)$$

Equation (16.2) also holds for the two- and one-dimensional case. In this case the volume V in equation 16.2 is replaced by the area S or the length L of the component. Correspondingly, in these cases, the volume number density of the flaws $\lambda(x, y, z)$ will become 'number of flaws per unit area' $\lambda(x, y)$ or 'number of flaws per unit length' $\lambda(x)$. Since

$$\bar{\lambda} = \frac{1}{V} \int_V \lambda(x, y, z) \, dv$$

is the expected (average) number density of flaws in the volume V, equation (16.2) can also be presented as

$$p_f = 1 - \exp(-\bar{\lambda} V F_c) \qquad (16.3)$$

A very important special case for the applications is obtained if the flaws locations follow a homogeneous Poisson process. In this case, the flaw number density is constant $\lambda(x, y, z) = \lambda = $ constant and the probability of failure in equation (16.3) becomes

$$p_f = 1 - \exp(-\lambda V F_c) \tag{16.4}$$

An upper bound of the probability of failure p_f can be obtained if *weak flaws* ($F_c \approx 1$) are assumed. This is a very conservative assumption, suitable in cases where the upper bound of the probability of failure is required.

Equation (16.3) can be generalised for multiple type of flaws. Thus, if M types of flaws are present, the probability that no failure will be initiated is

$$p^0 = \exp(-\bar{\lambda}_1 V F_{1c}) \times \cdots \times \exp(-\bar{\lambda}_M V F_{Mc}) = \exp\left(-V \sum_{i=1}^{M} \bar{\lambda}_i F_{ic}\right)$$

where $\bar{\lambda}_i$ and F_{ic} are the average flaw number density and the conditional individual probability of initiating failure characterising the ith type of flaws. This equation expresses the probability that no failure will be initiated by the first, the second, ..., the Mth type of flaws. The probability of failure then becomes

$$p_f = 1 - \exp\left(-V \sum_{i=1}^{M} \bar{\lambda}_i F_{ic}\right) \tag{16.5}$$

In order to distinguish between a complex stress state and a uniaxial stress state, for a volume V subjected to a uniaxial stress σ, the probability F_c in equation (16.4) will be denoted by $F(\sigma)$.

16.2 DETERMINING THE CONDITIONAL INDIVIDUAL PROBABILITY OF INITIATING FAILURE, CHARACTERISING A SINGLE FLAW

Suppose that a flaw resides somewhere in the volume V of the component. If the volume V is divided into infinitesimally small sub-volumes dV, $P(\text{flaw} \in dV \mid \text{flaw resides in } V)$ is the conditional probability that the flaw belongs to the elementary volume dV given that it resides in the volume V. This conditional probability can be determined from

$$P(\text{flaw} \in dV | \text{flaw resides in } V) = \frac{dV}{V} \frac{\lambda(x,\, y,\, z)}{\bar{\lambda}} \tag{16.6}$$

where

$$\bar{\lambda} = \frac{1}{V} \int_V \lambda(x,\, y,\, z)\, dv$$

is the mean number density of flaws in the volume V. Consequently, the locations of the flaw inside the volume V follow the conditional probability density $\lambda(x, y, z)/(\bar{\lambda}V)$. In the important special case where the flaw number density is constant ($\lambda = \text{constant}$), for a given number of flaws, their locations are uniformly distributed in the volume V.

In order to determine the conditional individual probability of initiating failure, for each random location of the flaw, a check is performed whether the flaw will be unstable (will initiate failure). In general, the failure criterion is a complex function of the local stress state specified by the stress tensor, the local properties of the matrix, the material properties of the flaw, the shape and size of the flaw and its orientation regarding the matrix. We assume that at each location x, y, z, the failure criterion $\Phi(x, y, z)$ can be calculated and $\Phi(x, y, z) \geq 0$ indicates failure (limit state), while $\Phi(x, y, z) < 0$ indicates that the flaw at location x, y, z will not initiate failure. Some of the controlling variables in the failure criterion will be random variables. Calculation of the failure criterion at a location x, y, z involves deterministic parameters characterising the properties of the matrix at location x, y, z and also, sampling from the joint distribution $\varphi(x_1, \ldots, x_m)$ of m random parameters in the failure criterion. Sampling from the joint distribution $\varphi(x_1, \ldots, x_m)$ can for example be done by using the Hastings–Methropolis algorithm (Ross, 1997). In cases where the random parameters are statistically independent: $\varphi(x_1, \ldots, x_m) = \varphi_1(x_1) \times \cdots \times \varphi_m(x_m)$, sampling from their joint distribution reduces to a sequential sampling from each of the marginal distributions $\varphi_i(x_i)$ characterising the separate random parameters. Sampling from the marginal distributions can for example be done by using the inverse transformation method described in Appendix A (Algorithm A2). All of the obtained realisations for the random variables are used to calculate the failure criterion $\Phi(x, y, z)$.

The conditional individual probability F_c of initiating failure characterising a single flaw can be estimated from the ratio

$$F_c \approx n_f / n \tag{16.7}$$

where n is the total number of generated flaw locations and n_f is the number of flaw locations which resulted in failures ($\Phi \geq 0$). In words, the conditional individual probability F_c of initiating failure characterising a single flaw has been estimated from a Monte Carlo simulation, by dividing the number of trials in which failure has been 'initiated' to the total number of simulation trials. Substituting the estimate F_c in equation (16.4) then yields the probability of failure of the stressed component/structure, *irrespective of its geometry, type of loading and flaw number density.* The algorithm in pseudo-code is given in Appendix 16.1.

In the case of a failure criterion dependent on the size of the flaws and their orientation, for each random location a random orientation is generated. Next, a random flaw size is generated by sampling the size distribution of the flaws. Given the specified location, orientation and size of the flaw, a failure criterion is applied to check whether the flaw will be unstable (will initiate failure).

The described approach is very flexible because it permits the conditional individual probability F_c of initiating failure to be estimated by using different methods. Indeed, the failure criterion is not restricted to fracture mechanics criteria only. It can also be determined from models based on the micromechanics of failure.

The efficiency of the algorithm can be increased significantly if the loaded component is divided into N sub-volumes. If a finite element solution is used, the sub-volumes are simply the finite elements which partition the volume of the component.

In the case of flaws following a homogeneous Poisson process, in order to generate a random flaw location, a finite element is randomly selected first, with probability proportional to its volume fraction (Fig. 16.3).

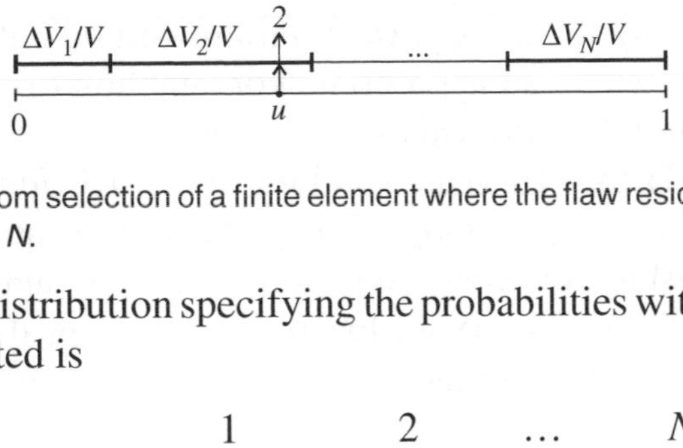

Figure 16.3 A random selection of a finite element where the flaw resides. The total number of finite elements is N.

The discrete distribution specifying the probabilities with which the finite element is selected is

$$
\begin{array}{ccccc}
X & 1 & 2 & \cdots & N \\
P(X=x) & \Delta V_1/V & \Delta V_2/V & \cdots & \Delta V_N/V
\end{array}
$$

where $X = 1, 2, \ldots, N$ is the index of the finite element, ΔV_X is its volume and V is the total volume of the component/structure. The probability with which the ith finite element is selected is proportional to its volume fraction $\Delta V_i/V$.

Usually the finite element solvers give the coordinates of all vertices belonging to a particular finite element. For the special case of a tetrahedron with coordinates of the vertices (x_1, y_1, z_1), (x_2, y_2, z_2), (x_3, y_3, z_3) and

(x_4, y_4, z_4), the volume is given by $V = (1/6)|D|$, where D is the value of the determinant (Rourke, 1994):

$$D = \begin{vmatrix} x_1 & y_1 & z_1 & 1 \\ x_2 & y_2 & z_2 & 1 \\ x_3 & y_3 & z_3 & 1 \\ x_4 & y_4 & z_4 & 1 \end{vmatrix}$$

The algorithm for selecting a random finite element is given in Appendix A (Algorithm A6).

Once a finite element has been selected, a defect location is generated inside and the principal stresses at this location are calculated by using interpolation. The calculation speed can be increased further at the expense of a slight decrease in the calculation precision if, instead of generating a random location for the flaw inside the randomly selected finite element, the principal stresses in the centre of the finite element are used to evaluate the failure criterion. Since most finite element solvers provide information regarding the three principal stresses in the centre of the finite elements, the speed of computation can be increased significantly.

Failure will be initiated most frequently in the highest-stressed regions where the conditions for a flaw instability will be met first during overloading. If in the highest-stressed region, no flaw of appropriate type, orientation and size for initiating failure is present, failure will be initiated in a region with lower stress, where an appropriate combination of stress, flaw type, flaw orientation and flaw size exists.

Equations (16.3) and (16.4) are valid for *arbitrarily loaded components and structures*, with *complex shape*. The power of the equations is in relating in a simple fashion the conditional individual probability of failure F_c characterising a single flaw (with locations following the specified flaw number density $\lambda(x, y, z)$) to the probability of failure p_f characterising a whole population of flaws.

Suppose that a direct Monte Carlo simulation was used to determine the probability of failure of the component. In this case, at each simulation trial, a large number of flaws need to be generated and for each flaw, and a check needs to be performed to determine whether there will be at least a single unstable flaw which initiates failure. If equation (16.3) is used to determine the probability of failure, only a single simulation trial involving a single act of generating flaws in the component volume is necessary. The aim is to collect statistical information from all parts of the component volume, locally stressed in different ways, which is necessary to estimate the conditional individual probability F_c. Once F_c has been estimated, it is simply plugged in equation (16.3) to determine the probability of failure of the component.

Because only a single simulation trial is involved instead of thousands or millions of trials, the calculation speed of the algorithm based on equation (16.3) is significantly greater than the calculation speed of an algorithm based on a direct simulation.

It is important to point out that F_c incorporates the influence of the particular local loading (stress) state throughout the entire volume of the component. If the loading is altered, F_c will be altered too despite that all locations, orientations and flaw sizes will remain the same. Another important feature of F_c which distinguishes it from the probability of failure p_f is that while p_f is an *absolute probability*, F_c is a *conditional probability*. It is the probability that a flaw will cause failure, *given* that it is already inside the volume of the stressed component. By 'moving' the flaw randomly inside the component and by simultaneously changing its shape and orientation, statistical information regarding the conditional probability F_c is gathered.

Equations (16.3) and (16.4) have been derived under the assumption that the flaws do not interact. This assumption is fulfilled in cases where the number density of the flaws guarantees that the distances between them exceed the distances at which the regions of local stress intensification associated with the flaws start interacting. In cases of dense populations of flaws however, this assumption is no longer valid. The distances between flaws are too small and the interactions of the regions of stress intensification can no longer be neglected. In this case, equations (16.3) and (16.4) should not be used for determining the probability of failure.

In effect, equations (16.3) and (16.4) constitute the core of a theory of failure initiated by flaws. The equations avoid conservative predictions related to the probability of failure and are a real alternative to existing approaches. The concept 'conditional individual probability of initiating failure' characterising a single flaw acknowledges that not all flaws in the material will initiate failure. Flaws initiate failure with certain probability.

16.3   IMPORTANT SPECIAL CASES RELATED TO THE CONDITIONAL INDIVIDUAL PROBABILITY OF INITIATING FAILURE

Let us consider a common special case in determining the conditional individual probability F_c, related to globular flaws whose locations follow a homogeneous Poisson process in the volume of the component. The flaws are characterised by a particular cumulative distribution $G(d)$ of their size. $G(d) \equiv P(D \le d)$ gives the probability that the flaw will have a diameter D not greater than a specified value d. The individual conditional probability of initiating brittle failure by the flaws can be determined under fairly general assumptions.

Failure of brittle materials (cast iron, high-strength steels, carbon steels at low temperatures, ceramics) is associated with almost no yielding. Failure occurs when the maximum principal tensile stress reaches a critical value. The maximum principal stress failure predictor (MPSFP) design rule (Samuel and Weir, 1999) states that if a component of brittle material is exposed to a multiaxial stress system, fracture will occur when the maximum principal stress anywhere in the component exceeds the local strength.

The likelihood of brittle fracture is increased significantly by the presence of inclusions in the steel (particularly oxides, which are characterised by tensile tessellation stresses). If inclusions are present, brittle fracture will be initiated by the principal tensile stress acting at some of the inclusions. Numerous observations of failure surfaces indicate that for cylindrical components made of high-strength steel, torsional overstress fracture typically results in a helicoidal fracture surface (plane γ in Fig. 16.4a) which indicates brittle fracture.

Consequently, if a torsional shock results in overstress failure, the shaft will fail by cleavage, caused by the maximum principal tensile stress σ_1 at an angle approximately 45° to the axis (Fig. 16.4b). This is the stress which controls brittle fracture, not the maximum shear stress. In combination with a surface defect the maximum tensile stress σ_1 will initiate brittle fracture at a smaller overstress load.

Suppose that the local maximum tensile stress in the component has been calculated for a large number n of locations, uniformly distributed into the

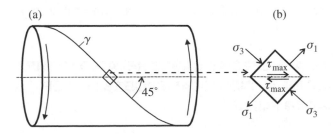

Figure 16.4 (a) Fracture surface of a brittle cylindrical component subjected to pure torsion and (b) stress state of an element at the surface.

volume of the component. Suppose that the local principal tensile stress has been determined for each of these locations (e.g. by the finite-element method) and an array $\sigma_t[n]$ of size n has been constructed containing the local principal tensile stress characterising each location. For a flaw with size D, let $\sigma_c(D)$ be the critical tensile stress which, if exceeded by the local tensile stress σ_t, will make the flaw unstable and initiate failure. Thus, for a flaw location with coordinates x, y, z, the failure criterion becomes

$$\sigma_t(x,\ y,\ z) \geq \sigma_c(D) \tag{16.8}$$

where the size of the flaw at location x, y, z is obtained from sampling the distribution of the flaw size. The algorithm for estimating the conditional individual probability F_c can now be formulated. It consists of the following steps.

Algorithm 16.1

$\sigma_t\ [n]$; /* *An array containing the maximum principal tensile stress for n uniformly distributed locations in the volume of the component* */

Failure_counter=0; // *A variable counting the number of failures*

For k=1 to Number_of_trials do
{
   • *A random finite element (location) X is selected, by generating a random number from 1 to n, according to algorithm A1 from Appendix A;*
   • *Simultaneously, a random size d_x for the flaw in the selected finite element is generated by sampling its size distribution, according to algorithm A2 described in Appendix A;*

   /* *The critical stress $\sigma_{c,x}$ necessary to initiate failure is calculated from the selected failure criterion* */
   $\sigma_{c,x} = \sigma_c(d_x)$;

   /* *The principal tensile stress $\sigma_{t,x}$ characterizing the selected finite element 'x' is compared with the critical stress $\sigma_{c,x}$ necessary to initiate failure. If $\sigma_{t,x} \geq \sigma_{c,x}$ is fulfilled, the flaw will be unstable and the failure counter is incremented* */

   If $(\sigma_t[x] \geq \sigma_{c,x})$ **then** Failure_counter= Failure_counter+1;
}
F_c = Failure_counter/Number_of_trials; // *Determining the conditional individual probability of initiating failure*

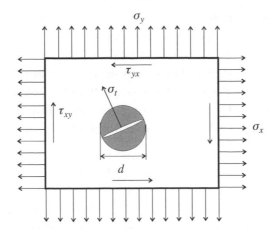

Figure 16.5 Failure of material with homogeneous matrix containing relatively weak globular second-phase particles which crack easily by the maximum principal tensile stress σ_t and produce penny-shaped microcracks.

The form of the failure criterion can vary depending on the material, the flaws, and most importantly, on the failure mechanism. Suppose that a material with homogeneous matrix contains globular second-phase particles (Fig. 16.5). These particles crack if subjected to stress, thereby producing penny-shaped microcracks which propagate in the matrix if the tensile stress is sufficiently large. The failure mechanism is similar to the failure mechanism of cleavage fracture in ferritic steels (Anderson, 2005).

In this model, the particles are assumed to be weak; in other words, the tensile stress necessary to produce microcracks in the particles is significantly smaller than the tensile stress necessary to propagate the microcracks through the matrix. In this case, the microcracks can be treated as Griffith cracks (Griffith, 1920). For a penny-shaped crack, the fracture stress is given by (Anderson, 2005):

$$\sigma_c = \left(\frac{\pi E \gamma_p}{(1 - v^2)D} \right)^{1/2} \tag{16.9}$$

where D is the diameter of the particle, γ_p is the plastic work required to create a unit area of fracture surface in the matrix, E and v are the modulus of elasticity and the Poisson's ratio for the matrix, correspondingly.

In reality, the local tensile stress necessary to crack the particle may be larger than the local tensile stress necessary to propagate the microcrack into the matrix. Furthermore, the particle may not crack at all. Failure may be initiated by the large tessellation stresses generated in the vicinity of the particle. In this case, different types of particles (e.g. oxide or sulphide inclusions) of the same size and in the same matrix will be characterised by different tessellation stresses and different critical tensile stress necessary to initiate failure.

The generic failure criterion used in the example to follow has the form

$$\sigma_c = \frac{K}{\sqrt{D}} \tag{16.10}$$

where σ_c is the critical tensile stress required to make the flaw unstable, D is the diameter of the flaw and K is a constant reflecting the type of the flaw, the properties of the matrix and its microstructure. As can be verified, for weak flaws which crack easily, this criterion reduces to the failure criterion (16.9) where

$$K = \left(\frac{\pi E \gamma_p}{(1 - v^2)} \right)^{1/2}$$

A failure criterion based on the maximum tensile stress is not the only failure criterion that can be used. For the special case of brittle fracture and flaws whose shape can be approximated well by penny-shaped cracks, a mixed-mode coplanar strain energy release rate criterion (Paris and Sih, 1965; Evans, 1978; Anderson, 2005) can be used. Fracture, according to this criterion occurs if the strain energy release rate G exceeds the critical strain energy release rate G_c for the material.

Using the algorithm described earlier, for different loading levels, the lower tail of the strength distribution for any loaded structure with internal flaws can be constructed. Important application areas of the discussed equations are (i) determining the lower tail of the strength distribution for components containing flaws and (ii) assessing the vulnerability of designs to failure initiated by flaws.

In the example to follow, the vulnerability of the double-cantilever structure in Fig. 16.6 is assessed, against overstress failure initiated by globular flaws whose locations follow a homogeneous Poisson process in the component. The purpose is to determine which type of loading is associated with the smallest probability of failure during overstress. For simplicity, only two different types of loading are considered here. Case 'a', where an overstress load $F = 30,000\,\text{N}$ subjects the structure to bending, shear and torsion and case 'b', where the overstress load $F = 30,000\,\text{N}$ subjects the structure to bending and shear (Fig. 16.6). The larger cantilever beam is 0.12 m long with a square cross-section with

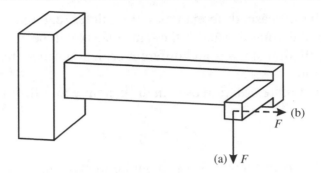

Figure 16.6 A double-cantilever structure loaded in (a) bending, shear and torsion and (b) bending and shear.

dimensions $0.02\,\mathrm{m} \times 0.02\,\mathrm{m}$. The smaller beam is $0.06\,\mathrm{m}$ long, with a square cross-section with dimensions $0.01\,\mathrm{m} \times 0.01\,\mathrm{m}$.

The constant K in the failure criterion given by equation (16.10), where D is in $\mu\mathrm{m}$, has been estimated to be $K = 32{,}700 \times 10^6$ and the diameter D of the flaws in $\mu\mathrm{m}$ has been assumed to be normally distributed, with mean $\mu_D = 300\,\mu\mathrm{m}$ and standard deviation $\sigma = 35\,\mu\mathrm{m}$. The volume of the structure has been divided into 32,535 equal-size tetrahedral finite elements and the maximum principal stress in the centre of each tetrahedral element has been produced by using a finite elements software package. By using Algorithm 16.1, the array containing the maximum principal stresses characterising all finite elements has been analysed.

Individual probabilities $F_{c,a} = 0.0018$ and $F_{c,b} = 0.0039$ corresponding to loading schemes 'a' and 'b' in Fig. 16.6, have been estimated from 1,000,000 Monte Carlo simulation trials. Assuming $\lambda V = 100$ expected number of flaws in the volume of the double-cantilever structure, from equation (16.4) we derive

$$p_{f,a} = 1 - \exp(-\lambda V F_c) = 1 - \exp(-100 \times 0.0018) \approx 0.16 \quad (16.11)$$

for the probability of failure related to loading scheme 'a' and

$$p_{f,b} = 1 - \exp(-\lambda V F_c) = 1 - \exp(-100 \times 0.0039) \approx 0.32 \quad (16.12)$$

for loading scheme 'b'. As a result, loading scheme 'a' in Fig. 16.6 is associated with smaller probability of overstress failure and should be preferred to loading scheme 'b' if overstress loading is likely to occur during the service history.

Equations (16.3) and (16.4) are very flexible and general because they permit the conditional individual probability F_c of initiating failure to be estimated by using different failure criteria. Indeed, the failure criterion is not restricted to fracture mechanics criteria only. It can also be based on other models related to the micromechanics of initiating failure. For the special case of brittle fracture and flaws whose shape can be approximated by penny-shaped cracks for example, a mixed-mode coplanar strain energy release rate criterion (Paris and Sih, 1965)

$$G = \frac{(1 - \nu^2)K_{\mathrm{I}}^2}{E} + \frac{(1 - \nu^2)K_{\mathrm{II}}^2}{E} + \frac{(1 + \nu)K_{\mathrm{III}}^2}{E}$$

can be used (Evans, 1978).

In this failure criterion, G is the strain energy release rate; K_{I}, K_{II} and K_{III} are the three stress-intensity factors corresponding to the three basic loading modes which are functions of the stress magnitude and crack geometry; E is the elastic modulus and ν is the Poisson ratio. Again, the local principal stresses calculated at the flaw's location act as remote stresses with respect to the flaw.

Fracture, according to this criterion occurs if the value of the strain energy release rate G exceeds the critical strain energy release rate G_c for the material. This criterion is based on the assumption that planar penny-shaped cracks propagate along their initial planes if $G > G_c$ is fulfilled.

16.4 DETERMINING THE LOWER TAIL OF THE STRENGTH DISTRIBUTION

By calculating the probability of failure at different loading levels, the lower tail of the strength distribution of any loaded component with internal flaws can be constructed. This is important because the lower tail of the strength distribution interacts with the upper tail of the load distribution and determines the reliability of the loaded structure. The lower tail of the strength is then plugged into a load–strength interference model (such as equation (13.3)) in order to determine the probability of failure of the structure. *In this way, the small probabilities of failure from the lower tail of the strength distribution are determined correctly, without the need for any conservative assumptions.*

This approach will be illustrated by an example related to a component with complex shape where the probability of component failure

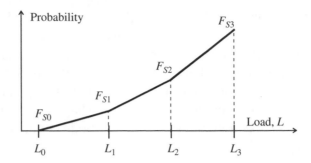

Figure 16.7 Building the lower tail of the strength distribution for different loads L_i.

has been calculated for n different magnitudes of the loading force $L_1 < L_2 < \cdots < L_n$, in ascending order.

At each load magnitude L_i, the probability of failure F_{si} of the component has been calculated according to the already presented method. The results are then plotted as shown in Fig. 16.7

If the points F_{si} are linked with segments (such as in Fig. 16.7), a conservative approximation of the lower tail of the strength distribution will be constructed. In this way, in the ith interval $L_{i-1}L_i$, $i = 1, \ldots, n$ for the load L, the lower tail of the strength distribution is approximated by the straight line $F_S = k_i L + a_i$ where

$$k_i = \frac{F_{Si} - F_{Si-1}}{L_i - L_{i-1}} \quad \text{and} \quad a_i = F_{Si-1} - L_{i-1} \frac{F_{Si} - F_{Si-1}}{L_i - L_{i-1}}$$

For a specified value of the load L in the sub-interval $L_{i-1}L_i$, $F_S = k_i L + a_i$ gives the probability of failure of the component at that load. The probability that the strength will be in the interval $L, L + dL$ in the ith load sub-interval $L_{i-1}L_i$ is given by $f_S(L)dL = k_i dL$ where $f_S(L) = dF_S(L)/dL = k_i$.

Suppose that the load also varies and $F_L(L)$ is its cumulative distribution function. The probability of failure in the region of the lower tail of the strength distribution can then be calculated by using the following method.

For each of the load intervals $L_{i-1}L_i$, the probability of no failure is calculated. Thus, the probability that the strength will be in the infinitesimal interval $L, L + dL$ belonging to the interval $L_{i-1}L_i$, and there will be no failure is given by $F_L(L)f_{Si}(L) dL = k_i F_L(L) dL$. The last relationship is a product of the probability $f_{Si}(L) dL$ that the strength will be in the interval $L, L + dL$ and the probability $F_L(L)$ that the load will be smaller than strength. Consequently, the probability that the strength will be in the

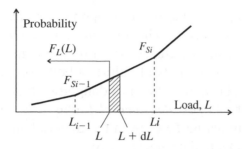

Figure 16.8 Determining the probability of failure in the region of the lower tail of the strength distribution.

interval $L_{i-1}L_i$ and there will be no failure is given by

$$P(\text{no\_failure in } L_{i-1}L_i) = \int_{L_{i-1}}^{L_i} k_i F_L(L)\,dL = k_i \overline{F}_{Li}(L_i - L_{i-1}) \quad (16.13)$$

where

$$\overline{F}_{Li} = \frac{1}{L_i - L_{i-1}} \int_{L_{i-1}}^{L_i} F_L(L)\,dL$$

is the average load in the interval $L_{i-1}L_i$ (Fig. 16.8).

Adding these probabilities yields

$$P(\text{no\_failure}) = \sum_{i=1}^{n} k_i \overline{F}_{Li}(L_i - L_{i-1}) \quad (16.14)$$

for the probability of no failure in the region of the lower tail of the strength distribution. Consequently,

$$P(\text{failure}) = 1 - \sum_{i=1}^{n} k_i \overline{F}_{Li}(P_i - P_{i-1}) \quad (16.15)$$

is obtained for the probability of failure.

16.5 STATISTICS OF FAILURE INITIATED BY FLAWS

The product $\lambda' = \lambda F_c$ in equation (16.4), which we refer to as *detrimental factor*, is an important parameter. Consider for example two components with identical material and geometry. One of the components is characterised by flaws with large number density λ_1 which initiate failure with small probability F_{c1} and the other component is characterised by flaws with small number density λ_2 which initiate failure with large probability F_{c2}. If both components are characterised by the same detrimental factors $(\lambda_1 F_{c1} = \lambda_2 F_{c2})$, the probability of failure of both components is the same.

Equation (16.5) shows that the most dangerous type of flaws is the one characterised by the largest *detrimental factor* $\overline{\lambda_i F_{ic}}$. Consequently, the efforts towards eliminating flaws from the material should concentrate on types of flaws with large detrimental factors.

For very weak flaws which initiate failure easily, the conditional individual probability of initiating failure can be assumed to be unity $F_c = 1$. In this case, the probability of failure $p_f = 1 - \exp(-\lambda V)$ of the stressed volume V equals the probability that at least one weak flaw will be present in it. In the general case however, the conditional individual probability F_c of initiating failure characterising a single flaw is a number between zero and unity. Consequently, equations (16.3) and (16.4) avoid overly conservative predictions regarding the probability of failure of components. From the equations it follows that the smaller the stressed volume V, the smaller the probability of failure. This is one of the reasons why between two similar components made of the same material, the larger component is weaker.

16.6 OPTIMISING DESIGNS BY DECREASING THEIR VULNERABILITY TO FAILURE INITIATED BY FLAWS

From equation (16.4) it is clear that given the volume of the component, the probability of failure p_f during overloading can be minimised by minimising the detrimental factor λF_c associated with the flaws. In the case of a large flaw number density λ, the probability of failure p_f is very sensitive to the conditional individual probability of failure F_c and relatively insensitive to the number density of the flaws λ. Consequently, a significant reduction of the probability of component failure can be achieved by decreasing the conditional individual probability of failure F_c. Conversely, in case of a large conditional individual probability of failure, the probability of failure becomes sensitive to the flaw number density and relatively insensitive to the conditional individual probability of failure. Consequently, an efficient reduction of the probability of component failure can be achieved by reducing the flaw number density. The decision about which method of reducing the probability of component failure should be preferred, depends also on the balance between the cost of investment and the risk reduction associated with it.

Given the volume of the component, the size distribution of the flaws and their number density, minimising the probability of failure requires minimising the conditional individual probability of failure F_c.

The application of equations (16.3) and (16.4) for decreasing the vulnerability of designs to failure caused by flaws, will be illustrated by a simple example. A solid bar with length L and constant cross-section S (Fig. 16.9(a)) contains flaws with locations in the volume of the bar following a homogeneous Poisson process, with constant flaw number density λ and size distribution according to Fig. 16.9(b). The bar is firmly supported (point A in Fig. 16.9(a)) at a distance x from its left end. There exists also a chance of excessive overload in axial direction. Given that overloading of the bar is present, there exists a chance measured by a probability q that a dynamic force of magnitude P_1 will overload the bar in tension right from the support, and by a probability $1 - q$ that a dynamic force of magnitude $P_2 < P_1$ will overload the bar in tension left from the support. Suppose that if the bar is overloaded in tension by a dynamic force of magnitude P_1, any flaw with size greater than the critical value d_1 (Fig. 16.9(b)) will trigger failure. The probability $P(D > d_1) = \alpha_1$ of having a flaw with size D greater than d_1 is equal to the area α_1 beneath the upper tail of the probability density distribution of the flaw size in Fig. 16.9b, located to the right of d_1. Accordingly, if the bar is overloaded in tension by the dynamic force P_2, any flaw with size D greater than the critical value d_2 ($d_2 > d_1$) will cause failure. The probability $P(D > d_2) = \alpha_2$ that a randomly selected flaw will have size greater than d_2, is equal to the area α_2 beneath the upper tail of the probability density distribution located to the right of d_2 in Fig. 16.9(b).

Given that overloading is present, according to the total probability theorem, the conditional individual probability of failure associated with a single flaw is

$$F_c = (x/L)(1 - q)\alpha_2 + (1 - x/L)q\,\alpha_1 \tag{16.16}$$

In equation (16.16), x/L is the probability that a single flaw with random location, existing with certainty in the volume of the bar, will be on the left side of the support; $(1 - x/L)$ is the probability that the flaw will be on the right side of the support. Given that overloading is present, the term $(x/L)(1 - q)\alpha_2$ in equation (16.16) is the probability that failure will be initiated left from the support and the term $(1 - x/L)q\,\alpha_1$ is the probability

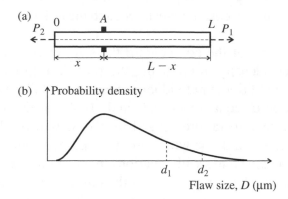

Figure 16.9 (a) A solid bar loaded in tension by dynamic forces and (b) size distribution of the flaws in the bar.

that failure will be initiated right from the support. After substituting these values in equation (16.4), the probability of failure of the bar given that overloading is present becomes

$$p_f = 1 - \exp(-\lambda LS\,[(x/L)\,(1 - q)\alpha_2 + (1 - x/L)\,q\,\alpha_1]) \qquad (16.17)$$

Now, let us select the distance x of the support in such a way, that the probability of failure triggered by flaws in case of overloading is minimised. Clearly, this is achieved if the conditional individual probability of failure F_c in equation (16.16) is minimised. Since F_c in equation (16.16) is a linear function of x, the minimum is attained if either $x = 0$ or $x = L$. Since $F_{c|x=0} = q\,\alpha_1$ and $F_{c|x=L} = (1 - q)\alpha_2$, if $q\alpha_1 < (1 - q)\,\alpha_2$ the support location minimising the probability of failure is at the left end of the bar. If $q\,\alpha_1 > (1 - q)\alpha_2$ the support location minimising the probability of failure is at the right end of the bar. Finally, if $q\,\alpha_1 = (1 - q)\,\alpha_2$, the support could be anywhere along the bar because, in this case, the conditional individual probability of failure F_c is the same. Interestingly, if $q\alpha_1 \neq (1 - q)\alpha_2$, the bar is least vulnerable to failure caused by flaws if the support is located at one of the ends, irrespective of the numerical values of the controlling parameters L, S, λ, α_1, α_2 and q.

The parameter λ and the size distribution in Fig. 16.9(b) can be determined by using *X-ray* or ultrasonic methods. It is not clear however how can the probability of failure be calculated by using equation (16.1). The expected number of critical flaws in equation (16.1) is not a measurable quantity. What is measured by using methods from the quantitative metallography is the actual number of flaws and the actual flaw size distribution.

16.7 DETERMINING THE SPATIAL DISTRIBUTION OF THE FAILURE INITIATION SITE

According to the failure criterion (16.8), failure occurs when the local maximum tensile stress exceeds the critical tensile stress σ_c and makes the flaw unstable. This criterion can also be presented as $\sigma_t - \sigma_c \geq 0$. Suppose that a component with complex shape is loaded by a force with magnitude P. Let us divide the stressed component in two regions: a high-stress region with volume ΔV and moderately stressed region with volume $V - \Delta V$. In the elastic region, the local tensile stress σ_t throughout the component is directly proportional to the magnitude of the load P. Consequently, during loading, when the magnitude of the load varies from 0 to P, failure will be initiated at a flaw for which the failure criterion $\sigma_t - \sigma_c = 0$ is first fulfilled. This flaw is not necessarily in the high-stress region. A Monte Carlo simulation can be developed where, at each trial, a number of random flaw locations are generated in the volume of the component. For each location, an additional check is performed regarding the smallest magnitude of the loading force P at which failure is initiated. Suppose that N random locations have been generated at a particular simulation trial. Determining the minimum magnitude of the loading force P at which failure will be initiated for each flaw, will produce a set of magnitudes $P_{1\,min}, P_{2\,min}, \ldots, P_{N\,min}$. The index k corresponding to the minimum magnitude among these magnitudes is essentially the index of the location where failure will actually be initiated:

$$P_{k\,min} = min\{P_{1\,min},\ P_{2\,min}, \ldots, P_{N\,min}\} \qquad (16.18)$$

Repeating this procedure will result in a map characterising the failure initiation sites.

For any particular sub-volume, dividing the number of trials for which failure has been initiated in this sub-volume to the total number of trials for which failure has been initiated in the component, gives the conditional probability of failure initiation characterising the sub-volume.

16.8 PROBABILITY OF LOCALLY INITIATED FAILURE IN A FINITE DOMAIN

The described method can be generalised for any system containing a random number of entities/events where failure is initiated locally, from a single entity/event.

Consider a case where random demands to a system in the time interval $(0, a)$, follow a non-homogeneous Poisson process with density $\lambda(t)$. The strength $S(t)$ of the system and the demand $L(t)$ are complex functions of the time. Failure occurs for a single demand where the demand/load exceeds the capacity/strength. In other words, the failure criterion is $\Phi \equiv L(t) - S(t) \geq 0$.

According to equation (16.3), the probability of failure is determined from the equation

$$p_f = 1 - \exp(-\overline{\lambda} t F_c) \tag{16.19}$$

where

$$\overline{\lambda} = \frac{1}{t} \int_0^t \lambda(v) \, dv$$

is the mean density of demands during the time interval $(0, t)$.

F_c is the conditional individual probability of failure given that a single demand exists in the time interval $(0, t)$. For the important special case of demands following a homogeneous Poisson process with constant density λ, equation (16.19) becomes

$$p_f = 1 - \exp(-\lambda t F_c) \tag{16.20}$$

Next, from equation (16.20) the reliability $R = 1 - p_f = \exp(-\lambda t F_c)$ associated with the time interval $(0, t)$ can be determined. In the case of time-dependent load and strength, F_c is determined by generating a uniformly distributed time t^* over the time interval $(0, t)$ (for a single demand) followed by a check whether $\Phi(t^*) \equiv L(t^*) - S(t^*) \geq 0$. The number of trials n_f for which $\Phi > 0$, divided by the total number of simulation trials yields an estimate of the conditional individual probability of failure $(F_c \approx n_f/n)$ which, if substituted in equation (16.20), yields the probability of failure.

If $\lambda(t)$ is the intensity of threats to a target in the finite time interval $(0, t)$ and F_c is the probability of damaging the target given that a threat arrives, the probability of damaging the target is again given by equation (16.20).

Another important application of the equation is in assessing the probability of system failure due to a fault. Suppose that the number of faults can be modelled by a random variable following a Poisson distribution with parameter μ, where μ is the average number of faults in the system. Let the probability of failure, given that the system contains a fault, be F_c.

The probability of system failure is then determined from

$$p_f = 1 - \exp(-\mu F_c) \tag{16.21}$$

In effect, equations (16.3) and (16.4) describe the probability of locally initiated failure by faults whose number is a random variable. The equations proposed avoid conservative predictions related to the probability of failure and are a real alternative to existing approaches based on the assumption that the probability of initiating failure in a particular section of the system is equal to the probability of existence of a fault in that section. The concept 'conditional individual probability of initiating failure' characterising a single fault acknowledges the fact that not all faults present in the system will initiate failure.

16.9 EQUATION RELATED TO THE FATIGUE LIFE DISTRIBUTION OF A COMPONENT CONTAINING DEFECTS

Equation (16.3) can also be generalised for determining the fatigue life distribution of a loaded component whose surface contains manufacturing defects or defects caused by a mechanical damage, with a specified number density, geometry and size distribution. The model is based on:

(i) the concept 'conditional individual probability that the fatigue life associated with a single defect will be smaller than a specified value *given* that the defect is on the stressed surface';

(ii) a model relating this conditional probability to the unconditional probability that the fatigue life of a component containing a population of defects will be smaller than a specified value;

(iii) the stress field of the loaded surface, determined by an analytical or numerical method.

Suppose that a component with complex geometry is fatigue loaded in arbitrary fashion, and contains non-interacting surface defects. It is assumed that the defects' locations on the surface of the component with total area S follow a non-homogeneous Poisson process. The variation of the defect number density on the surface of the component is described by the function $\lambda(x, y)$ which gives the defect number density in the infinitesimal surface element ds at a location with coordinates x, y.

Let $Q_c(n)$ denote the conditional individual probability (the index 'c' stands for 'conditional') that the fatigue life characterising a single defect with location following the defect number density $\lambda(x, y)$, will be smaller than n cycles, *given* that the defect resides on the surface. This probability is different from the probability $F(n)$ that the fatigue life of the component will be smaller than n cycles. The probability $F(n)$ is related to the whole population of defects and is not conditioned on the existence of a defect on the surface of the component. In other words, $F(n)$ is still meaningful even if no defects are present on the surface.

The probability $F(n)$ that the fatigue life of the component will be smaller than n loading cycles is equal to the probability that on the component's surface there will be at least a single defect with fatigue life smaller than n cycles. Similar to equation (16.2),

$$F(n) = 1 - \exp[- Q_c(n) \int_S \lambda(x, y)\, ds] \qquad (16.22)$$

Since

$$\overline{\lambda} = \frac{1}{S} \int_S \lambda(x, y)\, ds$$

is the expected (average) number density of defects on the surface S, equation (16.22) can also be presented as

$$F(n) = 1 - \exp(-\overline{\lambda} S Q_c(n)) \qquad (16.23)$$

An important special case of equation (16.22) can be obtained for defects whose locations follow a homogeneous Poisson process on the surface S. In this case, the defect number density is constant $\lambda(x, y) = \lambda = \text{constant}$ and the probability that the fatigue life of the component will be smaller than a specified number n of loading cycles becomes

$$F(n) = 1 - \exp(-\lambda S Q_c(n)) \qquad (16.24)$$

The conditional probability $Q_c(n)$ related to a single defect can be estimated by using a Monte Carlo simulation, similar to the way the conditional probability F_c in equation (16.3) was estimated.

In the special case, where the number of cycles expended on fatigue crack initiation is negligible, which approximately holds for shrinkage pores in highly stressed cast aluminium alloys (Fig. 16.10(a)), the conditional probability $Q_c(n)$ can be determined through a Monte Carlo simulation by

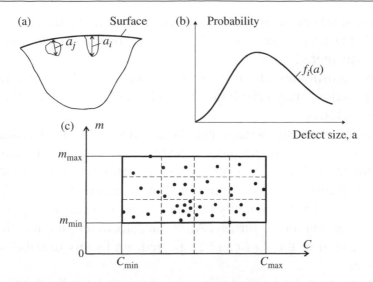

Figure 16.10 (a) Initial sizes a_i, a_j of the surface/subsurface shrinkage pores. (b and c) Input data required by the Monte Carlo model for determining the conditional probability $Q_c(n)$: (b) distribution of the size of the shrinkage pores and (c) joint distribution of the constants C and m in the Paris–Erdogan law.

integrating the Paris–Erdogan law (see equations (13.8) and (13.9)). As a result, the conditional probability that the fatigue life will be smaller than n cycles given that a shrinkage pore is present on the surface is given by

$$Q_c(n) = P\left(\int_{a_i}^{a_f} \frac{da}{C(\Delta K)^m} \leq n \right) \tag{16.25}$$

The size distribution of shrinkage pores (Fig. 16.10(b)) is sampled to determine the initial defect size a_i in the integral from (16.25) (Todinov, 1998b).

The empirical joint distribution of constants C and m, determined from a number of experiments, is also sampled to obtain realisations for C and m (Fig. 16.10(c)). For each location of the shrinkage pore, the stress-intensity factor range ΔK is calculated, followed by evaluating the Paris–Erdogan integral (13.9) from which the fatigue life is estimated.

A single defect with random orientation, shape, size and location following the specified number density $\lambda(x, y)$ is generated on the surface S. Next, for each generated location, orientation and size of the defect, the fatigue life is estimated. $Q_c(n)$ is obtained as a ratio of the number of defect locations for which the predicted fatigue life was not greater than n cycles and the total number of simulation trials. In this way, statistical information related to a single defect is collected first from different parts of the stressed surface. If the stress state is altered, the conditional individual probability $Q_c(n)$ is also altered.

Finally, substituting the estimated conditional individual probability $Q_c(n)$ in equation (16.24) yields the probability of fatigue failure $F(n)$ before n fatigue cycles.

The stress tensor, stress range and the mean stress characterising different locations of the flaw on the stressed surface can be obtained from a finite element analysis. The stressed surface can be partitioned into finite elements and instead of generating random locations for the defects, the finite elements can be randomly selected with probability proportional to their areal fraction on the surface. After the selection of a finite element, a random location of the flaw can be selected, uniformly distributed inside the element.

Similar to the overstress failure model, in case of a stress distribution on the surface obtained by using the method of finite elements, the calculation speed can be increased further at the expense of a slight decrease in the calculation precision if an approximation is used. Instead of generating a location for the defect in the randomly selected finite element and calculating the principal stresses at that location, the principal stresses in the centre of the finite element are used instead.

These are readily available from the file produced by the finite elements solver.

Parametric studies based on this stochastic model can be conducted to explore the influence of uncertainty associated with factors such as shape, size, number density of defects and residual stress fields, on the confidence levels of the fatigue life predictions. The stochastic model can be an excellent basis for specifying maximum acceptable levels for the defects number density which guarantee that the risk of fatigue failure remains below a maximum acceptable level.

Another important application of the model is for optimising designs and loading in order to minimise the probability of fatigue failure initiated by defects. In effect, this is a way to decrease the vulnerability of designs to fatigue failures initiated by surface flaws.

Similar to equation (16.3), equations (16.23) and (16.24) avoid overly conservative predictions related to the length of fatigue life. The reason is that the models recognise the circumstance that not all defects in the stressed region will evolve into propagating fatigue cracks. In other words, defects initiate propagating fatigue cracks with certain probability.

Calculating the probability of fatigue crack initiation for a particular combination of random defect size, orientation and location, characterised by a particular stress tensor, incorporates models and experimental data related to the micromechanics of initiating fatigue cracks (Jiang and Sehitoglu, 1999; Ringsberg et al., 2000; Wilkinson, 2001).

16.10   PROBABILITY OF FAILURE FROM TWO STATISTICALLY DEPENDENT FAILURE MODES

Suppose that failure can occur due to two failure modes: (i) due to individual defects triggering failure and (ii) due to clustering of flaws within a critical distance. The first failure mode has been discussed in the previous section. In many cases however, clustering of flaws within a critical distance is strongly correlated with the probability of failure, particularly for thin fibres and wires. Clustering of two or more flaws within a small critical distance s (Fig. 16.11) often decreases dangerously the load-bearing cross-section and increases the stress concentration which further decreases the load-bearing capacity. As a result, a configuration where two or more flaws cluster within a critical distance s, cannot be tolerated during loading, especially for thin fibres and wires.

Figure 16.11 Clustering within a critical distance s of two or more random flaws following a homogeneous Poisson process in a piece of thin wire with length L.

The two failure modes are not statistically independent. Indeed, the fact that there exists clustering in the length L affects the probability that there will exist flaws in the finite length L some of which may initiate failure. Let A_1 denote the event *no failure initiated from individual flaws* and A_2 denote the event *no failure initiated by clustering of two or more flaws within a critical distance s*. $A_1 \cap A_2$ is the event that *no failure will occur during loading at the stress level* σ. The probability of the intersection of events A_1 and A_2 (the probability of no failure) is given by

$$P(A_1 \cap A_2) = P(A_1) P(A_2|A_1) \tag{16.26}$$

where $P(A_2|A_1)$ in equation (16.26) is the conditional probability of no failure due to clustering of flaws within a critical distance given that 'no failure has been initiated by individual flaws'. According to equation (16.4), the probability $P(A_1)$ is given by $P(A_1) = \exp[-\lambda L F(\sigma)]$, where λ is the linear flaw number density and $F(\sigma)$ is the conditional individual probability of initiating failure at the specified loading (stress level) σ. This equation can also be rewritten as

$$P(A_1) = \exp(-\lambda' L) \tag{16.27}$$

where $\lambda' = \lambda F(\sigma)$. Equation (16.27) provides a convenient formalism which will be used in the subsequent derivations. Flaws with linear number density λ each of which initiates failure at a stress level σ with probability $F(\sigma)$ can be interpreted as an imaginary population of 'critical' flaws with linear number density $\lambda' = \lambda F(\sigma)$, initiating failure with certainty $F(\sigma) = 1$. According to this formalism, the probability of no failure is equal to the probability that no critical flaws will reside in the piece of length L.

According to an equation rigorously derived in (Todinov, 2004e), the probability of clustering p_c of two or more random flaws within a critical distance s is given by

$$p_c = 1 - \exp(-\lambda L) \left(1 + \lambda L + \frac{\lambda^2 (L-s)^2}{2!} + \cdots + \frac{\lambda^r [L - (r-1)s]^r}{r!} \right) \tag{16.28}$$

where r denotes the maximum number of flaws, with flaw-free gaps of length s between them, which can be accommodated into the finite length L ($r = [L/s] + 1$), where $[L/s]$ is the greatest integer which does not exceed the ratio L/s). The probability $P(A_2|A_1)$ can be determined from the following argument.

According to equation (16.28), the probability of no clustering, given that no critical flaws are present in the piece of length L, is given by

$$P(A_2|A_1) = \exp[-(\lambda - \lambda')L] \left(1 + (\lambda - \lambda')L + \frac{(\lambda - \lambda')^2 (L-s)^2}{2!} + \cdots \right.$$
$$\left. + \frac{(\lambda - \lambda')^r [L - (r-1)s]^r}{r!} \right) \tag{16.29}$$

The probability of no failure then becomes

$$P(A_1 \cap A_2) = P(A_1)\, P(A_2|A_1) = \exp(-\lambda'L) \times \exp[-(\lambda - \lambda')L]$$
$$\times \left(1 + (\lambda - \lambda')L + \frac{(\lambda - \lambda')^2(L - s)^2}{2!}\right.$$
$$\left. + \cdots + \frac{(\lambda - \lambda')^r[L - (r - 1)s]^r}{r!}\right)$$
$$= \exp(-\lambda L)\left(1 + (\lambda - \lambda')L + \frac{(\lambda - \lambda')^2(L - s)^2}{2!} + \cdots\right.$$
$$\left. + \frac{(\lambda - \lambda')^r[L - (r - 1)s]^r}{r!}\right) \tag{16.30}$$

Substituting $\lambda' = \lambda F(\sigma)$ in equation (16.30) finally results in

$$P(A_1 \cap A_2) = \exp(-\lambda L)\left(1 + \lambda[1 - F(\sigma)]L + \frac{\lambda^2[1 - F(\sigma)]^2(L - s)^2}{2!}\right.$$
$$\left. + \cdots + \frac{\lambda^r[1 - F(\sigma)]^r[L - (r - 1)s]^r}{r!}\right) \tag{16.31}$$

for the probability of no failure from individual flaws or from clustering of flaws.

Equation (16.31) has been verified by Monte Carlo simulations. In case of flaws initiating failure with certainty ($F(\sigma) = 1$), equation (16.31) transforms into $P(A_1 \cap A_2) = \exp(-\lambda L)$, which gives the probability of no flaws in the length L, as it should. The probability of failure is $p_f = 1 - P(A_1 \cap A_2)$ where $P(A_1 \cap A_2)$ is given by equation (16.31). Since, for the maximum acceptable probability of failure $p_{f\,max}$, the relationship $p_{f\,max} = K_{max}/C$ is fulfilled where K_{max} is the maximum tolerable level of risk and C is the cost given failure, combining with equation (16.31) results in

$$1 - \exp(-\lambda L)\left(1 + \lambda[1 - F(\sigma)]L + \frac{\lambda^2[1 - F(\sigma)]^2(L - s)^2}{2!}\right.$$
$$\left. + \cdots + \frac{\lambda^r[1 - F(\sigma)]^r[L - (r - 1)s]^r}{r!}\right) - \frac{K_{max}}{C} = 0 \tag{16.32}$$

which, if solved numerically with respect to λ, yields the upper bound of the flaw number density λ^* which guarantees that the risk K of failure initiated by flaws or due to clustering of flaws remains below the maximum acceptable value K_{max}. In short, if $\lambda \leq \lambda^*$ is satisfied for the linear number density of flaws, $K \leq K_{max}$ is satisfied for the risk of failure.

APPENDIX 16.1

An algorithm for evaluation of the probability of failure for a loaded structure with flaws whose locations follow a homogeneous Poisson process.

procedure **Calculate_stress_distribution()**
  {/ Calculates the distribution of stresses in the loaded structure by using a
    Finite Elements solution.*/}*

procedure **Select_a_random_finite_element()**
  {/* A random finite element is selected with probability proportional to its size */}

procedure **Select_a_random_location_in_the_element()**
  {/* A random, uniformly distributed location is selected in the finite element */.}

procedure **Interpolate_principal_stresses()**
  {/ Calculates the principal stresses associated with the random location
    in the selected finite element */}*

function **Generate_random_flaw_size()**
  {/* Samples the size distribution of flaws and returns a random flaw size */}

procedure **Generate_random_flaw_orientation()**
  {/ Generates the cosine directors of a randomly oriented flaw in space
    with respect to the directions of the principal normal stresses */}*

procedure **Generate_random_flaw_location()**
  {/ Generates a point with uniformly distributed coordinates (x,y,z) in the volume of
    the structure */}*

function **Check_for_failure_initiation()**
  {/ Uses a failure criterion to check whether the flaw is unstable and returns TRUE
    if the flaw with the selected location, size and orientation initiates failure */}*

Failure_counter = 0;
Calculate_stress_distribution();

For $i = 1$ **to** Number_of_trials **do**
```
{
    Generate_random_flaw_size ();
    Generate_random_flaw_orientation();
    Select_a_random_finite_element();
    Select_a_random_location_in_the_element();
    Interpolate_principal_stresses();

    Unstable = Check_for_failure_initiation();
    If (Unstable) then Failure_Counter = Failure_Counter+1;
}
```

$F_c = $ Failure_counter / Number_of_trials;

Probability_of_failure $= 1 - \exp(-\lambda V F_c)$.

Appendix

MONTE CARLO SIMULATION ROUTINES USED IN THE ALGORITHMS FOR RISK-BASED RELIABILITY ANALYSIS

A1. SIMULATION OF AN UNIFORMLY DISTRIBUTED RANDOM VARIABLE

By using the linear transformation $x_i = a + (b - a)u_i$, where u_i is a random number uniformly distributed in the interval $(0, 1)$, an uniformly distributed random value x_i in any specified interval (a,b) can be generated. Uniformly distributed integer numbers in the range $(0, n - 1)$, with equal probability of selecting any of the numbers $0, 1, 2, \ldots, n - 1$ can be obtained by using the expression

$$x_i = [nu_i]$$

where $[nu_i]$ denotes the greatest integer which does not exceed nu_i. Consequently, the formula

$$x_i = [nu_i] + 1$$

will generate with equal probability the integer numbers $1, 2, \ldots, n$. On the basis of the last equation, a function *Rand(k)* can be constructed which selects with the same probability $1/k$ one object out of k objects. The algorithm in pseudo-code is given below.

```
function Rand(k)
{
 u=u_random(); /* Generates a uniformly distributed random value in the
     interval 0, 1 */
 x= Int (k x u)+1;/* Generates a uniformly distributed integer value x in the
     interval 1, ..., k */
 return x;
}
```

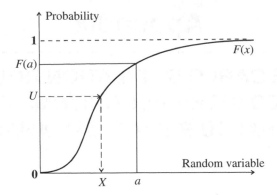

Figure A1 Inverse transformation method for generating random numbers.

Function **Int**(k × u) returns the greatest integer value which does not exceed the product ku. The function u_random() generates a uniformly distributed random value in the interval (0, 1) and can be found in most of the numerical-oriented computer packages. A good algorithm for generating uniformly distributed pseudo-random numbers is the multiplicative pseudo-random number generator suggested by Lehmer (1951).

A2.   INVERSE TRANSFORMATION METHOD FOR SIMULATING CONTINUOUS RANDOM VARIABLES

Let U be a random variable following an uniform distribution in the interval (0, 1) (Fig. A1). For any continuous distribution function $F(x)$, if a random variable X is defined by $X = F^{-1}(U)$, where F^{-1} denotes the inverse function of $F(x)$, the random variable X has a cumulative distribution function $F(x)$.

Indeed, because the cumulative distribution function is monotonically increasing (Fig. A1) the following chain of equalities holds

$$P(X \leq a) = P(U \leq F(a)) = F(a)$$

Consequently, $P(X \leq a) = F(a)$, which means that the random variable X has a cumulative distribution $F(x)$.

A3.   SIMULATION OF A RANDOM VARIABLE FOLLOWING THE NEGATIVE EXPONENTIAL DISTRIBUTION

The cumulative distribution function of the negative exponential distribution is $F(x) = 1 - \exp(-\lambda x)$ whose inverse is $x = -(1/\lambda) \ln(1 - F)$. Replacing $F(x)$ with U, which is a uniformly distributed random variable in the interval $(0, 1)$, gives

$$x = -\frac{1}{\lambda} \ln(1 - U)$$

which follows the negative exponential distribution. A small improvement of the efficiency of the algorithm can be obtained by noticing that $1 - U$ is also a uniform random variable in the range $(0, 1)$ and therefore $x = -(1/\lambda) \ln(1 - U)$ has the same distribution as $x = -(1/\lambda) \ln U$. Finally, generating a uniformly distributed random variable u_i in the interval $(0, 1)$ and substituting it in

$$x_i = -\frac{1}{\lambda} \ln(u_i)$$

results in a random variable x_i following the negative exponential distribution with parameter λ.

During the simulation, the uniformly distributed random values u_i are obtained either from a standard or from a specifically designed pseudo-random number generator.

A4.   SIMULATION OF A RANDOM VARIABLE FOLLOWING THE THREE-PARAMETER WEIBULL DISTRIBUTION

Since the Weibull cumulative distribution function is

$$F(x) = 1 - \exp\left(-\left(\frac{x - x_0}{\eta}\right)^m\right),$$

the first step is to construct its inverse

$$x = x_0 + \eta \left(\ln\left(\frac{1}{1 - F(x)}\right)\right)^{1/m}.$$

Next, $F(x)$ is replaced with U, which is a uniformly distributed random variable in the interval $(0, 1)$. As a result, the expression

$$x = x_0 + \eta \left(\ln\left(\frac{1}{1 - U}\right)\right)^{1/m}$$

is obtained. Generating uniformly distributed random values u_i in the interval (0, 1) and substituting them in

$$x_i = x_0 + \eta \, (-\ln(u_i))^{1/m}$$

yields random values x_i following the three-parameter Weibull distribution.

A5.   SIMULATION OF A RANDOM VARIABLE FOLLOWING A HOMOGENEOUS POISSON PROCESS IN A FINITE INTERVAL

Random variables following a homogeneous Poisson process in a finite interval with length a can be generated in the following way. Successive, exponentially distributed random numbers $x_i = -(1/\lambda) \ln(u_i)$, are generated according to the inverse transformation method, where u_i are uniformly distributed random numbers in the interval (0, 1). Subsequent realisations t_i following a homogeneous Poisson process with intensity λ can be obtained from: $t_1 = x_1, t_2 = t_1 + x_2, \ldots, t_n = t_{n-1} + x_n$ $(t_n \leq a)$. The number of variables n, following a homogeneous Poisson process in the finite interval, equals the number of generated values t_i smaller than the length a of the interval.

The nth generated value $t_n = -(1/\lambda) \ln(u_1) - (1/\lambda) \ln(u_2) - \cdots - (1/\lambda) \ln(u_n)$ can also be presented as $t_n = (-1/\lambda) \ln(u_1, u_2, \ldots, u_n)$. Generating uniformly distributed random numbers $u_1, \ldots, u_i$ continues while $t_i = (-1/\lambda) \ln(u_1, u_2, \ldots, u_i) \leq a$ and stops immediately if $t_i > a$. Because the condition $(-1/\lambda) \ln(u_1, u_2, \ldots, u_i) \leq a$ is equivalent to the condition

$$u_1, u_2, \ldots, u_i \geq \exp(-\lambda a)$$

the algorithm in pseudo-code for simulating a variable following a homogeneous Poisson process in a finite interval a becomes.

Limit = exp $(-\lambda a)$; S = **u_random()**; k=0;
While (S $\geq$ Limit) **do** { S=S* **u_random()**; k=k+1; }

At the end, the generated random variable following a homogeneous Poisson process remains in the variable k. Simulating a number of random failures characterised by a constant hazard rate λ in a finite time interval with length a can also be done by using the described algorithm.

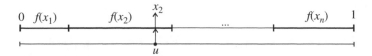

Figure A2 Simulating a random variable with a specified discrete distribution.

A6. SIMULATION OF A DISCRETE RANDOM VARIABLE WITH A SPECIFIED DISTRIBUTION

A discrete random variable X takes on only discrete values $X = x_1, x_2, \ldots, x_n$, with probabilities $p_1 = f(x_1)$, $p_2 = f(x_2), \ldots, p_n = f(x_n)$ and no other value.

X	x_1	x_2	$\ldots$	x_n
$P(X = x)$	$f(x_1)$	$f(x_2)$	$\ldots$	$f(x_n)$

where $f(x) \equiv P(X = x)$ is the probability (mass) function of the random variable: $\sum_{i=1}^{n} f(x_i) = 1$.

The algorithm for generating a random variable with the specified distribution consists of the following steps.

Algorithm

1. Construct the cumulative distribution $P(X \leq x_k) \equiv F(x_k) = \sum_{i \leq k} f(x_i)$ of the random variable.
2. Generate a uniformly distributed random number u in the interval $(0, 1)$.
3. If $u \leq F(x_1)$, the simulated random value is x_1, else if $F(x_{k-1}) < u \leq F(x_k)$ the simulated random value is x_k (Fig. A2).

A7. SIMULATION OF A GAUSSIAN RANDOM VARIABLE

A standard normal variable can be generated easily by using the *Central Limit Theorem* applied to a sum X of n random variables U_i, uniformly distributed in the interval $(0,1)$. According to the *Central Limit Theorem*, with increasing n, the sum $X = U_1 + U_2 + \cdots + U_n$ approaches a normal distribution with mean

$$E(X) = E(U_1) + \cdots + E(U_n) = n/2$$

and variance

$$V(X) = V(U_1) + \cdots + V(U_n) = n \times (1/12).$$

Selecting $n = 12$ uniformly distributed random variables U_i gives a reasonably good approximation for many practical applications. The random variable

$$X = U_1 + U_2 + \cdots + U_{12} - 6$$

is approximately normally distributed with mean $E(X) = 12 \times (1/2) - 6 = 0$ and variance $V(X) = 12 \times (1/12) = 1$, or, in other words, the random variable X follows the standard normal distribution (Rubinstein, 1981).

Another method for generating a standard normal variable is the Box–Muller method (Box and Muller, 1958). A pair of independent standard normal variables x and y are generated by generating a pair u_1, u_2 of statistically independent, uniformly distributed random numbers in the interval $(0,1)$. Random variables following the standard normal distribution are obtained from

$$x = \sqrt{-2 \ln u_1} \cos (2\pi u_2)$$

$$y = \sqrt{-2 \ln u_1} \sin (2\pi u_2)$$

From the generated standard normal variable $N(0, 1)$ with mean zero and standard deviation unity, a normally distributed random variable $N(\mu, \sigma)$ with mean μ and standard deviation σ can be obtained by applying the linear transformation

$$N(\mu, \sigma) = \sigma N(0, 1) + \mu$$

A8. SIMULATION OF A LOG-NORMAL RANDOM VARIABLE

A random variable follows a log-normal distribution if its logarithm follows a normal distribution. Suppose that the mean and the standard deviation of the logarithms of a log-normal variable X are $\mu_{\ln}$ and $\sigma_{\ln}$, correspondingly. A log-normal random variable can be obtained by first generating a normally distributed random variable Y with mean $\mu_{\ln}$ and standard deviation $\sigma_{\ln}$ from

$$Y = \sigma_{\ln} N(0, 1) + \mu_{\ln}$$

where $N(0, 1)$ is a standard normal variable generated by using Algorithm A7. The log-normal variable X is obtained by exponentiating the normal random variable Y:

$$X = e^Y$$

REFERENCES

Abernethy R.B., *The New Weibull handbook*, Ed. Robert B. Abernethy, North Palm Beach (1994).

Abramowitz M. and Stegun I.A., *Handbook of Mathematical Functions/with Formulas, Graphs, and Mathematical Tables*, edited by, Dover Publications (1972).

Altshuller G.S., *The Algorithm of Invention, Narodna Mladej* (Translation from Russian) (1974).

Anderson T.L., *Fracture Mechanics: Fundamentals and Applications*, Taylor & Francis (2005).

Andrews J.D. and Moss T.R., *Reliability and Risk Assessment*, Professional Engineering Publishing (2002).

Ang A.H.S. and Tang W.H., *Probability Concepts in Engineering Planning and Design, Vol. 1*, Basic principles, John Wiley & Sons (1975).

Arnold G., *Corporate Financial Management*, 3rd edn, Prentice Hall (2005).

Ascher H. and Feingold H., *Repairable Systems Reliability*, Marcel Dekker Inc. (1984).

Ashby M.F. and Jones, D.R.H. *Engineering Materials 1: An Introduction to Their Properties & Applications*, Butterworth-Heinemann (2000).

ASM, American Society for Metals, *Metals Handbook 9th ed.: Mechanical testing*, Vol. 8, Metals Park, Ohio, (1985).

Astley R.J., *Finite Elements in Solids and Structures*, Chapman & Hall (1992).

Aven T., *Foundations of Risk Analysis: A Knowledge and Decision-Oriented Perspective*, John Wiley & Sons (2003).

Barlow R.E. and Proschan F., *Mathematical Theory of Reliability*, John Wiley & Sons, Inc. (1965).

Barlow R.E. and Proschan F., *Statistical Theory of Reliability and Life Testing*, Rinehart and Winston, Inc. (1975).

Batdorf S.B. and Crose J.G., A statistical theory for the fracture of brittle structures subjected to non-uniform polyaxial stresses, **41**, *Journal of Applied Mechanics*, 459–464 (1974).

Bazovsky I., *Reliability Theory and Practice*, Prentice-Hall, Inc. (1961).

Beasley M., *Reliability for Engineers: An Introduction*, Macmillan Education Ltd (1991).

Bedford T. and Cooke R., *Probabilistic Risk Analysis, Foundations and Methods*, Cambridge University Press (2001).

Bessis J., *Risk Management in Banking*, 2nd edn, John Wiley & Sons Ltd. (2002).

Billinton R. and Allan R.N., *Reliability Evaluation of Engineering Systems*, 2nd edn, Plenum Press (1992).

Bird G.C. and Saynor D., The effect of peening shot size on the performance of carbon-steel springs, *Journal of Mechanical Working Technology*, **10** (2), 175–185 (1984).

Blake I.F., *An Introduction to Applied Probability*, John Wiley and Sons (1979).

Blischke W.R. and Murthy D.N., *Reliability: Modelling, Prediction, and Optimisation*, John Wiley and Sons (2000).

Booker J.D., Raines M. and Swift K.G., *Designing Capable and Reliable Products*, Butterworth (2001).

Box G.E.P. and Muller M.E., A note on the generation of random normal deviates, *Annals of Mathematical Statistics*, **28**, 610–611, (1958).

Budinski K.G., *Engineering Materials: Properties and Selection*, 5th edn, Prentice-Hall, Inc. (1996).

Burrell N.K., Controlled shot peening of automotive components, *SAE Transactions*, Section 3, Vol. 94, paper 850365 (1985).

Carter A.D.S., *Mechanical Reliability*, Macmillan Education Ltd. (1986).

Carter A.D.S., *Mechanical Reliability and Design*, Macmillan Press Ltd (1997).

Chatfield C., *Problem Solving*: A Statistician's Guide, Chapman & Hall (1998).

Cheney W. and Kincaid D., *Numerical Mathematics and Computing*, 4th edn, Brooks/Cole Publishing Company (1999).

Chernykh Yu.M., Determination of the depth of the defective surface layer of silicon steel springs, *Metal Science and Heat treatment*, **33** (9–10), 792–793 (1991).

Chiang D.T. and Niu S.C., Reliability of consecutive k-out-of-n F system, *IEEE Transactions on Reliability R-30*, 87–89 (1981).

Christensen P. and Baker M., *Structural Reliability Theory and Its Applications*, Springer-Verlag (1982).

Collins J.A., *Mechanical Design of Machine Elements and Machines*, John Wiley & Sons (2003).

Crouhy M., Galai D. and Mark R., *Risk Management*, McGraw-Hill (2001).

Crouhy M., Galai D. and Mark R., *The Essentials of Risk Management*, McGraw-Hill (2006).

Cullity B.D., *Elements of X-ray Diffraction*, 2nd edn, Addison-Wesley (1978).

Curry D.A. and Knott J.F., Effect of microstructure on cleavage fracture toughness of quenched and tempered steels, *Metal Science*, **7**, 341–345 (1979).

Dasgupta A. and Pecht M., Material failure mechanisms and damage models, *IEEE Transactions on Reliability*, **40** (5), 531–536 (1991).

Danzer R. and Lube T., New fracture statistics for brittle materials, *Fracture mechanics of ceramics*, **11**, 425–439, (1996).

DeGroot M., *Probability and Statistics*, Addison-Wesley (1989).

Dieter G.E., *Mechanical Metallurgy*, McGraw-Hill (1986).

Dhillon B.S. and Singh C., *Engineering Reliability: New Techniques and Applications*, John Wiley & Sons (1981).

Dowling N.E., *Mechanical Behaviour of Materials*, Prentice Hall (1999).

Drenick R.F., The failure low of complex equipment, *Journal of the Society for Industrial and Applied Mathematics*, **8**, 680–690 (1960).

Ebeling, C.E., *An Introduction to Reliability and Maintainability Engineering*, McGraw-Hill (1997).

Elegbede A.O.C., Chu C., Adjallah K.H. and Yalaoui F., Reliability allocation through cost minimisation, *IEEE Transactions on Reliability*, **52** (1) 106–111 (2003).

Elsayed A.E., *Reliability Engineering*, Addison Wesley Longman, Inc. (1996).

Evans A.G., A general approach for the statistical analysis of multiaxial fracture, *Journal of the American Ceramic Society*, **61**, 302–308 (1978).

Fishman G.S., *Principles of Discrete Event Simulation*, John Wiley & Sons (1978).

French M., *Conceptual Design for Engineers*, 3rd edn, Springer-Verlag London Ltd (1999).

Freudenthal, A.M., Safety and the probability of structural failure, *American Society of Civil Engineers Transactions*, paper No. 2843, 1337–1397 (1954).

Gere J. and Timoshenko, *Mechanics of Materials*, 4th edn, Stanley Thornes Ltd. (1999).

Gildersleeve M.J., Relationship between decarburisation and fatigue strength of through hardened and carburising steels, *Materials Science and Technology*, **7**, 307–310 (1991).

Gnedenko B.V., *The Theory of Probability*, Chelsea Publishing company (1962).

Grosh D.L., *A Primer of Reliability Theory*, John Wiley (1989).

Griffith A.A., *Philosophical Transactions of the Royal Society of London Series A*, **221**, 163 (1920).

Haddon Jr. W., Energy damage and the ten counter-measure strategies, *Human Factors Journal*, August, 1973.

Hale A., Goossens L., Ale B., Bellamy L., Post J., Oh J. and Papazoglou I.A., Managing safety barriers and controls at the workplace, *Proceedings of PSAM7-ESREL'2004 conference*, 14–18 June, Berlin, 609–613 (2004).

Harry, M.J. and Lawson J.R., *Six Sigma Producibility Analysis and Process Characterisation*, Addison-Wesley (1992).

Haugen E.B. *Probabilistic Mechanical Design*, John Wiley & Sons (1980).

Hecht H., *Systems Reliability and Failure Prevention*, Artech house 2004.

Heitmann W.E., Oakwood T.G. and Krauss G., Continuous heat treatment of automotive suspension spring steels, In: *Fundamentals and Applications of Microalloying Forging Steels*, Proceeding of a symposium sponsored by the Ferrous Metallurgy Committee of TMS, Eds. by Chester J. Van Tyne et al. (1996).

Heldman K., *Project Manager's Spotlight on Risk Management*, Harbor Light Press (2005).

Henley E.J. and Kumamoto H., *Reliability Engineering and Risk Assessment*, Prentice-Hall Inc. (1981).

Heron R.A., *System Quantity/Quality Assessment – The Quasi-steady State Monitoring of Inputs and Outputs'* in *Handbook of Condition Monitoring: Techniques and Methodology*, 1st edn, Ed. A. Davis, Chapman and Hall (1998).

Hertzberg R.W., *Deformation and Fracture Mechanics of Engineering Materials*, 4th edn, John Wiley & Sons, Inc. (1996).

Hobbs G.K., *Accelerated Reliability Engineering, HALT and HASS*, John Wiley & Sons Ltd (2000).

Horowitz E. and Sahni S., *Computer Algorithms*, Computer Science Press (1997).

Hoyland A. and Rausand M., *System Reliability Theory*, John Wiley and Sons (1994).

Hussain A. and M.T. Todinov, Optimisation of system reliability by using genetic algorithms, to be published.

IEC (International Electro-technical Commission), *International Vocabulary*, Chapter 191: *Dependability and Quality of Service*, IEC 50 (191) (1991).

Inglis C.E., Stresses in a plate due to the presence of cracks and sharp corners, *Transactions of the Institute of Naval Architects*, **55**, 219–241 (1913).

Jiang Y. and Sehitoglu H., A model for rolling contact failure, *Wear*, **224**, 38–49 (1999).

Jensen F., *Electronic Component Reliability*, Wiley (1995).

Kalpakjian S. and Schmid S.R., *Manufacturing Engineering and Technology*, 4th edn, Prentice Hall (2001).

King J.L., *Operational Risk: Measurement and Modelling*, John Willey & Sons Ltd, (2001).

Kuo W. and Prasad R., An annotated overview of system-reliability optimisation, *IEEE Transactions on Reliability*, **49** (2), 176–187 (2000).

Larousse, *Dictionary of Science and Technology*, Ed. P.M.B. Walker, Larousse plc (1995).

Leadbetter M.R., On basic results of point process theory. *Proceedings of the 6th Berkely Simposium on Mathematical Statistics and Probability*. University of California Press, Berkeley, 449–462 (1970).

Lee P.M., *Bayesian Statistics, an Introduction*, Arnold, 1997.

Lehmer D.H., Mathematical methods in large-scale computing units. *Proceedings of the 2nd Annual Symposium on Large-Scale Digital Computing Machinery*. Harvard University Press, 141–145 (1951).

Lewis E.E., *Introduction to Reliability engineering*, John Wiley & Sons, Inc. (1996).

Mattson E., *Basic Corrosion Technology for Scientists and Engineers*, Wiley (Halsted Press) (1989).

Mc Laney E. and Atrill P., *Accounting: An Introduction*, 3rd edn, Prentice Hall (2005).

Meeker W.Q. and Escobar L.A., *Statistical Methods for Reliability Data*, John Wiley & Sons, Inc. (1998).

Mepham M.J., *Accounting Models*, Polytech Publishers Ltd. (1980).

MIL-HDBK-217F, *Reliability prediction of electronic equipment*, US Department of Defence, Washington, DC (1991).

MIL-STD-1629A, *US Department of Defence procedure* for performing a Failure Mode and Effects Analysis, (1977).

Miller K.J., *Materials Science and Technology*, **9**, 453–462 (1993).

Miller I. and Miller M., *John E.Freund's Mathematical Statistics*, 6th edn, Prentice Hall International, Inc. (1999).

Mitrani I., *Simulation Techniques for Discrete Event Systems*, Cambridge University Press (1982).

Montgomery D.C., Runger G.C., and Hubele N.F., *Engineering Statistics*, 2nd edn, John Wiley & Sons (2001).

Nakov P. and Dobrikov P., *Programming = ++ Algorithms*, 2nd edn, TopTeam (2003).

Niku-Lari A., Shot-peening, In: The *First International Conference on Shot Peening*, 14–17 September, Paris, 1–27, Pergamon Press (1981).

O'Connor P.D.T., *Practical Reliability Engineering*, 4th edn, John Wiley & Sons (2003).

Ohring M., *Engineering Materials Science*, Academic Press, Inc. (1995).

OREDA-1992 (Offshore Reliability Data) 1992. DNV Technica (1992).

Orlov P.I., Osnovi construirovania, Vol. 1., Moskva, Mashinostroenie (in Russian) (1988).

Ortiz K. and Kiremidjian A., Stochastic modelling of fatigue crack growth, *Engineering Fracture Mechanics*, **29** (3), 317–334 (1988).

Paris P.C. and Erdogan F., *Journal of Basic Engineering*, **85**, 528–534 (1963).

Paris P.C. and Sih G.C., Stress analysis of cracks. In: *Fracture Toughness Testing and Its Application*, American Society for Testing and Materials, Annual Meeting, 67th, Chicago, 21–26 June 1964 (1965).

Park S.K. and Miller K.W., Random number generators: good ones are hard to find, *Communications of the ACM*, **31** (10), 1192–1201 (1988).

Parzen E., *Modern Theory of Probability and its Applications*, John Wiley & Sons, Inc., New York (1960).

Phadke M.S., *Quality Engineering Using Robust Design*, Prentice Hall, Englewood Cliffs (1989).

Pham H., Reliability of systems with multiple failure modes, In: *Handbook of Reliability Engineering*, Ed. H. Pham, Springer (2003).

Pickford J. (Ed.), *Mastering Risk*, Vol. 1: Concepts, Pearson Education Ltd (2001).

Press W.H., Teukolsky S.A., Vetterling W.T. and Flannery B.P., *Numerical Recipes in C*, Cambridge University Press (1992).

Ramakumar R., *Engineering Reliability, Fundamentals and Applications*, Prentice Hall (1993).

Ravi V., Reddy P.J. and Hans-Jürgen Zimmermann, Fuzzy global optimisation of complex system reliability, *IEEE Transactions on Fuzzy Systems*, **8**, (3), 241–248 (2000).

Rigaku Corporation, *Instruction Manual for Strainflex MSF-2M/PSF-2M* (1994).

Ringsberg J.W., Loo-Morrey M., Josefson B.L., Kapoor A. and Beynon J.H., Prediction of fatigue crack initiation for rolling contact fatigue, *International Journal of Fatigue*, **22**, 205–215 (2000).

Roberts F.S., *Measurement Theory with Applications to Decision-Making, Utility and the Social Sciences*, Addison-Wesley publishing company (1979).

Ross P.J., *Taguchi Techniques for Quality Engineering*, McGraw-Hill (1988).

Ross S.M., *Simulation*, 2nd edn, Harcourt Academic Press (1997).

Ross S.M., *Introduction to Probability Models*, 7th edn, A Harcourt Science and Technology Company (2000).

Rourke J., *Computational Geometry in C*, Cambridge University Press (1994).

Rubinstein R.Y., *Simulation and the Monte-Carlo method*, John Wiley & Sons, New York (1981).

SAE J936, *Report of Iron and Steel Technical Committee* (1965).

Samuel A. and Weir J., *Introduction to Engineering Design*, Elsevier Butterworth-Heinemann (1999).

Sedgewick R., *Algorithms in C++*, Addison-Wesley Publishing Company (1992).

Sedricks J., *Corrosion of Stainless Steels*, Wiley (1979).

Sherwin D.J. and Bossche A., *The Reliability, Availability and Productiveness of Systems*, Chapman and Hall (1993).

Smith C.O., *Introduction to Reliability in Design*, McGraw-Hill Inc. (1976).

Smith D.J., *Reliability Maintainability and Risk*, Butterworth (2001).

Sobol I.M., *A Primer for the Monte-Carlo Method*, CRC Press (1994).

Sundararajan C. (Raj), *Guide to Reliability Engineering: Data Analysis, Applications, Implementations and Management*, Van Nostrand Reinold (1991).

Sung C.S. and Cho Y.K., Reliability optimisation of a series system with multiple choice and budget constraints, *European Journal of Operational Research* **127**, 159–171 (2000).

Sutton I.S., *Process Reliability and Risk Management*, Van Nostrand Reinhold (1992).

Teall J.L. and Hasan I., *Quantitative Methods for Finance and Investments*, Blackwell Publishing (2002).

Thompson G., *Improving Maintainability and Reliability Through Design*, Professional Engineering Publishing Ltd (1999).

Thompson W.A., *Point Process Models with Applications to Safety and Reliability*, Chapman & Hall (1988).

Tillman F.A., Hwang F.A. and Kuo W., *Optimisation of Systems Reliability*, Marcel Dekker (1985).

Ting J.C. and Lawrence F.V., Modelling the long-life fatigue behaviour of a cast aluminium alloy, *Fatigue and Fracture of Engineering Materials and Structures*, **16** (6), 631–647 (1993).

Todinov M.T., Mechanism for the formation of the residual stresses from quenching, *Modelling and Simulation in Materials Science and Engineering*, **6**, 273–291 (1998a).

Todinov M.T., A probabilistic method for predicting fatigue life controlled by defects, *Materials Science and Engineering A*, **A255**, 117–123 (1998b).

Todinov M.T., Influence of some parameters on the residual stresses from quenching, *Modelling and Simulation in Materials Science and Engineering*, **7**, 25–41 (1999a).

Todinov M.T., Maximum principal tensile stress and fatigue crack origin for compression springs, *International Journal of Mechanical Sciences*, **41**, 357–370 (1999b).

Todinov M.T., Probability of fracture initiated by defects, *Materials Science and Engineering A*, **A276**, 39–47 (2000).

Todinov M.T., Probability of fracture initiated by defects, *Materials Science and Engineering A*, **A276**, 39–47 (2000a).

Todinov M.T., Residual stresses at the surface of automotive suspension springs, *Journal of Materials Science,* **35**, 3313–3320 (2000b).

Todinov M.T., Necessary and sufficient condition for additivity in the sense of the Palmgren–Miner rule, *Computational Materials Science*, **21**, 101–110 (2001a).

Todinov M.T., Probability distribution of fatigue life controlled by defects, *Computers and Structures*, **79**, 313–318 (2001b).

Todinov M.T., Distribution mixtures from sampling of inhomogeneous microstructures: variance and probability bounds of the properties, *Nuclear Engineering and Design*, **214**, 195–204 (2002).

Todinov M.T., Modelling consequences from failure and material properties by distribution mixtures, *Nuclear Engineering and Design*, **224**, 233–244 (2003).

Todinov M.T., Setting reliability requirements based on minimum failure-free operating periods, *Quality and Reliability Engineering International*, **20**, 273–287 (2004a).

Todinov M.T., A new reliability measure based on specified minimum distances before the locations of random variables in a finite interval, *Reliability Engineering and System Safety*, **86**, 95–103 (2004b).

Todinov M.T., Reliability analysis and setting reliability requirements based on the cost of failure, *International Journal of Reliability, Quality and Safety Engineering*, **11** (3), 1–27 (2004c).

Todinov M.T., Uncertainty and risk associated with the Charpy impact energy of multi-run welds, *Nuclear Engineering and Design*, **231**, 27–38 (2004d).

Todinov M.T., Reliability governed by the relative locations of random variables in a finite interval, *IEEE Transactions on Reliability*, **53** (2), 226–237 (2004e).

Todinov M.T., *Reliability and Risk Models: Setting Quantitative Reliability Requirements*, Wiley (2005a).

Todinov M.T., Limiting the probability of failure for components containing flaws, *Computational Materials Science*, **32**, 156–166 (2005b).

Todinov M.T., Equations and a fast algorithm for determining the probability of failure initiated by flaws, *International Journal of Solids and Structures*, **43**, 5182–5195 (2006a).

Todinov M.T., Reliability analysis based on the losses from failures, *Risk Analysis* **26** (2), 311–335 (2006b).

Todinov M.T., Reliability value analysis of complex production systems based on the losses from failures, *International Journal of Quality and Reliability Management*, **23** (6), 696–718 (2006c).

Todinov M.T., Reliability analysis of complex systems based on the losses from failures, *International Journal of Reliability, Quality and Safety Engineering*, **13** (2), 127–148 (2006d).

Trivedi K.S., *Probability and Statistics with Reliability, Queuing and Computer Science Applications*, 2nd edn, John Wiley (2002).

Tryon R.G. and Cruse T.A., Probabilistic mesomechanical fatigue crack nucleation model, *Journal of Engineering Materials and Technology*, **119**, 65–70 (1997).

Tuckwell H.C., *Elementary Applications of Probability Theory*, Chapman & Hall (1988).

Ullman D.G., *The Mechanical Design Process*, 3rd edn, McGraw Hill (2003).

Villemeur A., *Reliability, Availability, Maintainability and Safety Assessment*, **2**, Wiley (1992).

Vose D., *Risk Analysis: A Quantitative Guide*, 2nd edn, John Wiley & Sons, Ltd. (2000).

Wallin K., Saario T., Torronen K., *Metal Science*, **18**, 13–16 (1984).

Wattanapongsakorn N. and Levitan S.P., Reliability optimisation models for embedded systems with multiple applications, *IEEE Transactions on Reliability*, **53** (3), 406–416 (2004).

Weibull W., A statistical distribution of wide applicability, *Journal of Applied mechanics*, **18**, 293–297 (1951).

Wilkinson A.J., Modelling the effects of texture on the statistics of stage I fatigue crack growth, *Philosophical Magazine A*, **81** (4), 841–855 (2001).

Wright M.G., Discounted Cash Flow, 2nd edn, McGraw-Hill Book Company (1973).

Xu Z., Kuo W. and Lin H.H., Optimisation limits in improving system reliability, *IEEE Transactions on Reliability*, **39** (1), 51–60 (1990).

Zahavi E. and Torbilo V., *Fatigue Design, Life Expectancy of Machine Parts*, CRC Press (1996).

INDEX